Altering Nature

Philosophy and Medicine

VOLUME 98

Founding Co-Editor
Stuart F. Spicker

Senior Editor

H. Tristram Engelhardt, Jr., *Department of Philosophy, Rice University, and Baylor College of Medicine, Houston, Texas*

Associate Editor

Lisa M. Rasmussen, *Department of Philosophy, University of North Carolina at Charlotte, Charlotte*

Editorial Board

George J. Agich, *Department of Philosophy, Bowling Green State University, Bowling Green, Ohio*
Nicholas Capaldi, *College of Business Administration, Loyola University New Orleans, New Orleans, Louisiana*
Edmund Erde, *University of Medicine and Dentistry of New Jersey, Stratford, New Jersey*
Christopher Tollefsen, *Department of Philosophy, University of South Carolina, Columbia, South Carolina*
Kevin Wm. Wildes, S.J., *President Loyola University, New Orleans, New Orleans, Louisiana*

For other titles published in this series, go to
www.springer.com/series/6414

B. Andrew Lustig • Baruch A. Brody
Gerald P. McKenny

Editors

Altering Nature

Volume Two: Religion, Biotechnology,
and Public Policy

Editors:
B. Andrew LustigBaruch A. Brody
Davidson CollegeBaylor College of Medicine

Gerald P. McKenny
University of Notre Dame

ISBN 978-1-4020-6922-2e-ISBN 978-1-4020-6923-9

Library of Congress Control Number: 2008925365

© 2008 Springer Science + Business Media B.V.
No part of this work may be reproduced, stored in a retrieval system, or transmitted in any form or by any means, electronic, mechanical, photocopying, microfilming, recording or otherwise, without written permission from the Publisher, with the exception of any material supplied specifically for the purpose of being entered and executed on a computer system, for exclusive use by the purchaser of the work.

Printed on acid-free paper

9 8 7 6 5 4 3 2 1

springer.com

Contents

Contributors vii

Introduction 1
B. Andrew Lustig, Baruch A. Brody, and Gerald P. McKenny

1 **Compatible Contradictions: Religion and the
  Naturalization of Assisted Reproduction** 15
  Cristina Traina, Eugenia Georges, Marcia Inhorn,
  Susan Kahn, and Maura A. Ryan

2 **Religion, Conceptions of Nature, and Assisted
  Reproductive Technology Policy** 87
  John H. Evans

3 **Religious Traditions and Genetic Enhancement** 109
  Ted Peters, Estuardo Aguilar-Cordova, Cromwell Crawford,
  and Karen Lebacqz

4 **How Bioethics Can Inform Policy Decisions
  About Genetic Enhancement** 161
  Robert Cook-Deegan, Kathleen N. Lohr, and Julie Gage Palmer

5 **The Machine in the Body: Ethical and Religious
  Issues in the Bodily Incorporation of Mechanical Devices** 199
  Courtney S. Campbell, James F. Keenan, David R. Loy,
  Kathleen Matthews, Terry Winograd, and Laurie Zoloth

6 **Medical Devices Policy and the Humanities:
  Examining Implantable Cardiac Devices** 259
  Jeremy Sugarman, Courtney S. Campbell, Paul Citron,
  Susan Bartlett Foote, and Nancy M. P. King

7	**Biodiversity and Biotechnology**	285
	Nicholas Agar, David M. Lodge, Gerald P. McKenny, and LaReesa Wolfenbarger	
8	**Swimming Upstream: Regulating Genetically Modified Salmon**	321
	Paul A. Lombardo and Ann Bostrom	
Index		337

Contributors

Nicholas Agar, Ph.D., is a Reader in the School of History, Philosophy, Political Science and International Relations at Victoria University in Wellington, New Zealand.

Estuardo Aguilar-Cordova, M.D., Ph.D., is Founder and CEO of Advantagene, Inc.

Ann Bostrom, Ph.D., MBA, is Associate Dean of Research and Associate Professor in the Daniel J. Evans School of Public Affairs at the University of Washington, Seattle.

Baruch A. Brody, Ph.D., is Leon Jaworski Professor of Biomedical Ethics and Director of the Center for Medical Ethics and Health Policy at Baylor College of Medicine. He is also the Andrew Mellow Professor of Humanities in the Department of Philosophy at Rice University.

Courtney S. Campbell, Ph.D., is a Professor of Philosophy and the Director of the Program for Ethics, Science, and the Environment at Oregon State University.

Paul Citron, MSEE, retired from Medtronic. Inc., a medical device company, after 32 years in senior R&D positions. He is currently on the faculty in the Department of Bioengineering at the University of California, San Diego.

Robert Cook-Deegan, M.D., directs the Center for Genome Ethics, Law & Policy, Institute for Genome Sciences & Policy at Duke University.

Cromwell Crawford, D.Lit., Ph.D., is an Emeritus Professor in the Department of Religion at the University of Hawaii at Manoa.

John H. Evans, Ph.D., is an Associate Professor in the Department of Sociology, University of California, San Diego.

Susan Bartlett Foote, JD, is a Professor in the Division of Health Policy and Management at the University of Minnesota School of Public Health.

Eugenia Georges, Ph.D., is an Associate Professor in the Department of Anthropology, Rice University.

Marcia Inhorn, Ph.D., MPH, is a Professor at the University of Michigan, with joint appointments in the Department of Health Behavior and Health Education (School of Public Health), the Department of Anthropology, Program in Women's Studies, and Center for Middle Eastern and North African Studies.

Susan Kahn, Ph.D., is a Lecturer on Near Eastern Languages and Civilizations at Harvard University and Associate Director of the Center for Middle Eastern Studies at Harvard University.

James F. Keenan, S.J., STL, STD, is a Professor in the Department of Theology at Boston College.

Nancy M. P. King, JD, is a Professor in the Department of Social Sciences and Health Policy and Director of the Program in Bioethics, Health, and Society at Wake Forest University.

Karen Lebacqz, Ph.D., is Robert Gordon Sproul Professor of Theological Ethics, Emerita, at Pacific School of Religion in the Graduate Theological Union, Berkeley, California.

David M. Lodge, D.Phil., is a Professor in the Department of Biological Sciences and Director of the Center for Aquatic Conservation at the University of Notre Dame.

Kathleen N. Lohr, Ph.D., is a Distinguished Fellow (Health Services Research) in Social and Statistical Sciences at RTI International, Research Triangle Park, North Carolina.

Paul A. Lombardo, Ph.D., JD, is a Professor of Law at Georgia State University's College of Law.

David R. Loy, Ph.D., is the Besl Family Chair of Ethics/Religion and Society at Xavier University.

B. Andrew Lustig, Ph.D., is the Holmes Rolston III Professor of Religion and Science at Davidson College.

Kathleen Matthews, Ph.D., is Dean of the Wiess School of Natural Sciences and the Stewart Memorial Professor in the Department of Biochemistry & Cell Biology at Rice University.

Contributors

Gerald P. McKenny, Ph.D., is an Associate Professor in the Department of Theology at the University of Notre Dame.

Julie Gage Palmer, JD, is Lecturer in Law at the University of Chicago School of Law.

Ted Peters, Ph.D., is a Professor of Systematic Theology at Pacific Lutheran Theological Seminary and the Graduate Theological Union (GTU) in Berkeley, California.

Maura A. Ryan, Ph.D., is John Cardinal O'Hara, C.S.C. Associate Professor of Theology at the University of Notre Dame.

Jeremy Sugarman, M.D., MPH, MA, is the Harvey M. Meyerhoff Professor of Bioethics and Medicine in the Berman Institute of Bioethics and Department of Medicine at Johns Hopkins University.

Cristina Traina, Ph.D., is an Associate Professor in the Department of Religion at Northwestern University.

Terry Winograd, Ph.D., is a Professor of Computer Science at Stanford University.

LaReesa Wolfenbarger, Ph.D., is an Associate Professor in the Department of Biology at the University of Nebraska at Omaha.

Laurie Zoloth, Ph.D., is a Professor of Medical Ethics and Humanities at the Feinberg School of Medicine and a Professor of Religion at the Weinberg College of Arts and Sciences at Northwestern University.

Introduction

B. Andrew Lustig, Baruch A. Brody, and Gerald P. McKenny

In this second volume of the "Altering Nature" project, we situate specific religious and policy discussions of four broad areas of biotechnology within the context of our interdisciplinary research on concepts of nature and the natural in the first volume (*Altering Nature, Concepts of Nature and the Natural in Biotechnology Debates*). In the first volume, we invited five groups of scholars to explore the diverse conceptions of nature and the natural that shape moral judgments about human alterations of nature, as especially exemplified by recent developments in biotechnology. A careful reading of such developments reveals that assessments of them—whether positive or negative—are often informed by different conceptual interpretations of nature and the natural, with differing implications for judgments about the appropriateness of particular alterations of nature. These varying interpretations of nature and the natural often result from the distinctive perspectives that characterize various scholarly disciplines. Therefore, in an effort to explore the variety of meanings that attend discussions of the concepts of nature and the natural, the contributors to the first volume of *Altering Nature* addressed those concepts from five different disciplinary vantages.

A first group of scholars analyzed a range of religious and spiritual perspectives on concepts of nature and the natural. Their research highlighted the thematic, historical, and methodological touchstones in those traditions that shape their perspectives on nature. They were especially concerned with the ways in which concepts of nature function both to authorize and to constrain judgments about divine purposes for the natural order of creation (in theistic traditions) and for judgments about appropriate human responsibility vis-à-vis that order (in all spiritual traditions).

A second group of scholars analyzed philosophical perspectives on concepts of nature and the natural. They focused particularly on developments since the Scientific Revolution of the seventeenth century and discussed the alleged "death of nature" associated with the mechanization of the world picture. They also discussed a number of alternatives to reductionistic attitudes toward nature that have persisted across the modern era, including organicism in the life sciences, the emphasis on functional explanations, and normative naturalism in such recent developments as deep ecology.

A third group of scholars explored scientific and medical interpretations of nature and the natural through the prism of three case studies. Their first case study reviewed discussions in the early modern period about food, animals, and plants as "pure" or "impure" and considered the relevance of those historical precedents to recent debates about genetically modified foods and organisms. The second case study focused on medical interventions to correct "deficiencies" in nature, drew on the writings of David Hume and John Gregory to develop an account of the "observably normative" in nature, and considered the relevance of such secular accounts of the normativity of nature to current discussions of biotechnology. The third case study traced the shifts in historical conceptions of aging as "natural" or as "pathological," with implications for interventions to forestall or reverse aging, depending on which conception is invoked.

A fourth group of scholars researched legal and economic understandings of nature, focusing their discussion on the legal and economic structures that govern the appropriation of nature for private ends. In the biotechnology arena, issues concerning the patenting of nature have emerged as a central locus of conflict over the appropriation of nature for private ends. The group discussed DNA patenting as a primary case study of issues that arise at the intersection of biotechnology, legal and economic concepts of nature as property, and religion. In their wide-ranging survey of religious perspectives on nature as property, the group identified two core issues in current debates: concerns about the commodification of nature and concerns about injustices in the distribution of the benefits of the patented inventions.

Finally, a fifth group of scholars contributing to the first volume explored aesthetical and representational perspectives on nature and the natural. The group examined how different sensibilities toward nature and the natural are reflected in recent visual works that incorporate DNA and other biomaterials and in images of the body and of body parts that are employed in popular culture and advertising. In the course of their examination, they explored parallels between religion and art as two ways of dealing with boundaries between the sacred and the secular or between nature and culture.

In light of the richly interdisciplinary conversation described above, Volume Two of *Altering Nature, Religion, Biotechnology, and Public Policy*, situates religious and policy discussions of four broad areas of biotechnology within the context of the analysis of concepts of nature and the natural set forth in Volume One. In the present volume, our authors review religious, ethical, and policy discussions in four areas: biotechnology and assisted reproduction, biotechnology and genetic enhancement, biotechnology and human-machine incorporation, and biotechnology and biodiversity. We have selected these four areas as especially important because they form a continuum of perspectives on human nature and the natural world. The first two areas involve reflections on human nature as a biological "given." The third area invites reflections on the amalgamation of "given" biological functions or processes with "non-human" mechanical, chemical, or electrical constituents, including bionics and nanotechnology. The fourth areas invites reflections on the relations between human nature and the larger natural order. While many developments in biotechnology emerge as worthy areas of research, the four areas we have chosen are especially rich examples of the ways that various interpretations of nature and the natural shape the range of religious responses to particular

Introduction   3

issues in biotechnology. The aim of this volume, then, is to provide information vital to fostering scholarly, public, and policy dialogue among major religious, ethical, and biotechnology perspectives.

Each of our four subject areas is the focus of two interrelated chapters. In one chapter, our authors review and analyze current discussions of their topic in light of core theological and ethical themes and perspectives. In an accompanying chapter, a second group of authors considers the implications of those recent religious and ethical discussions for public policy deliberations and choices. In particular, these second groups suggest various ways that policy choices in these areas may be illumined by paying much greater attention to religious perspectives.

1 Assisted Reproduction Chapters

The treatment of assisted reproduction in this volume consists of two related essays. The first, "Compatible Contradictions: Religion and the Naturalization of Assisted Reproduction," examines the attitudes of various religious communities and their members to a wide variety of reproductive technologies, with a special emphasis on the processes whereby these communities do or do not adapt their understanding of the nature of family, of reproduction and parenthood, and of sexuality in order to incorporate these reproductive technologies into the life and practices of these communities. At the end, its authors discuss the implications of their findings for policy makers struggling to develop public policies that are sensitive to the range of attitudes of different communities. This leads, quite naturally, to the second essay, "Religion, Conceptions of Nature, and Assisted Reproductive Technology Policy," which examines the possibility of developing public policies that are sensitive to these communal attitudes. The essay analyzes the ways in which moral/religious conceptions of nature may or may not be incorporated into policy discourse. The argument is cross-cultural, with reasons being given to explain why it might be more possible in Europe than in the United States.

One overarching point made by the first essay is that it is necessary to combine a religious-ethical inquiry, focusing on official religious teachings, with an anthropological inquiry, focusing on the thinking and behavior of the ordinary members of the community for whom these teachings were created and by whom they are ratified or modified. This has important implications for the use of these religious teachings by policy makers desiring to develop community-sensitive public policies; they need to focus on more than the official teachings of the communities in question by examining the ways in which such communities actually live their faiths. This is nicely illustrated by a contrast the authors draw between the ways in which infertile American Roman Catholic infertile couples and Egyptian and Lebanese Muslim infertile couples can situate themselves in relation to the official teachings of their communities. For the former, they can seek support from the official teachings of their community, which stress that if a couple remains infertile after having been open to having children, they should practice hospitality and generosity in other ways. Official community teachings provide a model for their way of life. By contrast, although official Muslim community teachings allow (and perhaps even support) men who choose to divorce their wives or who choose to marry a second wife, this option is often rejected and the couples structure their lives around intimacy, love and fidelity.

Another overarching point made by the essay is that the concept of the natural in the moral/religious reflections of these communities is not static. In the attempt to incorporate reproductive technologies into the life and practices of their communities, communities may change their conceptions of the natural. But these changes are rarely completely revolutionary. Rather, they exhibit "subversive continuity," preservation of some traditional conceptions of the natural combined with transformation of others. In this way, technological advances do not necessarily weaken the grip of religious thought on the morality of these communities; instead, communities use traditional modes of analysis to understand the compatibilities of these new technologies with their traditional understandings of morality and nature. The essay supports the view that conceptions of the natural family are most often preserved, while conceptions of natural biological processes are most often transformed. This is nicely illustrated by recent developments related to egg donation in Shi'ite Iran. Under the desire to legitimate this form of assisted reproduction, leading Shiite clerics dealt with the problem of adultery by restricting the donors to single women who were willing to enter into a temporary marriage with the men in question. This works, of course, in a society which accepts polygamy and has a tradition of temporary marriages. Since polyandry is not accepted, the same strategy will not work for sperm donation. The crucial point is that this revolutionary new approach is allowed not by breaking with the traditions of the community but by emphasizing that society's traditional concepts of natural relations (polygamy and temporary marriage) at the cost of traditional conceptions of natural biological processes. This approach was then officially adopted into Iranian law.

A final overarching point made by the essay is that these changes often lead to unintended consequences. One that the authors are particularly concerned about is the possibility that the availability and legitimization of these technologies may put pressure on couples to use these expensive technologies even when the chances of success are modest. Policy makers, they argue, need to keep issues about social patterns and about justice as part of the formation of public policy. In this respect they cite the example of the results of the passage of the Israeli surrogacy law. According to that law, which was intended to preserve genetic links with the rearing parents, the sperm is to be taken from the father-to-be and the surrogate must be single and not the source of the ovum. In practice, this legalization of single women serving as the surrogate mother has made it easier for this technique to be used by single women to have their own children.

The second essay begins by reviewing the current state of regulation of assisted reproduction in the United States, both by the government and by the professional societies. The author's survey indicates that while regulations related to safety and efficacy exist, there are none related to the additional ethical concerns of various communities which have been reviewed in the first essay. Why?

The answer begins with the distinction between thin and thick moral discourse. The former assumes the ends to be pursued and examines the choice of the best means, while the latter is an examination of the ends to be pursued. The author observes that invocations of the natural by various religious communities are attempts to engage in thick moral discourse, and that it is difficult to engage in such discourse in the United States in many policy-setting venues. Elected officials are more able to do so, because their conception of the ends to be pursued have been validated by the citizens who elected them. Others, who cannot invoke this validation, confine themselves to thin moral discourse. Courts and commissions are, from his perspective, in-between cases. But even in those venues in which thick moral discourse is possible, the community's discussion of ends must be translated into a public language while still maintaining its lived meaning, and this has proven to be a difficult task.

The author notes that few of the Western European countries have followed the lead of the United States; many of them, to varying degrees, restrict the technologies because of thick moral appeals. After presenting a brief account of these stricter regulations, he goes on to postulate that different attitudes towards the legitimate authority of regulative authorities and the collectivist nature of European society make possible these regulations based upon a thicker moral discourse.

2 Genetic Enhancement Chapters

The treatment of genetic enhancement in this volume consists of two essays. The first, "Religious Traditions and Genetic Enhancement," considers the treatment of genetic enhancement in many of the world's religions. The second, "How Bioethics Can Inform Policy Decisions About Genetic Enhancement," discusses policy development concerning enhancement, and reflects on the role that religious/moral thinking might have in policy development. For both essays, questions of social justice are crucial.

The authors of the first essay begin with a number of crucial points: (a) any discussion of enhancement will have to come to grips with the legacy of the eugenics movement; (b) for the sake of their discussion, they distinguish enhancement from therapy on the grounds that the latter is aimed at bringing the individual to some average level of functioning while the former is aimed at producing a level of excellence beyond the average; and (c) there are many different meanings to nature and to the natural, and the moral status of altering the natural through enhancements may depend upon which concept of the natural is being invoked.

To give concreteness to their discussion, they present four scenarios by which height might be promoted through gene therapy, rather than directly through the injection of growth hormone. The first involves a child with abnormally low levels of growth hormone who is estimated to reach a height of 3.5 feet. The goal of the gene therapy would be for the child to reach a height that is at the lower range of normality. The second involves a child with a normal growth rate but who is estimated to grow to no more than 5 feet. The goal of the gene therapy would be for the child to reach an average height.

The third involves a child with a normal growth rate who is estimated to grow to an average height. But he would be much shorter than his family members who are unusually tall and the goal of gene therapy would be for the child to reach a height similar to the members of his family. The final case involves a child who is at the 97th percentile in the growth chart who is predicted to grow to a height greater than 6 feet. The parents want him to be even taller, and the goal of gene therapy would be to satisfy their desires. Different accounts of natural functioning might lead to different classifications of these four cases. Their goal then is to analyze the implications of different moral/religious traditions for the administration of gene therapy in these four cases.

The authors note that within the naturalist position, which is often equated with scientism and the support for anything that science can do, some have argued against the use of techniques of gene therapy for enhancement purposes. This may be based upon some appeal to latent essences, but it may also be based upon a feeling that enhancement leads to fantasies about perfection and to the neglect of human finitude. They also note that there are similar tensions within another secular tradition, feminism, since its emphasis on individual autonomy may conflict with its concerns about the impact of such practices on gender relations.

The authors then examine the issue of enhancement from the perspectives of a variety of religious traditions. These include the Roman Catholic, Protestant, Jewish, Muslim, Hindu and Buddhist perspectives. The authors once more stress the existence of conflicting strands within these traditions, but feel that they can postulate certain predominant conclusions.

They feel that the Roman Catholic view will be structured by its affirmation of human dignity, of the inviolability of life from the moment of conception, and of the development of virtues in response to challenges, even if that involves some suffering. This leads them to the conclusion that the Catholic tradition will support gene therapy in the first case, but not in the others.

In their discussion of Protestant thought, the authors note at least five tensions between views that might have differing implications for enhancements. Two deserve special notice. The first is the tension between the conception of human stewardship of nature (that might support non-interference with nature) and the concept of human co-creation with God (that might support more extensive human interventions). The second is the tension between human freedom (that might support the right to choose enhancements) and justice (that might be concerned about enhancements heightening already existing disparities). They therefore conclude that Protestants are likely to disagree about issues of enhancement.

Their discussion of Jewish thought emphasizes its support of interventions as a way of fulfilling the divine mandate to heal and to improve nature. Wherever there is suffering, there is a mandate to intervene. They therefore postulate that Jewish thought would support the genetic interventions in the first three cases. Concerns about justice might, however, lead them to oppose their use in the fourth case, where the parents are seeking special advantages for their child.

With less by way of direct Islamic discussions to draw upon, the authors forecast that Islamic views will differ between those who have a dualistic anthropology (engineered changes leave the soul uncompromised) and those who have a more

Aristotelian anthropology (who may identify human dignity, at least in part, with the inviolability of bodily essence). They note an interesting parallelism with differing Muslim views about brain death and organ transplantation.

The fundamental Hindu principle that the authors appeal to is the principle of *ahimsa* (the avoidance of harm). They provide us with a subtle account of potential harms, including the issue of social fairness, and argue that Hinduism would express extreme caution about genetic alterations, supporting it in the first case but not in the others.

Finally, they provide an analysis of these issues from the Buddhist perspective. Buddhists do not have moral concerns about altering nature, so they would have no problem with genetic alterations in any of the cases. But they would be hesitant with an excessive concern about height, since that would produce the very sort of attachment which is the root of suffering according to Buddhism.

The authors of the second essay begin their analysis by pointing out how bioethical reflection played a major role in the introduction of human DNA transfer. The very decision to devote special attention to this issue reflected, at least in part, concerns expressed by major religious denominations about this technology, expressed in an open letter to President Carter. These concerns led to the 1982 *Splicing Life* report and a 2004 OTA report. In the end, the body charged with overseeing such research (the RAC) issued a "Points to Consider" document. These bioethical and religious developments led, the authors conclude, to three major benefits: (a) there was greater public awareness of these advantages; (b) questions of long-range implications were addressed directly; and (c) policy debates incorporated questions of fairness and of sensitivity to moral/religious pluralism.

The authors turn to the example of genetic enhancements of memory emerging from medical treatments that turn out to have enhancement potential. These will surely raise issues of the ethics of research and of innovative therapy, but they will also raise issues of social justice and of the naturalness of these techniques. The authors focus on the issue of public resources being used to pay for these enhancements, and suggest that various theories of justice in health care suggest that they should not be so used. They also note that this raises the question of the fairness of some having access while others do not. All of this raises further policy questions about quality and costs of care.

The authors end with a discussion of the need for any useful bioethical discussion to focus on the formation of policy. This leads them to contrast the *Splicing Life* report with the more recent *Beyond Therapy* report. They note that the latter, unlike the former, paid little attention to the implications of its discussion for policy formulation. They end with an enthusiastic call for bioethics to enter into this type of policy relevant analysis.

3 Human-Machine Incorporation Chapters

The treatment of human-machine incorporation technologies in this volume consists of two essays. The first, "The Machine in the Body: Ethical and Religious Issues in the Bodily Incorporation of Mechanical Devices," analyzes the discussion of incorporation themes in a number of religious and philosophical traditions. The second, "Altering Nature: Medical Devices Policy and the Humanities: Examining Implantable Cardiac Devices," focuses on the history of implantable cardiac devices as a fruitful example of regulatory and policy discussion and highlights the contributions that perspectives from the humanities can make to policy debates.

In the first chapter, our authors place recent discussions of incorporation technologies in broader context by observing that human interactions with a mechanized world presuppose the ability to differentiate between our selves (in our organic and bounded embodiment) and the "other" we encounter as a technological artifact. The recent literature of religious studies, anthropology, and cultural studies has paid particular attention to interpretations of the body as boundary between self and other, between human and nonhuman, and between human and divine. The first emphasis—the boundary between the self and the other—is especially important to our authors. This boundary, which in actuality has been permeable for several centuries, may be dissolving further as new mechanical devices are introduced in biomedicine and incorporated in the body. For example, the authors point out, contemporary medicine is rife with examples of innovative technologies that restore, repair, rehabilitate, and, in rare cases, enhance our physical and psychological capacities.

In their discussion, our authors develop a useful taxonomy by which to identify four core dimensions of normative discussions of incorporation technologies. At the level of *systems*, the body itself may be interpreted primarily as a mechanical system, as an organismic whole, as a "collection of sensory channels," or as the "corporeal home" for mental life. Incorporation technologies can then be distinguished according to their *method of* intervention: chemical, mechanical, or electrical. Assessments of incorporation technologies are also influenced by the *degree of impingement on the body* (for example, whether such devices are worn, attached, penetrating, or embedded). Finally, incorporation technologies can be assessed in light of the *objectives* of the intervention, with different judgments made depending on whether particular interventions are seen as primarily therapeutic, enhancing, or hedonistic in their aims.

Our authors then focus on the question of whether the prospects for ever-greater "convergence" or bodily incorporation" of such technologies constitute actual or possible violations of religious understandings of natural boundaries or the natural order. Many mechanical or electrical restorations of function (for example, restoring mobility to impaired soldiers) are ethically uncontroversial and can be easily justified. Other interventions (for example, brain implants for neurocognitive enhancement) are far more controversial, because, as our authors characterize them, they "present improvements over our natural bodily endowments or pose implications for human nature, identity, and destiny."

The authors propose three patterns of normative response from the world's classic religions—appropriation, ambivalence, and resistance. On the basis of that taxonomy, the authors survey representative examples of current and prospective incorporation technologies, including include implanted medical electronics (e.g., cochlear implants, research on retinal implants, pacemakers and ventricular assist devices for the heart); neurally controlled external devices (e.g., advanced prosthetics, direct neural stimulation, defense research on "cyber soldiers"); and ongoing research in nanomedicine. The authors also review far more speculative discussions of downloaded consciousness to external devices.

Our authors next turn to specific religious responses to such developments according to the three patterns identified. The first strategy is that of appropriation. For example, despite specific prohibitions on consuming pork, both Judaism and Islam allow the use of porcine heart valves because of the "compelling imperative" in both traditions to save or preserve life. A second pattern, one of ambivalence, is found particularly in traditions such as Buddhism and Hinduism. Therein, incorporation technologies are not judged to be violations of an intrinsic bodily order, but may well be seen as unduly focused on the body at the expense of spiritual wisdom and enlightenment. The third pattern, resistance, our authors find most readily among certain evangelical Protestant perspectives which emphasize therapy, in keeping with a biblical emphasis on healing, but discourage efforts at enhancement.

The authors consider the central functions of metaphors and casuistry in various religious traditions as important elements in their assessments of incorporation technologies. For example, discussions of the chimera and the golem invite rich religious speculations on the nature of self and identity in light of new possibilities for incorporating devices. Moreover, our authors affirm the applicability of traditional casuistry, notably Roman Catholicism and Judaism, to such new developments. As our authors observe, methods of casuistry fulfill two important purposes: first, casuistry allows the extension of characteristic norms and values to new and unfamiliar cases, while second, in the reasoning prompted by new applications, traditional norms may themselves be enriched, amended, or revised.

The authors then review specific traditions and draw tentative conclusions about their likely reactions to incorporation technologies. Among the Asian traditions of Hinduism, Buddhism, and Daoism, the authors conclude that all three perspectives affirm the value of technological interventions "if they can reduce suffering," but if they are likely to increase our attachment to bodily desires and to interfere with spiritual pursuits, they will be judged negatively.

Among the Abrahamic traditions, various positions are taken. Despite its conservatism on natural law in other contexts, our authors conclude that Roman Catholicism, with its emphasis on the Incarnation, provides theological warrant for a range of incorporation technologies, both therapeutic and enhancing, because "[t]he body's ability to incorporate the divinity as it does in Jesus implies it is quite able to incorporate the inanimate as well." Our authors therefore conclude that there are "no warrants *in se* for prohibiting possible enhancements, although questions of social justice and the common good will be central values in such discussions." In Judaism, nature is not deemed normative; moreover, the central Jewish duties to heal the body and to preserve life have led to "an enthusiastic use" of technologies that would heal or enhance human welfare. In Islam, recent debates about organ transplantation might suggest concerns about bodily integrity relevant to judgments about incorporation technologies. The two cases, however, remain distinct. Concerns regarding brain death criteria and the status of donor bodies arise with the former but not the latter. Our authors conclude, therefore, that the Abrahamic traditions are, to varying degrees, able to provide positive warrants for the appropriation of incorporation technologies.

The authors also discuss transhumanism, a movement which seeks to correct, improve upon, and even overcome our natural biological limits through the use of technology. The religious response to the vision of transhumanism has been especially strong, and decidedly critical, among several Protestant commentators. The biblical language of humans as made in the "image of God" suggests, on this reading, that the dignity of humans as imagers of God "should be affirmed" against transhumanism's express desire to remake ourselves into "post-humans." On quite distinctive grounds, as the authors point out, Buddhism and Daoism also critique the ambitions of transhumanism: overcoming "finitude" is not the problem, and fixation on technology is more likely to increase, rather than decrease, the habits and attachments that are the basis of all suffering. The central issue under debate is whether human improvement is "simply a matter of ... scientific technology" or primarily a matter of character, virtue, and spiritual practice.

In the second essay on incorporation technologies, our authors explore the recent history of developments related to enhancing or replacing impaired cardiac function by means of devices. The authors consider three categories of devices: devices that correct abnormal heart rhythms, devices designed to improve cardiac blood flow (prosthetic heart valves and stents), and devices designed to enhance or replace the mechanical function of the heart muscle. They provide a thorough review of the issues associated with the development, testing, approval, and use of such devices, as well as of the ethical questions which arise. They emphasize three important aspects of that history. First, cardiac devices offer a vivid example of how the rapid pace of technological innovation gives rise to the need for creative regulatory responses. Second, novel aspects of some devices (for example, devices that may involve zenotransplantation) underscore the need for broader input from communities whose religious and cultural values may be implicated in judgments about the licitness of such techniques. Third, standard measurement tools for assessing new technologies (e.g., cost-effectiveness analysis, cost-benefit analysis, quality-adjusted life years, and cost utility analysis) need to be amplified by broader non-utilitarian considerations, especially concerns about distributive justice.

In their conclusion, the authors examine how the humanities, including perspectives from religious studies, can play an important role in judgments about the development and dissemination of new technologies. Ordinarily, the central criteria for clinical research and regulatory approval of devices are safety and efficacy. Perspectives from the humanities, by emphasizing non-consequentialist values—especially compassion for those marginalized by current health care distributive patterns—may serve to challenge conventional judgments about proceeding with new technologies, especially when such challenges occur at the preregulatory phase of policy discussions.

4 Biodiversity Chapters

The treatment of biodiversity issues in this volume consists of two essays. The first, "Biodiversity and Biotechnology," discusses the meanings of biodiversity as a concept and applies those understandings to issues raised by transgenic organisms, using the case of Bt maize as a primary example. The essay develops a model of risk analysis that more fully reflects the range of factors, including cultural and religious values, that are relevant to choices about transgenics, The second essay, "Swimming Upstream: Regulating Genetically Modified Salmon," analyzes recent discussions of genetically modified salmon as its focus for the interface of religious values and policy concerns.

In the first essay on transgenic organisms, our authors initially focus on two key questions. The first is an empirical or factual one: what is the current and potential impact of transgenic organisms on biodiversity? Uncertainty about the nature and magnitude of such effects makes measurement and prediction difficult, but our authors seek to answer this question by discussing the most likely foreseeable developments in transgenics. The second set of questions is a normative one: what is ethically at stake in appealing to the "value" of biodiversity? How should we assess that value in relation to others, including the interests of those who introduce transgenic organisms into various ecosystems and those affected by such interventions?

Our authors discuss various conceptions of biodiversity, which can be defined in terms of genetic variety, "organismal" diversity (the dynamics of species and populations in changing environments), and "ecosystem" diversity (which include broader features of habitat and ecosystems). Genetically engineered organisms are likely to have both positive and negative effects for all three sorts of biodiversity. While arguments are often made about likely negative effects of transgenics on biodiversity, biotechnology may also be a positive tool for conservation biology, in cases when particular biotechnology products offer environmentally sound alternatives to current products that are environmentally harmful. Given the range of possible effects, our authors emphasize the need to avoid blanket generalizations and to assess particular applications on a case-by-case basis.

Our authors then assess the value of biodiversity in light of claims about nature and the natural in various religious and cultural traditions. The chapter draws on a basic distinction in environmental ethics between anthropocentrism and nonanthropocentrism. Anthropocentric theories interpret the value of non-human nature entirely in light of current or potential human interests. Nonanthropocentric theories claim an innate or intrinsic value to non-human nature independent of or in addition to human interests. (There are also "mixed theories" that seek to respect and combine both sorts of value.)

Our authors assess the characteristic weaknesses of both anthropocentric and nonanthropocentric approaches. While anthropocentrism may include a clear emphasis on preserving biodiversity as a form of enlightened instrumentalism, the less easily specified values associated with maintaining biodiversity may tend to be underweighted relative to measurable economic gains associated with specific transgenic applications. But when nonanthropocentric theorists affirm the intrinsic value of the "natural" state of an ecosystem or respect for the evolutionary process, they face challenges: first, the philosophical challenge of deriving a prescriptive norm from a descriptive fact; and second, the scientific difficulty of determining what actually constitutes a "natural" condition to be respected and preserved. For example, what *counts* as an "ecosystem" in the first place? Moreover, the notion that ecosystems are characterized by relative stability rather than radical flux has been significantly challenged in recent environmental literature.

Our authors then discuss the proposal to introduce genetically engineered maize (Bt maize) into Mexico as a rich case study of the central ethical issues involved. Maize, as one of the world's three major crops, has an extensive history of modification and domestication. However, the introduction of transgenic maize "could alter the species composition and structure of Mexican agroecosystems," with the overall effect of causing local indigenous groups to bear the risks." Local species may well have religious or cultural significance, so the transfer of transgenes "would constitute a harm." Moreover, maize agriculture in Mexico reflects dramatically different social patterns, with the majority of farming there involving peasants using traditional farming methods.

Our authors' discussion of Bt maize illustrates the difficulty of assessing the risks associated with transgenics in terms rich enough to "do justice" to the range of relevant considerations. First, which factors qualify as legitimate aspects of assessing risks, harms, and benefits? While risk analysis often assumes the priority of direct and easily quantifiable variables, a comprehensive set of risks and trade-offs is difficult to specify even on those narrow terms. Second, when components appear to be incommensurable, can one proceed (and if so, how) in the absence of a common metric? How, in such circumstances, should noneconomic factors—such as long-term impacts on cultural and religious meanings of agriculture—figure in one's judgments?

Given such uncertainties, many theorists invoke the so-called Precautionary Principle, where the burden of proof is borne by the proponent of an activity rather than the public. The authors critically assess that principle, noting a number of practical and conceptual problems with its rationale and use, including the tendency of many of its supporters to be inattentive to the real costs of inaction. At the same time, our authors are sensitive to the factor of novelty with transgenic organisms. The never-before-seen character of such organisms makes extrapolations from past patterns of species-environment interaction far more problematic.

In light of the complexity of such deliberations, our authors emphasize a *via media* between those who would interpret risk in merely technical terms and those who would deny the relevance of scientific metrics to such judgments. Instead, our authors discuss an illuminating example from the Maori culture of New Zealand to suggest the need to appreciate that "different peoples may stand in different positions in respect of the potential costs and benefits of environmental changes on the introduction of transgenics." The Maori invoke a concept called "tapu" to describe their appropriate relationship with the natural environment. The concept, while sometimes translated as "forbidden," "restricted," "sacred," or "set apart," cannot be easily captured by such terms, Rather, our authors identify a richly holistic character to its meaning, with implications for interventions into nature that never reduce such actions to merely instrumental uses. The authors conclude their essay chapter by criticizing narrow models of risk that serve to exclude factors, (especially cultural/religious ones) that defy easy quantification. The "unavoidability of normative judgments combined with their irreducibility to scientific judgments" leads our authors to recommend a much broader vision of the voices and values that should be included in assessing the risks and benefits in particular cases.

The second essay on biodiversity addresses the public policy dimensions of debates about genetically modified organisms. The authors consider the case of genetically modified salmon as an especially rich example of the legal, ethical, and cultural aspects of such environmental discussions. For several indigenous cultures, salmon is an icon that "symbolizes ties to the land and the water that are in danger of being permanently severed." The salmon holds a central place in the cultural mythologies of these peoples and offers another example of the broader cultural and religious values relevant to public policy choices.

Despite the seriousness of debates about possible impacts of genetically modified salmon on the environment, on the economy, and on indigenous cultures, the authors note that the United States lacks a unified policy on salmon aquaculture. The chapter therefore discusses how existing policies on biotechnology might be employed in the North American context. Our authors review the broad domestic and international regulatory framework that provides the general context for discussing genetically modified foodstuffs. At the international level, the authors discuss several important documents, including the *Biosafety Protocol*, the *Convention on Biodiversity*, the *United Nations Law of the Sea*, and a number of statements from the United Nations Conference on Environment and Development. Domestically, the chapter also surveys relevant state legislation and a number of judicial decisions.

Policy debates about salmon aquaculture illustrate at least four different understandings of nature and the natural at work in such discussions. First, nature may be defined as the preservation of existing species and ecosystems. The major policy implication of this interpretation would be to avoid technologies that may increase the likelihood of species extinction or threaten current balances among species. Second, nature may be linked to respect for the belief systems of indigenous people. The major policy implications of this interpretation would be that we protect extant fish stocks and habitat, that we prohibit the introduction of genetically modified stocks that may threaten indigenous economies and cultures, and perhaps that traditional indigenous aquaculture be subsidized. Third, nature is viewed by the *Convention on Biodiversity* as coterminous with existing biosystems. The major policy implications of this interpretation include protection of existing species and habitats, and regulation meant to minimize risks to genetic diversity and the environment. Fourth, nature may be seen in somewhat Hobbesian terms as a state of enduring competition. The policy implications of this interpretation would include an emphasis on market-based competition among ideas and products, minimal regulation, and "unfettered consumer choice."

The authors conclude that debates about genetically modified salmon in the United States will turn on whether the interpretation of nature implied in international agreements such as the *Convention on Biodiversity* is likely to prove decisive.

5 Conclusion

In this two-volume project, we have sought to achieve a comprehensive multidisciplinary, cross-traditional, and cross-temporal analysis of major religious understandings of nature and to suggest the relevance of those insights to scholarly and public discussions of four major areas of biotechnology. We are confident that several fundamental conclusions have been persuasively drawn. First, the range of religious responses to current or prospective developments in biotechnology reflects the ways that various concepts of, and appeals to, nature function in religious traditions, both substantively and methodologically. Second, it is important to attend to the variety of such religious interpretations both within and across traditions, especially for the ways that particular traditions maintain the integrity of their convictions by different of strategies, including appropriation, accommodation, and reinterpretation and creative extension of traditional commitments. Finally, greater attention to the variety of religious interpretations of nature and their implications for judgments on particular biotechnologies is crucial to understanding the possibilities for both cooperation and conflict between religious and other perspectives in theoretical and policy debates about biotechnology.

As we close this introduction to the second volume of *Altering Nature*, several specific acknowledgements are in order. First, we wish to express our gratitude to the Ford Foundation for their generous funding of our research, with a special word of thanks to Constance Buchanan, for many years the Senior Program Officer for Religion, Society, and Culture in Ford's "Knowledge, Creativity and Freedom Program." Ms. Buchanan has been a generous and constructive partner to the project from its inception, and we are deeply indebted to her. Second, we wish to thank the many scholars who participated in the project during the last five-and-a-half years and who have been colleagues in a fascinating multidisciplinary conversation. We trust that our common efforts will make a useful contribution to ongoing debates about the conceptual and policy dimensions of issues posed by new developments in biotechnology. Finally, we wish to thank Dr. Lisa Rasmussen, who has served as managing editor for both volumes of *Altering Nature*. With characteristic diligence and grace, she has proved invaluable in helping us to ready the project for publication.

Chapter 1
Compatible Contradictions: Religion and the Naturalization of Assisted Reproduction[1]

Cristina Traina, Eugenia Georges, Marcia Inhorn, Susan Kahn, and Maura A. Ryan

1.1 Introduction

In first-world countries the popular autobiography of assisted reproduction organizes itself largely around the paradigm of childless marriage, which most of the world regards as a social problem, or at least as a social disvalue. Biotechnology enters the equation when culture transmutes childlessness into infertility, a disease to which one can apply medical therapy.[2] Reproductive technologies rarely cure the "illness," which affects 8% to 12% of heterosexual couples worldwide (Inhorn, 2003a, 1837), but they often provide a way around the symptoms, enabling wives to conceive and bear children. Hence the stereotypical, sympathetic American portrait of infertility patients: loving, otherwise healthy, childless (or childless but for the intervention of reproductive technology), white, economically stable, married women and men.

These married couples raise the first-level questions about "nature" and "naturalness" that have dominated Euro-American bioethical debates about assisted reproduction: do its methods resemble unassisted marital reproduction closely enough to be acceptable? Is assisted reproduction being used as an excuse for an objectionable level of tampering, so that it becomes a means of back-door genetic enhancement? These concerns in turn reveal the degree to which "nature" (supposedly mere matter that can be dissected and manipulated) and "culture" (human

---

[1] We refer to Sarah Franklin's argument that "the point of much social scientific analysis of IVF… is not that it offers yet further proof of the inherent irrationality of either people or scientific thought, but rather that *its rationalities are fully compatible with others that may contradict them*" (Franklin, 2006, 550; italics in original).

The authors owe a debt of gratitude to Meghan Courtney, Hayley Glaholt, Joseph Moser, Michal Raucher, and Tobin Miller Shearer, all of whom provided invaluable assistance with research or editing. We also thank the Ford Foundation for additional funding for research assistance.

[2] Paxson (2004, 220). Adopting the "illness" metaphor creates a more sympathetic image, one under which assistance, generosity, and compassion are appropriate responses. Adopting the metaphor of "consumer," which would be perfectly appropriate given the financial investment infertility treatment typically involves, calls up quite different sentiments. See, e.g., Kahn (2000, 170) and Becker (2000, 27).

constructs in which meaning and value are created) in fact remain intertwined in contemporary experience and moral language.

The picture becomes even more complicated when we acknowledge the patriarchal social context of many childless marriages. In many cultures treatment is motivated by a drive to parenthood, which is considered a nearly ubiquitous condition for marital happiness and security, which in turn is often also a woman's primary hope for financial stability. These are often the same settings in which rapid rates of population growth, high rates of unwanted and dangerous pregnancy, inadequate health care, compromises of women's rights, and precarious social status for women are the rule; in which reproductive tract infections render disturbing numbers of women infertile; and in which exposure to unregulated pollution compromises male fertility. These conditions raise pressing questions about the welfare, rights, and freedoms of married women and their children, but also increasingly of their husbands.

In addition, the true story of assisted reproduction extends far beyond the borders of heterosexual childless marriage. People increasingly employ assisted reproductive technologies outside this context. A close look at true uses of assisted reproduction raises further questions (sometimes in addition to the previously-mentioned ones, sometimes instead of them). In the United States and other first-world countries fertile single women and men, gay couples, and lesbian couples employ reproductive technologies in the hope of raising genetically related offspring from infancy. Likewise, fertile married couples make use of them in order to avoid genetic defects and, in some cases, even simply to determine the sex of their offspring.[3]

In each of these cases new challenges arise, nearly all connected to "nature" and "the natural": Does nature, through sexual differentiation, determine who may reproduce, and if so, to what degree? What is the natural shape of the family? What is the natural meaning of marriage? What relationship to nature does infertility bear when it clearly results from disease or pollution arising from human decisions, or when it seems arbitrary? What is "natural" or "unnatural" about the patriarchal family structures that fertility treatment reinforces in many cultures? How does self-conscious adherence to a concrete religious tradition function to certify these convictions and to dictate which will hold sway when they appear to conflict?

We argue that narrow approaches to these questions are of limited use. We choose instead to combine anthropological and religious-ethical inquiry in the hope of providing a sense not only of official religious teaching on assisted reproductive technologies (ARTs) but also of the thinking and behavior of ordinary people for whom (and ideally by whom) policy is created and ratified.[4] Consequently our coverage will not be absolutely even. We will revisit a few religious communities and cultures

---

[3] Gender selection through ART procedures is illegal in many nations and considered immoral in others, but it is clearly highly desirable even in the United States. Under the banner "NEW, LOWER PGD SEX SELECTION PRICING" one clinic offers "the chance of obtaining a desired pregnancy gender outcome that ranges from excellent to virtually GUARANTEED." See http://www.fertility-docs.com/fertility_gender.phtml (accessed August 11, 2004).

[4] On this tension, see Inhorn and Sargent (2006); on the values, contexts, and priorities that shape this reasoning, see Ryan (2004).

several times in order to elaborate the connections among their politics, theologies, and social practices, with the hope that this exercise will enable the reader better to discern such connections in communities we have not discussed. In addition, we intentionally attend at some moments to the ways in which religious individuals interpret ARTs and at others to the debates and declarations of recognized representatives of religious communities. At some points we will analyze a religious community's official documents; when these are lacking, we will report majority opinions and typical approaches. As we hope to show, these data and strategies of interpretation exhibit considerable resonance. But here again our intent is not to present a comprehensive picture or to make universal claims but to suggest approaches that might be used to test our hypotheses in communities and circumstances we do not cover.

We introduce our topic with a discussion of the significance of recent debates over nature and kinship for evaluation of ARTs. We articulate our own method and introduce the technologies that give rise to the questions. We then turn to an extensive discussion of the "natures" that ARTs enforce or transgress, paying close attention to the way in which religious belief—taking shape in a particular political, economic, and communal setting—affects these judgments about naturalness. We explore ideas of the natural family and the place of marriage and childbearing that they entail; we analyze reactions to third-party involvement in family-formation through adoption, gamete donation, and surrogacy; we discuss ideas about the nature and status of unused embryos; we examine ideas about natural gender roles and identities and the ways in which infertility and ARTs erode or preserve them; we examine the normative status of the so-called new families ARTs make possible: genetically connected gay and lesbian partnered families, and intentionally single motherhood; we analyze intercourse and conception as a single or divisible natural process; and we reflect briefly on the "naturalness" of using ARTs to select for or against disease, gender, or traits. Then we step back from these distinct but related issues to ask two larger questions. First, how does pronatalism of various kinds interact with a specific set of religious beliefs and practices in a particular cultural setting, and what are the sometimes surprising practical consequences of this interaction? Second, how do religious people incorporate the idea of ARTs into their functional religious universes, whether they ultimately accept or reject them? How must they alter their understandings of authority, power, law, divine will, or the cosmos in order to accept ARTs as natural? We close with recommendations that policymakers pay attention to religious people, and not merely to religious statements and religious leaders, if they desire an accurate understanding of religious people's responses to ARTs and also of the pressing social issues ARTs raise but that religious discussions of them often ignore.

1.1.1 *The Boundaries of "Religion"*

If "religion" is notoriously hard to define, "religious people" is nearly impossible. Within the field of religious studies, the test is merely that a person claims to be religious. However, for the purposes of this article, which is concerned with broad global trends that may affect policy and law, religious people are people who self-consciously bind themselves to one of the established, communal "great traditions" (Buddhism, Hinduism, Christianity, Judaism, or Islam). Religious people suffuse significant dimensions of human activity with religious meaning and subject important decisions about the shape of life (marriage, parenthood, etc.) to religious reflection, guided by their community's particular tradition of arraying religious authority. In addition, theistic religious people typically see divine power and intention at work in their daily activities, successes, and trials.

1.1.2 Nature's Polymorphism

Scholars have noted that questions about "nature" have two important consequences for our evaluations of biotechnology: "'nature' often continues to serve, as it did prior to the Enlightenment, as a moral touchstone" (Lock and Kaufert, 1998, 19–20), and science and religion, rather than being assigned to different domains of life and intellect, are instead "often actively [woven] together"[5] in a way that confounds Western predictions about which fertility practices will be acceptable or popular in a given culture.

We find Marilyn Strathern's work helpful for articulating and analyzing these connections from a Western point of view. Strathern's first helpful insight is that as the mystery of nature—seen in the West as the machine that humanity can neither fully understand nor overpower—shrinks in the face of science and technology, our sense of living within a vital cultural tradition inventively built within nature's limits weakens as well. In short, a weakened nature means a weaker, and not a stronger, culture (Strathern, 1992a, b). Religion—relying on both a strong sense of the natural and on a highly developed, detailed cultural tradition—would therefore be expected to loosen its grip on our moral vision. Strathern argues that meaning-making in the West is already becoming mostly a matter of self-consciously self-creative choice, a choice that—thanks to the lack of a default set of agreed communal meanings—is bleakly mandatory.[6] In such a setting, religious concerns would seem to have little influence over reproductive choices.

---

[5] Linda Layne's response to "Divine Interventions and Sacred Conceptions: Religion in the Global Practice of IVF," American Anthropological Association, November 23, 2003, 2; see also Layne (1992).

[6] For instance, it is no longer possible for a couple—or for a woman as always-and-everywhere potential mother—simply to fall into childlessness by default. One must now actively choose whether to pursue technological means of becoming a parent or mother. On this pressure, see Lauritzen (1993, 39–40); Franklin (1998, 107–113, *passim*); Handwerker (1998); and Paxson (2004, 31–32). Handwerker relates one Chinese woman's comment on Chinese reproductive policy: "The one child policy is really the 'you must have one-child policy'" (1998, 183).

The discussion that follows will contradict this conventional wisdom. We argue that, on the contrary, all over the world communal religion continues to exist in living, meaning-generating, holistic forms; in most cultures its strength never ebbed, but even where it might seem to have been in decline, it is reasserting itself. The concern to conform reproduction to religious tradition is common, and in many places ascendant. In most of the world questions of religion and nature are vital both to ART practitioners and patients and to those who opt not to perform or use ARTs. Religious people tend to believe that only a credible account of ARTs' compatibility with an upright life generally—and with a religious view of nature's relationship to the divine or to the structure of the cosmos in particular—can justify their use. Therefore the religious questions—What are the spiritual and theological meanings of infertility? Of parenthood? In what way is the divine active in daily life, and how can its presence best be acknowledged and respected? What is the divine will, or the shape of a holy path?[7] In what way does the divine or the power of the cosmos inhabit or use persons and their bodies?—are in fact paramount. Consequently, anyone who wishes to craft or study ART policy must attend carefully to religious customs and arguments.[8]

However, we are not therefore embracing the typical arguments of contemporary constructive religious bioethics. This literature tends to assume that each biotechnological development presents an entirely new, never-before-encountered threat to religious ways of thinking and behaving and therefore to religion itself. Thus ARTs are said to pose unprecedented challenges to religious concepts of nature. We agree that they pose new questions, but we contend that they do not pose new kinds of questions. Technology, kinship, and the meaning of nature are recurring challenges for religious traditions.

We argue rather that many religious traditions are resisting the diminution of both culture and nature by employing methods of reasoning and response that they have developed over centuries. Religious communities tend to naturalize technology, incorporating it pragmatically but artfully into a religious vision of the world that has a definite place for nature-as-sacred-creation. An extreme but familiar example might be the Amish. As a community they long ago rejected the use of electricity, motorized vehicles, and telephones. These are still forbidden in cases of what might be considered personal convenience, but where business with the "English" world requires it, they are acceptable. Hence electricity might be used to meet refrigeration requirements for a dairy farm but not for light or refrigeration in the home; a pay telephone might be used to transact business, but phones are not found in Amish kitchens; owning a car is "worldly," but use of a taxi service to take landless young Amish to paying jobs, or a sick person to a distant doctor or

---

[7] From a Christian or Muslim perspective, it makes sense to speak of God's will for humanity. Jews and Buddhists would not use the language of divine will. As a way of indicating the outlines of a devout or righteous life across traditions we have chosen the phrase "holy path."

[8] Language about "god" and "divinity" is inappropriate to Buddhism, which is non-theistic but nevertheless possesses a strong sense of order, function, and relationships in the universe.

hospital, is acceptable. Technology is selectively accommodated precisely for the purpose of preserving health and maintaining the community's independence and separation from the highly technological culture in which it must function. Similarly, often religious people—sometimes authoritative leaders, sometimes only laypeople—embrace ARTs specifically in order to fulfill devotional, vocational ends that are integral to their religious traditions as they understand and practice them, even when the ARTs apparently also transgress some existing religious laws or theological claims.

Thus when religious people are deciding how to react to ARTs they are not typically deciding between flouting nature and uncritically obeying it. Rather, they are drawing on traditional understandings of nature—making use of precedents and analogies that already have meaning within their traditions—in order to give meaning and place to the new technologies, to whatever degree they may embrace them. As Aditya Bharadwaj has written about Indian Hindu clinical practice, Western secular assumptions about the opposition between science and religion do not fit religious people's understanding of the world: "assisted conception conjoins seemingly disparate domains of the traditional and the modern, the sacred and the profane, the human and the superhuman, science and religion, working in tandem to produce human life" (Bharadwaj, 2006). Each acceptance of new methods (and each thoughtful rejection) is a transformation of the tradition's view of nature, inevitably, but on the tradition's own terms. There is development but rarely a sharp discontinuity.

This is the second point at which Marilyn Strathern's analysis is helpful to our argument. Strathern has described how this process of naturalization functions with regard to the category of kin relations, a process that is essential to any analysis of ARTs. She argues that views of "natural" kin relations are not static. The "traditional" Euro-American natural family (in which "real" kin relations are biogenetic) is actually a relatively recent, post-Darwinian social construction that prides itself on its basis in modern, scientific fact. ARTs simultaneously unsettle this vision of the nature of things by interfering in and complicating kin relations (which in people's minds remain dependent on genetic relationship) and reinforce the vision by enabling genetically related families to exist where formerly they could not. One consequence of the reinforcement is the normalization or naturalization of ARTs as alternate means to the accepted end, the "natural" biogenetic family. The other consequence is awareness that what seemed to have been an immutable or infallible vision of nature (biogenetic families produced through heterosexual coitus) is in fact open to revision and therefore may be an uncertain basis for "reality." This awareness in turn reduces nature's justificatory power (Strathern, 1992a, 43, 47, 52–53, 177–178, 195; 1992b, chs. 1–3; see also Franklin, 1991). Strathern sees Westerners as caught between two alternatives: abandoning the modern project of natural scientific rationality altogether for some non-scientific, immutable vision, or replacing "immutable nature" with a new gold standard, our evolving, possibly novel scientific knowledge of nature.

We accept the validity of this dilemma for some Westerners but argue that, in most circumstances worldwide, religious people regard themselves as rejecting Strathern's dilemma and taking a third road. They do not believe that new discoveries can dislodge nature, for nature itself is (depending on the tradition) either a divine creation, or at least the given system in within which all existent beings can and must find their meaning. Religious people therefore appropriate only scientific descriptions of the world that can be deployed within existing religious visions of it (often even employing scientific language to elaborate these religious visions). Similarly, they make use of ARTs only when and if they believe ARTs help them adhere to approved, existing ideas of the divinely-intended natural family.[9]

In this process of naturalization the ideas of "nature" and the "natural" family evolve somewhat even when traditional ideals are upheld. Strathern (1992b) and Charis Thompson (Thompson, 2005, ch. 5; see also Lewin, 1998, 164–177) have implied that as a result ARTs, even when employed in support of traditional family forms, inevitably subvert them, heralding the end of the "natural" Western, hegemonic, genetically related, heterosexual nuclear family and the beginning of new, if undetermined norms that are themselves biological/cultural amalgams. To some extent the debate can be framed as a matter of emphasis (is an affirmation of a traditional form that nonetheless leads to that form's transformation fundamentally conservative, subversive, or both?[10]). But for us the more important point is that all use of ARTs involves naturalizing technology by reifying unquestioned "natural" ideals and then incorporating technology within them. In Thompson's words, naturalization in this sense is

> the means by which what sometimes gets referred to as 'bedrock' is established and maintained. Its examination invites an analysis of the role of specific configurations of bedrock in establishing the moral, epistemic, and technical taken-for-granteds essential to the practice of infertility medicine (Thompson, 1998, 66).

In the West, for example, this means adhering to nature by appropriating ARTs under the banner of supporting the reigning vision of the genetically related "natural" family rather than under the banner of interfering in (disrupting) it. Indeed, Europeans and Great Britons have attempted to anticipate and cut off possibilities for radical uses of these technologies. For instance, the 1978 Warnock report's limitation of ARTs to stable, cohabiting, heterosexual couples was endorsed by liberals and conservatives alike. Research conducted in Finland reveals a sense that ARTs are justified by an inevitable, biologically-rooted human desire for genetically related children (and that adoptive parents—and by extension any parents of genetically unrelated children—are performing a risky act of supererogation).[11] For similar

---

[9] For non-theistic traditions traditions like Buddhism, use of terms like "God" and "divine" is clearly problematic. We have attempted to avoid the term "God" when speaking of religions in this broadest sense; in places we have settled on "divine" as the easiest concise indicator of a foundational "natural" reality with structure, logic, and purpose.

[10] For an example of this tension between intent and result, see Mulkay (1994, 712–713); for examples of parenthood that both conform and subvert, see Lewin (1998).

[11] In England supporters of embryo research argued "that it would help to make the conventional family more widely available." See Mulkay (1994, 708, see also 710, 711) on "normal," disease-free children as part of the new definition of the conventional family. On Finland see Malin (2002).

reasons the Greek Orthodox communion, and other communions that endorse limited use of ARTs, tend to permit homologous procedures but frown on the introduction of third-party gametes (Sheean, 2004).

We will address the religious naturalization, or "divinization," of technologies in greater detail later in this chapter. For the moment, it seems important to say that some traditions—Roman Catholicism, for instance—give explicit theological justifications for this process, arguing that reason, which is part of human nature, does and should analyze and utilize material nature for human ends. Not surprisingly, the most interesting debates within traditions erupt in the tensions between "nature" as an essence or characteristic to be realized (for instance, motherhood, seen as a dimension of each woman's potential and a goal toward which she strives) and "nature" as a pattern laid down in biological events uninterrupted by technology (for instance, "natural" heterosexual fertilization and conception of a child). Quite frequently both patients and practitioners of ARTs see themselves to be subordinating one vision of nature to the other and also feel this subordination as a loss or sacrifice that, however necessary to the pursuit of the other vision of nature, is genuinely regrettable. So, for instance, in Confucianism, accepting surgical intervention (giving up "natural" heterosexual intercourse as a path to fertilization) is justified by its potential to fulfill the "natural" goal of parenthood (Qiu, 2002, 78). Conversely, in Roman Catholicism, declining the "natural" end of parenthood is justified, however regretfully, by respect for the duty not to transgress the boundaries of the "natural" process of conception (Congregation for the Doctrine of the Faith, accessed online, 8, 11). Thus both intervention in material nature and refusal to intervene in it are naturalized by their ultimate support for regnant visions of nature (see Becker, 2000, 6; Strathern, 1992b, 56).

Yet, as we will also show, not all cultures root family in immediate genetic relationships in quite the same way Euro-Americans do (Franklin, 1997, 97–99). In other places the tension that naturalization must resolve might not be the tension between immutable biological nature (or matter) and artificial technology (or mind) but another: group identity versus foreignness; motherly selflessness versus modern self-possession; self-interested individualism versus communal concern; or the sense of God as personal companion versus God as guardian of a religious institution (Kahn, 2000, 159–160; Paxson, 2004; Lock and Kaufert, 1998; Roberts, 2006). In these situations ARTs do not necessarily symbolize an encroachment by one side of the tension on the other, and so—while they present challenges to thinking about kinship—they do not always disrupt kinship's foundations and may even reinforce them (Kahn, 2000).

Given this variety of responses, we propose that ART debates are an opportunity not only to compare the traditions' views of nature but also to examine how the traditions' strategies for dealing with technology differ. Obviously, we also reject claims that ARTs can destroy traditional religion, Western civilization, or the idea of nature in general; traditions and cultures survive precisely because they have well-honed strategies for dealing with just such challenges. Thus the question is not only how ARTs alter the *concept* of nature in each tradition, but also, what are the institutional and personal *strategies* for dealing with technological innovation in each tradition? And what do these strategies, when applied to the case of ARTs, further reveal about the *existing idea of nature* in each tradition, in each cultural context?

1.1.3 Method

In order to answer these questions satisfactorily, we have departed from the method normally used to present religious thought to policymakers. Most such writing on bioethics and religion begins with a survey of the official positions of religious groups on an issue and, possibly, of the opinions and arguments of their major exponents and critics. While this kind of investigation is certainly necessary and responsible, it has two important disadvantages. First, it unavoidably privileges traditions that organize themselves bureaucratically, develop coherent ethical systems, have mechanisms for issuing statements, and value theological and ethical unanimity highly, leaving the religious opinions of large proportions of the population unrepresented. Second, potential practitioners and users of ARTs are not by and large theologians but laypeople attempting to live their lives faithfully in an imperfect world and—perhaps even more important—in a concrete, complex cultural, social, political, and economic setting. These real lives are lived in the nexus not only of competing cultural and religious values and practical needs but, even more importantly, of religious values that are in practice in tension with each other. The methods that religious people employ in order to resolve all of these tensions are likewise complex and can be reduced neither to official statements nor to statistics about how people actually behave.

For these reasons we have chosen to follow Arthur Kleinman's lead, combining cultural anthropology and bioethics to investigate "the local moral worlds in which patients and practitioners live, worlds that involve unjust distributions of power, entitlements, and resources" (Kleinman, 1995, 48). Kleinman argues that to the degree that bioethical argument (on which many religious statements are modeled) is abstract, universal, and dependent on assumptions about relatively easy access to resources, it is a "charming romance" recommending "utopian virtues" in situations of overwhelming suffering and existential and practical conflict (Kleinman, 1998, 48). He recommends that bioethics begin not with abstract systems but with "the lived flow of interpersonal experience in an intensely particular local world," a flow that includes the commitments, uncertainties, and contradictions endemic to life in that community and time, intentionally embracing "the murky indeterminacy of real lives and the messy uncertainty of real conditions" (Kleinman, 1998, 54). Although we will frequently refer to religious bioethical statements, our intention is to alert policymakers to the ways in which such statements are actually received and employed in these local moral worlds, and to the tensions the statements attempt to resolve. Public bioethical debate tends to be either utterly realist (about the possible) or utterly idealist (about theology or philosophy, thinly veiled) but rarely combines the two in any finely-textured discussion of real moral experience.

In order fully to grasp what religious people think about ARTs and nature, then, it is essential to look carefully at how they respond to official statements in concrete circumstances or—where statements are lacking—how they generate their own working understandings and guidelines. Qualitative anthropological research is critical for unearthing these fundamentally theological moves, as it is for understanding how a person's context affects the strategies she employs in making them and the conclusions she draws from them. Our choice to combine anthropological and theological inquiry will necessarily result in some loose ends, but we hope the richness of the texture that this approach gains will make up for this unavoidable loss of tidiness.

Finally, our analysis is informed by feminist reflection on infertility and ARTs. Charis Thompson's artful review of recent feminist fertility literature points out that trends in feminism are charted perfectly in the evolution of feminist concerns over infertility. Early second-wave feminism, eager to disrupt ideologies and institutions that enforce gendered roles and identities within a system of racial and social stratification, viewed all unhappily infertile women as victims of patriarchal ideology that had brainwashed them into considering motherhood as a need. As a result, some women who desired to become mothers felt unjustly condemned. More recent feminist writing, expressing compassion for what turns out to be a particularly female heartache that transcends class, has embraced ARTs as instruments of reproductive choice but has also lobbied for policy justice and procedural safety and against exploitation of women surrogates and donors (Thompson, 2002; see also Thompson, 2005, ch. 2). We hold the latter position but will argue that, worldwide, patriarchal practices and racial and social stratification converge to produce a situation in which most women's decisions about reproduction are still far from free. Patriarchal cultures outside the United States receive somewhat more attention than American patriarchy, but this is because less familiar cultural and religious settings are often fresher—and therefore more instructive—examples of the convergence of multiple factors to create patriarchal social structures. In addition, the inordinate focus on women and infertility has obscured the personal heartache and social stigma of male infertility, a problem whose significance and resolution are just beginning to receive attention (see for example Inhorn, 2002, 2003b, 2004, 2006a, 2007). It is not possible to explore all dimensions of these questions of gender, race, and class explicitly in the argument that follows, but they inform our thinking at every point.

1.2 Points of Ambiguity and Resolution

1.2.1 Themes and Variations

Lurking among the social and religious factors that influence people's thinking about ARTs are several that cannot be reduced to the standard questions but have a great deal to do with how the standard questions are resolved. These factors are strongly connected with the practical exigencies of daily life in a region and with ways in which practical structures and decisions of life are interpreted. We shall reflect on these more systematically later in this essay, but they should be kept in mind throughout the discussion.

- Explicit or implicit pronatalism at any of several levels (governments, religious institutions, cultural traditions, economies, and local communities) can strongly influence views of ARTs, as can an atmosphere of anti-natalism, delayed child-bearing, or strict family limitation. Messages that conflict at various levels produce especially interesting cases.
- The methods by which people navigate "upstream" against their traditions, or deal with genuinely conflicting strands within traditions, vary greatly. They depend in turn upon the locus of religious and moral authority (priest or rabbi? text? medical practitioner?) and upon the means that religious groups use to determine which of a number of competing values "trumps" the others.
- In discussions of kinship it matters whether nature is seen as identical to society (elevating adherence to social structures to a direct religious obligation), as society's foundation (providing a critical standard according to which social structures can be judged and reformed), or as society's antithesis (representing a condition of chaos, disorganization, or sin that society must control or overcome).
- The degree to which the divine will or cosmic structure is understood to be evident in universal physical structures, traditional social arrangements, technology, religious leaders, and portentous events in individuals' lives is of profound significance for their vision of religious life.
- A community's eagerness or reluctance to use "nature" as an interpretive category is telling but not determinative. Reluctance to use it is sometimes an intentional resistance to other, apparently hegemonic religious or to secular discourses that employ "nature" as a source of moral authority. Acceptance might imply any of several related themes: for example, nature as well-designed physical creation, perhaps left to its own devices, or perhaps inhabited by a power that occasionally gives it a little boost; and nature as the divinely intended character, purpose, and end of a thing. In either case, human reason may have a role to play in assisting the existing systems to function or in improving realization of existing ends.

In some cases these variations are determined by official theology or law; but whether they possess this kind of authority or not, they have immediate authority also at the level of ethos, of customary assumptions and behaviors. This latter, more immediate authority can be more compelling.

1.2.2 The Technologies

Put simply, any conception that does not follow from heterosexual intercourse, unassisted and uninterrupted by third parties, counts as assisted reproduction. In its most basic and unproblematic form, assisted reproduction involves optimizing the chances of conception by drawing upon increasingly sophisticated knowledge of the human reproductive system to adjust the timing and conditions of intercourse, possibly employing fertility-enhancing drugs. Conception under these circumstances is by all accounts still considered "natural," for no external process alters the structure of heterosexual intercourse or guides the paths of the gametes or the process of fertilization or implantation. This is instructive, as it implies a nearly unanimous, lowest common denominator description of "natural" reproduction. Controversy arises over the following progressively more complex interventions:

- **Artificial insemination (AI)**: In artificial insemination, sperm are introduced into a woman's uterus during the fertile part of her menstrual cycle. In some cases the sperm are collected from the woman's husband or male partner; he may ejaculate into a container after masturbation or into a perforated condom during intercourse. In this case the procedure is known as **artificial insemination-homologous (sometimes "husband," AIH)**. If a woman has no male partner, or her partner lacks adequate sperm, donor (or heterologous) sperm may be used; in this case the procedure is labeled **artificial insemination-donor (AID)** or **donor insemination (DI)**. Especially among lesbians AI has often been performed at home, with anonymous or known donor sperm, but concerns about disease and paternity rights have led more and more women to pursue AI in clinics under medical supervision.
- **Gamete intrafallopian transfer (GIFT)**: A woman with intact fallopian tubes takes drugs to induce superovulation; a doctor removes ova, mixes them with ejaculated homologous or donor sperm, and inserts the gametes into her fallopian tubes, the normal environment for fertilization. In a variation, **zygote intrafallopian transfer (ZIFT)**: the doctor transfers already-fertilized ova into the fallopian tubes. In either case, zygotes may develop into embryos, descend into the uterus, and implant there.
- **In vitro fertilization (IVF)**: Commonly known as the "test-tube baby" procedure, in vitro fertilization is particularly helpful for women with blocked fallopian tubes. It too begins with a drug regimen to induce superovulation; in this case, however, the harvested ova are mixed with homologous or donor sperm and cultured for 48–72 hours. At the end of this period, the physician removes a small number of embryos from the culture dish and transfers them to the woman's uterus through a catheter, where one or several may implant and grow normally. Unused viable embryos may be frozen for future use or destroyed, and unused ova may be donated or discarded. This procedure also allows for the use of **donor ova** and/or **surrogate mothers**, since fertilization occurs in the laboratory.
- **Surrogacy**: In some countries, when no viable womb is available—either when a woman cannot carry a child, or when a single man or gay couple wishes to conceive—aspiring parents may contract with fertile women to undergo IVF and carry a pregnancy to term for them.
- **Intracytoplasmic sperm injection (ICSI)**: In this procedure designed to overcome very low sperm count or low sperm motility, a doctor removes ova from a

woman just as in other procedures, but also—often surgically—removes sperm from the man. Single sperm are then injected into single ova, forcing fertilization of particular ova by particular sperm. After 48–72 hours, a small number of embryos are transferred to the woman's uterus through a catheter. As in standard IVF, donor ova may be used, although donor sperm rarely are, as the procedure was developed to facilitate fertilization by nearly-infertile men.

- **Ooplasm donation**: Designed to overcome the weaknesses of "elderly" eggs, this technique involves introducing donor ooplasm (which contains another, younger woman's mitochondrial genetic material) into an older woman's ovum in order to facilitate fertilization; then ICSI is performed with the older woman's husband's (or possibly even a donor's) sperm. The resulting embryo then contains genetic material from three gametes, although it contains nuclear material from only two.[12]
- **Microsorting**: Applicable to any form of ART, microsorting sperm by weight allows sperm to be separated by sex before fertilization, either to avoid sex-linked diseases or simply to increase the chances that a child will be of the desired sex.
- **Preimplantation genetic diagnosis (PGD)**: This procedure may be used with any form of IVF. Single cells are removed from eight-cell zygotes for genetic testing. Zygotes that match desired traits (preferred gender or freedom from genetic disease) are selectively implanted in the woman's uterus.

Although each technique invites its own questions, IVF and its variants, as we will see below, raise issues that cover nearly the whole range of concerns about "nature" and "the natural." For this reason the discussion will concentrate primarily on IVF, with the hope that implications for other procedures will be clear.

1.2.3 Processes and Nature: The Sticking Points

In work with American Jewish, Christian, and Muslim thought, Baruch Brody has noted that the following areas of concern, rather than the traditional bioethical categories of harms, rights, or injustices, preoccupy religious leaders:

1. The new reproductive techniques disrupt the connection between unitive conjugal intimacy and procreative potential that is required by morality.
2. The new reproductive techniques often introduce third parties into the process of reproduction, and this is morally illicit.
3. The new reproductive techniques often result in a morally illicit confusion of lineage, since children are often unaware of their biological parents.
4. Some new reproductive techniques (IVF) often involve a failure to implant fertilized eggs. This is a form of early abortion and is therefore morally illicit.

---

[12] According to Charis Thompson this procedure was forbidden in the United States in 2001 but nevertheless has "wide clinical application" here (Thompson, 2005, 272–273).

5. The new reproductive technologies often involve a dehumanization of the reproductive process and are therefore morally illicit.
6. Some new reproductive techniques (especially surrogacy) involve commercialization and exploitation that makes them morally illicit (Brody, 1990, 46–47).[13]

As we show below, questions about ARTs and nature closely parallel these concerns. In addition, at each step the moral objections often amount to a judgment about "unnaturalness." However, particular religious cultures often choose to question the "naturalness" of very different reproductive techniques and processes or have very different reasons for finding the same procedures problematic.

1.2.3.1 The Natural Family

A significant proportion of the questions that ARTs raise has to do with the shape of the "natural family": What are its limits? To what degree can they be stretched, and for what purposes? Significantly, in our judgment ARTs typically are employed in the service of (and extension of) existing notions and ideals of family rather than in order to replace them (Inhorn, 2003c, 120; Becker, 2000, 35; Khanna, 1999). A familiar American example of extension is lesbian and gay couples' use of ARTs to raise biologically related children in households that closely resemble those of "normal" heterosexual married couples. In India and China ARTs may even be providing unprecedented means to transform a preference into a near mandate. Before, bearing sons rather than daughters either was mainly a matter of luck or required willingness to commit infanticide; now a combination of diagnostic testing, abortion, and ARTs is being used to support the traditional preference for male children. Because failure to produce male children often impairs women's status and economic stability, and in patrilocal rural societies can spell poverty for the entire household, women make use of these services in large numbers, however reluctantly.[14]

---

[13] At times the language of harm, right, or injustice is used, but it often follows on judgments about nature, so that harm or injustice is defined as a transgression of "the way things are meant to work."

[14] On China, see Handwerker (2002). On India, see Sen (1990). Emily Oster has suggested that at least part of the "missing women" phenomenon may be attributable to hepatitis B infections in mothers, which significantly increase the chance of giving birth to a boy (Dubner and Leavitt, 2005). In India diagnostic techniques are being used to abort girls in huge numbers and also to exploit women's fear of carrying a girl, sometimes resulting in abortions of boys by unscrupulous clinics out to make a profit. These abortions also tend to lead to reproductive tract infections because diagnostic and surgical procedures by clinics vying to stay low-cost and profitable are often unsanitary or otherwise sloppy; infection in turn raises the infertility rate. See Khanna (1999). Note that religion is often an argument used by unscrupulous, often unqualified clinic staff to reinforce a woman's/couple's desire for a boy. Khanna implies that ARTs may be fanning an existing Indian preference for boys into a frenzy, therefore magnifying the tendency by providing a way to act on it. See also Meseguer et al. (2002); Schenker (2002); Allahbadia (2002); Mori and Watanabe (2002); Chan et al. (2002); Hanson et al. (2002); Sills and Palermo (2002); Renteln (1992).

Other questions abound. If the family's given nature includes transfer of wealth, education, status, and other advantages according to a strict system, then transgression of that system becomes a matter of justice not just to one family's children but to divine plans for human society broadly. On this view ARTs could, if not carefully regulated, cause suffering, confusion, and social collapse. In addition, in some accounts there is an assumption that blood relations "naturally" produce more constant, perceptive parents than chosen relations, throwing donor procedures into doubt.

There is also the question of genetics and identity: do genes dictate the moment of human existence, i.e., the moment after which a zygote is an existentially privileged being and not merely a [collection of] cell[s]? Do genes dictate the character of relations between persons legally, instinctively, emotionally? The key test is adoption: Is adoption at root a second-best way to make up for parentlessness and childlessness, and thus to complete the true natures of adults and children whom chance has deprived of a family? An act of benevolence rooted more in compassion and a sense of justice than in the desire to reproduce the model of the natural family? Or a recognition that commitment and social practice, not physical relatedness, bond parents and children, no matter what their genetic connection? ART arguments can follow adoption arguments unless another criterion is determinative in the speaker's vision of "nature."

Far more commonly, the question is simply whether ARTs realize or transgress the structure and intent of what religious people typically hold to be the "natural" standard, heterosexual, married procreation. For married Jews, the biblical command to be fruitful and multiply falls—depending on their affiliation within Judaism and their cultural setting—somewhere on the spectrum between mandate and permission.[15] Thus seeking assistance for infertility falls along the same spectrum. For most Muslims and Hindus, similar convictions hold. For them as for Jews the question is, when does the treatment transgress the boundaries of the marriage or the divine initiative? For Protestant and Roman Catholic Christian marriages, surprisingly, children are hoped for but never mandated theologically; one crucial indication of this is that childlessness is not considered proper grounds for divorce.[16] As a result some

---

[15] In some opinions the mandate, which is directed to men only, is a command to continue to engage in procreative marital intercourse until the requisite number of children has been born. In this view a man can fulfill it simply by having sex with his wife during her fertile periods, whether or not children result; conception is the goal but is not the measure of obedience. See Bleich (1997a, 147–148).

[16] Paxson reports that Greek Orthodoxy has long permitted divorce for childless couples on the grounds that (1) parenthood is a divinely ordained vocation of marriage; (2) children are born (or not) at God's will; and (3) it is parenthood, and not merely the marriage liturgy, that cements the permanent bond of marriage. However, this opinion may reflect Greek national pronatalism. One highly revered early Orthodox theologian argued that because immortality is assured, couples need not reproduce in order ensure their own vicarious presence on earth; in marriage it is enough to provide each other spiritual and emotional support and sexual comfort (St. John Chrysostom, 1997, 85–86).

Christian communions (the Roman Catholic in particular) allow little intervention for the infertile. A related distinction may be made between traditions that begin by asking, "What is the (natural) ideal or proper setting for reproduction?" and those that begin with the query, "Who is (naturally) obligated or permitted to reproduce?" In the first case, the originating concern is how and where children can best be born and raised; in the second case, the originating concern is defining and assisting adults in accomplishing their religious vocations. In both cases the originating concern eventually includes the secondary, but it can make a difference where one begins, especially when one contemplates variety in family structure.

The picture becomes even more complex because the cultural and devotional traditions in many places either make childbearing nearly mandatory or declare sex and reproduction to be purely a matter of parental choice; either approach can soften *de facto* or even *de jure* objections to ARTs.

From a Western viewpoint, Buddhism presents perhaps the most intriguing example of "natural" family structure and reproduction. Because all attachments to temporal things—sexual pleasure, child, spouse, or even self—discourage enlightenment, a member of the *sangha* would neither marry nor desire to have children. For spiritually serious people, sex and reproduction are thus suspect on several counts. First, and most obviously, sexual desire and the desire for progeny are particularly vivid instances of attachment and desire and so are unhelpful to one's own spiritual progress.[17] Second, birth—for which sex is a precondition—is ambivalent because, although it provides imperfect souls life (a genuine good) and new opportunities to attain enlightenment, it also reintroduces them to the world of contingency and suffering.[18] In addition, as infertility is thought to be a consequence of sins committed in one's last life,[19] there may be a sense that working to overcome infertility defies spiritual laws in a way that is ultimately injurious, a belief that in turn creates psychological and spiritual barriers to use of ARTs; this reluctance reflects a larger Buddhist predisposition to work with rather than upon or against nature, bolstering an attitude of non-intervention.[20] Finally, the uncritical desire to reproduce in the face of ecosystematic strain contradicts Buddhist responsibility to promote life in its interdependency (Gross, 1997). This complete absence of theological pronatalism throws doubts on the theory that religious people automatically see the "natural" family as a good to be pursued.

Yet Buddhism does not utterly reject worldly life. Buddhist belief in reincarnation assumes that only the purest souls are cut out for the ascetic life, and for this reason although the lay life, marriage, and parenthood fall short of the ideal, they also are not frowned upon. And even the enlightened are believed to be able to

---

[17] Keown (1995, 67, 126); Keown (2005, 55–58); Williams (1991, 101–102); Satha-Anand (2001); Gross (1997, 302–304); Qiu (2002, 75); Fagley (1965, 333–334).

[18] Keown (1995, 66–67, 128); Keown (2005, 54).

[19] Qiu (2002, 77).

[20] Qiu (2002, 78); Lock (1998, 214). But note that compassion, which involves an obligation to reduce suffering, justifies many cases of medical intervention. For exceptions see Keown (1995, 58–64).

participate fully in the life of sense and relationship without spiritual danger; their sexual fluids may continue to flow even in the absence of a couple relationship, attachment, or desire.[21] Still, Buddhism's failure to prescribe marriage or enjoin childbearing means that only local custom, with its secondary devotional and moral traditions, can create a religious, moral impetus for marriage and a strong connection between marriage and childbearing.

The tradition to which East Asian Buddhists most often look for such moral guidance is Confucianism, which presents an equally ambiguous picture. According to Ren Zong Qiu, the Confucian emphasis on family and communal identity yields a variety of filial piety that commands the continuation of the male ancestral line. The purpose of sex, then, is primarily reproduction; health (preservation of the yin-yang balance in both men and women) and pleasure are secondary and tertiary goods (Qiu, 2002, 76, 78). In addition, patriarchal complementarity rules gender relations; as infertility has traditionally been seen as a woman's problem, infertile wives are, not surprisingly, at risk for divorce. These problems could be alleviated quietly by ARTs. However, Confucianism also erects two significant barriers to ARTs: concern about introduction of extra-familial gametes (this, if the family learns about it, can raise conflicts out of concern for the ancestral line) and the belief that one should avoid donating sperm—both because a man should not harm himself by wasting the vital forces in his sperm, and because sperm stored outside a man's body may lose its vital force and produce a weak or defective child (Qiu, 2002, 78–79). Even disturbing the natural process of conception in general (e.g., AIH) is suspect. But as Qiu succinctly concludes, these suspicions are surmountable: according to Confucianism "disturbing the *dao* of nature is more acceptable than being without an offspring" (Qiu, 2002, 78).

Political pressures also temper Buddhist traditions. In China, many moral arguments about ART implementation center on the future health of the population and the right of parents to reproduce ("Should Surrogate Motherhood Be Banned?", 2001, 28–29); for example, the one-child policy increases parents' desires for a single, healthy, male child, creating a market for methods that increase parents' chances of this outcome.

In the face of this complexity, we might best envision a spectrum of positions, one end of which is held down by Orthodox Judaism, Indian Hinduism, and Middle Eastern Islam, in which marriage and procreation of genetically related children is nearly universal and practically mandatory (for related but not identical reasons), and the other end of which is held down by the less syncretistic strands of the Buddhist tradition, which recommends that childbearing occur within marriage but is not particularly eager to promote either marriage or procreation for its most serious practitioners. Somewhere in the middle might fall Athenian Greek Orthodoxy, most Protestant denominations, and more liberal Jews. Members of many religious communions—notably the Roman Catholic—in fact span the whole spectrum despite being united

---

[21] Satha-Anand (2001, 120); westerners will find a parallel in Augustine's vision of sex in paradise, not in Plato's account of Socrates' "take it or leave it" attitude toward sexual relationships.

under an apparently univocal official policy, making it important to examine relationships of local belief and practice to religious authority very closely. Although all these traditions agree that moral goodness entails following nature in some sense, they do not agree on whether "nature" mandates, permits, or only tolerates marriage and childbearing. One can imagine a great willingness to use ARTs to fulfill the narrow requirements of family structure at one end of the range and a great degree of flexibility in family configuration, either with or without acceptance of ARTs, at the other.

Some examples may help to illustrate this range. In the Middle Eastern Muslim world, marriage is highly valued, and in most Middle Eastern countries nearly all adults marry if they can. Middle Eastern societies are also pronatalist: they highly value children for numerous reasons and expect all marriages to produce them. Children are usually desired from the beginning of marriage and loved and cherished once they are born. The notion of a married couple living happily without children is unthinkable. In fact Islamic personal status laws consider a wife's barrenness to be a major ground for divorce. As a result, childless couples are often under tremendous social pressure to conceive, and infertile women in particular often live in fear that their marriages will "collapse."[22] Put over-simply, all normal adults marry, in part in order to have children. In this setting, the bond of the marriage actually depends—*de facto* and *de jure*—on fertility; the "natural" act of reproducing legitimizes or ratifies the "natural" union of marriage, and not the reverse. And, as we will see, kinship is biological-genetic; it is normally believed that children must be the direct biological descendants of both parents. These patterns are cemented by Islamic inheritance laws, which stipulate that children can inherit only from their biological (interpreted as genetic) parents. A child of uncertain parentage lacks inheritance rights and therefore the social position from which to launch his or her own marriage and life. Consequently use of homologous ARTs is nearly mandatory for Middle Eastern Muslim women and men in infertile marriages.[23]

Indian Hindu culture is similarly concerned with marriage and procreation, but—as Cromwell Crawford and others point out—for slightly different reasons. First, the āśramas, or traditional ideal stages of life for Hindu men, call for a period of single studenthood followed by marriage and parenting, forest dwelling, and hermitic life. Married parenthood has traditionally been seen as the superior stage, for it not only fulfills a social function but also creates a school of virtue for adults (Crawford, 2003, 16, 116–119). In addition, it produces sons who can carry on the family responsibility to venerate ancestors, delivering their father from hell: "The sonless man was born to no end, and he who does not propagate himself is godless (*adharmika*); for to carry on the blood is the highest duty and virtue" (Crawford,

---

[22] Although Islam also allows women to divorce if male infertility can be proven, a woman's initiation of divorce continues to be so stigmatizing that women rarely choose this option unless their marriages are truly unbearable.

[23] Sunni communities (the dominant form of Islam, comprising 80–90% of the world's 1.3 billion Muslims (Inhorn, 2006a)) typically permit only AIH and homologous IVF, for all the reasons outlined above. Iranian and Lebanese Shi'a communities are beginning to permit donor procedures under particular conditions; see below.

2003, 116; see also Bharadwaj, 2003, 1869–1870). Finally, it creates an opportunity for souls to reenter the world through reincarnation (Crawford, 2003, 123). Crawford notes that, thanks to its emphasis on begetting, Hinduism weights preserving nature-as-end (procreation) more heavily than preserving nature-as-means (uninterrupted marital intercourse): "It is this end that makes *vivaha* [marriage] obligatory and justifies all *emergency means* to attain that end, including means that would ordinarily be deemed immoral. The morality of such situations resides in the *intention* and *outcome* of the act (progeny), and not in the act (intercourse) itself" (Crawford, 2003, 119). Thus in ancient times *niyoga* (in which a widow or the wife of an infertile man was impregnated by her husband's male relative), and in contemporary times limited use of ARTs, seem less distasteful to most Hindus than the prospect of childlessness (Crawford, 2003, 116–119; see also discussion below).

It is worth noting that in both Arab Muslim and Indian Hindu culture, substantial proportions of couples of child-bearing age (over 60% of infertile couples in one Indian study, for example) live in extended family households (Bharadwaj, 2003, 1869). There is substantial anecdotal evidence that this living situation increases pressures on couples to reproduce and to pursue fertility therapy and (in patrilocal settings) may also make the wife more vulnerable to her husband's family's disfavor and therefore to divorce (Inhorn, 1996). It also makes surreptitious use of third-party gametes or adoption more difficult.

Nearer the center of the spectrum lie official Roman Catholic Church teachings, which in many respects echo Islamic and Hindu positions on the "natural" family but reach somewhat different conclusions. For, in a tradition in which singleness and vowed celibacy have long been considered valid—even superior—vocations, neither marriage nor childbearing is a prerequisite to a faithful adult life, nor is childbearing in theory necessary to a legitimate marriage (though openness to childbearing is).

Roman Catholic statements on assisted reproductive technologies assume a natural law ethical framework. Their explicit appeals to nature function normatively; they are not simply observations about human reproductive experience, but also conclusions about what is inherently important or valuable in human reproduction. In its comprehensive 1987 letter *The Instruction on Respect for Human Life in Its Origin and on The Dignity of Procreation* (*Donum Vitae*), the Vatican Congregation for the Doctrine of the Faith invokes a "natural language of the body," which sets the parameters for legitimate reproduction and gives a rationale for the Catholic Church's insistence on the inseparability of sex, reproduction and marriage. Because the most basic ethical question posed by assisted reproduction is whether conception must occur within sexual intercourse, *Donum Vitae*'s analysis of various ARTs begins with an assertion about the meaning of human reproduction: "'The transmission of human life is entrusted by nature to a personal and conscious act and as such is subject to the all-holy laws of God: immutable and inviolable laws which must be recognized and observed.'"[24]

---

[24] Congregation for the Doctrine of the Faith, p. 3, quoting Pope John XXIII (1961, 447).

Several intersecting appeals to nature are at work in *Donum Vitae*'s description of sexual reproduction as a human act. First, there is an appeal to the nature of human persons as rational subjects, able and obligated to participate intentionally in God's creative activity through the act of transmitting human life. In addition, there is an appeal to the nature of sexual intercourse as generative as well as capable of expressing intimacy. It refers to sexual intercourse as "most closely uniting husband and wife [and capacitating] them for the generation of new lives, *according to laws inscribed in the very being of man and of woman*" (Congregation for the Doctrine of the Faith, 1987, 9; reference to Pope Paul VI, 1968, 488–489; emphasis added). Thus the physical union of sexual partners has a "natural generosity" that is normative and material and can never legitimately be subverted. Finally, marriage (assumed to be the permanent union of a man and a woman) is said to be "ordered by its nature" to a dual end: the union of the spouses and the generation of offspring (Congregation for the Doctrine of the Faith, 1987, 9). Here, then, is the argument for the "natural" family. These various appeals come together in the Catholic Church's opposition to barrier, surgical, or pharmacological forms of contraception and any form of assisted reproduction that "replaces the conjugal act": "In order to respect the language of their bodies and their natural generosity, the conjugal union must take place with respect for its openness to procreation; and the procreation of a person must be the fruit and result of married love" (Congregation for the Doctrine of the Faith, 9).[25]

Donum Vitae's objections to interventions such as IVF do not turn on concerns that medical assistance renders reproduction *artificial*. The Roman Catholic tradition in general welcomes developments in science and technology as potentially valuable means for human beings to participate in divine activity in the world. Rather, constraints on reproduction take the form of "objective and inalienable properties" of human beings and of sexual intercourse and marriage. Thus, to alter or violate embodied nature in the context of the Roman Catholic moral tradition is to fail to attend to the way things are and ought to be, that is, to the distinctive character of reproduction as a human act set, by God's design, within the enduring and intimate marital partnership. Roman Catholic moral theology founds procreation on marital sexual union, implying that all people are to limit sexual gratification to marital intercourse, that married couples are to be spiritually and emotionally open to parenthood in general, and that they are to leave each act of intercourse physically open to procreation. Conversely, they are forbidden to conceive other than by marital intercourse, although they are welcome to create or expand their families through adoption. Thus the vision of the ideal family is identical to the Sunni Muslim ideal, but the theology behind it both weakens the mandate to procreate and elevates the link between sexual relations and procreation, making it difficult to justify use of ARTs. In practice, however, many infertile Roman Catholic couples depart from this reasoning; valuing procreation and parenting more highly than avoidance of artificial means, they quietly pursue ARTs.

---

[25] Congregation for the Doctrine of the Faith, 9. The possible exception is GIFT performed after sexual relations in which a perforated condom has been used. On the arguments of the foregoing two paragraphs, see also Lauritzen (1993, 6).

The Greek Orthodox Church, part of the worldwide Orthodox Christian communion, proceeds on similar premises but accepts limited use of ARTs because its understanding of the relationship between intercourse and procreation diverges from the Roman Catholic view at two key points. The Greek Orthodox Church recognizes four purposes of marriage: (1) the birth and care of children; (2) the mutual aid of the couple; (3) satisfaction of the sexual drive; and (4) growth in mutuality and oneness, i.e., love (Harakas, 1980). According to the Orthodox Church in America, although "the procreation of children in marriage is...a blessing of God...[and] the natural result of the act of sexual intercourse in marriage," sexual relations and conception may be separated: "married couples may use medical means to enhance conception of their common children" if they use their own gametes and avoid procedures that result in the destruction of embryos (Orthodox Church in America, 1992, confirmed in Sheean, 2004). Children resulting from AIH are still products of the "one flesh" joined by God through marriage. The concern with AID and one of the concerns with surrogacy are not lineage (adopted children are welcomed) but the adulterous involvement of a third party in what is normally an intimate marital act. Furthermore—again in departure from the Roman Catholic view—the prohibition against masturbation can be lifted in the interest of obtaining a semen sample for AIH. A forbidden "unnatural" act becomes tolerated in the greater interest of procreation, a "natural" desirable outcome of marriage. This reasoning is consistent with the Greek Orthodox Church's principle of *ekonomia*, which accepts a minor wrongdoing committed in the process of realizing God's plan for salvation of the world.[26]

Toward the other end of the spectrum we find, among others, most Protestants. Here the question of "nature" might be regarded as a concern not about inheritance or genetic connection but discerning the criteria that make the "natural" family a typically superior environment for raising children and protecting the dignity of children who are conceived. First, although when Protestant denominations make statements on assisted reproduction they rarely treat the connections between genetic relation and parenthood as absolute values, marriage typically functions as the "fitting" or "faithful" setting for procreation and thus as the ideal against which alternatives are to be judged. Second, Protestant denominational statements on assisted reproduction sometimes invoke the distinction popularized by Gilbert Meilaender between "begetting" and "making" a child (Meilaender, 1997; see also O'Donovan, 1984). Drawn from creedal descriptions of the relationship between the aspects of God as Creator and as Redeemer, the image of the child as "begotten" suggests divine agency: "When a new human life—which can be regarded as the 'created word' of the living God—comes to be through the marital act it is in truth 'begotten, not made,' just as God's eternal and uncreated Word, who became man (a 'created word') for our sake...'begotten, not made'" (May, online). "Made," in contrast, suggests the relationship of a human agent to a product, i.e., a relationship of alienation, subordination, and instrumentalization. The extreme, cloning, is often

---

[26] On *ekonomia* see for example Paxson (2004, 61).

presented as the culmination of the technological transformation of procreation into reproduction (of begetting into making): here natural humanity itself is treated as a raw material for "constructing a form of life that is *not* natural humanity but is an artificial development *out of* humanity" (Campbell, 1997, D13, quoting O'Donovan, 1984, 16; italics in original).

The distinction between "begetting" and "making" provides a helpful lens for describing uses of "nature" in Protestant analyses of reproductive technologies. Because Protestant ethics looks to the Bible for moral guidance, we do not often find appeals to normative patterns of existence or "the laws of nature." Indeed, "nature" is frequently used to describe the state of humankind as "unredeemed," analogous to "the flesh" as contrasted with "the Spirit." It is God's word, not nature, that establishes the boundaries for the exercise of human freedom. Thus, the moral task is to discern what is fitting with respect to God's intentions for procreation as known through Scripture. In the hands of conservative interpreters, this yields an interpretation very similar to the Roman Catholic one: Hebrew Bible dicta and models of family life are taken as the permanent norms, modified by New Testament mandates for monogamous marriage.[27] In the hands of more liberal interpreters, the rationales for biblical models of family life understood within their historical and cultural contexts carry more weight than the models themselves. So, for instance, a fertile heterosexual marriage might be seen as "naturally ideal" not in itself but *because* it combines factors ideal for a non-instrumental view of children and for children's flourishing: loving sexual reproduction, the emotional and financial security of a two-adult household, the intimate presence of both male and female adults, the long-term commitment of marriage, and religious, social, and political approbation (Waters, 2001). From this perspective fulfilling as many of these criteria as possible, and not simply living within the institution that usually produces them, is what makes a household worthy of children. Contemporary Greek Orthodox women often employ a similar criterion: do ARTs like IVF enable "proper parenting" (Paxson, 2004)? These kinds of reasoning can open the door to carefully considered ART use by married couples, intentionally single parenthood, and parenthood by committed gay and lesbian partners.

These examples show that although the vision of the "natural" family—including fertile, married heterosexual parents and their children—may be virtually universal, its significance for the "naturalness" of ARTs cannot be understood except in the context of a particular religious tradition as it unfolds in a concrete cultural setting. Indeed, as we have seen with Buddhism and Roman Catholicism, the "natural" family may not even be seen as the only or the ideal context for a holy life.

Finally, it is important to recognize that accepting infertility, either after or without infertility treatment, can lead couples though a crisis of identity to a change in their own religious understandings of "natural" marriage and family. Maura Ryan notes that if they are to lead healthy, well-rounded lives, infertile American Roman

---

[27] For a useful recent statement, see "The Natural Family: A Manifesto," available at http://www.sutherlandinstitute.org/images/FamilyManifesto.pdf (accessed July 6, 2006).

Catholic couples must divest themselves of a culturally Catholic expectation (a marriage must produce children and is meaningless without them) and adopt something closer to the official Church teaching on family (a married couple must be open to children but must practice hospitality and generosity in other ways if they are not forthcoming) (Ryan, 2001, 58–60). Marcia Inhorn reports that in Egypt and Lebanon, Muslim men in long infertile marriages tend not to avail themselves of their legal opportunities to divorce and remarry or to take second wives; implicitly, these men and their wives give up the traditional Muslim ideal of a procreative union for a companionate vision of marriage in which their intimacy, love, and fidelity alone, not offspring, justify their continued union (Inhorn, 2006a).

1.2.3.2 Third-Party Donation

As the foregoing discussion suggests, different religious and cultural groups regard various moments and processes as normatively natural, leading to divergent evaluations of their acceptability. One telling sticking point is third-party donation. For example, when they have addressed the question of assisted reproduction, both mainline and conservative Protestant communities have in general found separating procreation from sexual union acceptable *for the sake of marital reproduction*.[28] Likewise, on criteria of promoting life and providing opportunities for rebirth (after all, no child can be born unless there is a soul awaiting rebirth), Keown surmises that AIH might be acceptable to Buddhists.[29] Thus interventions such as artificial insemination and in vitro fertilization, which use a married couple's own gametes, are more widely acceptable. Donor methods, which risk severing what is often seen as a divinely-willed connection between a married couple and their offspring, are more controversial.

Even if the concern is unanimous, the reasons given are varied, revealing further differences in groups' varied understandings of nature. For some the question of nature and third-party donation is purely and simply a question of *physical genetic identity*: have both parents put their genetic stamp upon the child?[30] If not, the connection between the non-genetically-contributing parent and the child is compromised, even if (for example) a woman who conceives with donor ova then gestates and raises the child. In some cases the argument is that non-genetically-related parents cannot feel adequately devoted to children to form proper bonds with them; in other cases it is that the child's behavioral, emotional, and intellectual patterns will be hard for its parent to comprehend and support; in still others it is the idea that

---

[28] American Protestant denominations holding this position include, among others, the United Methodist Church, the Evangelical Lutheran Church in America, the Episcopal Church, the Presbyterian Church (U.S.A.) and the United Church of Christ.

[29] Keown (1995, 135–136).

[30] Marilyn Strathern calls this the replacement of social relationships with biological or genetic relationships. See Strathern (1992a, 52–53, 178).

immortality is not eternal life but self-perpetuation through genetic temporal reproduction (see for example Lock and Kaufert, 1998; Kass, 2002, 153–157). Interestingly, although some religious communities place great value on homologous fertilization, the people most likely to phrase the question of third-party donation in these ways are biological determinists. Max Charlesworth has pointed out that such extreme reductionism is unsupported by experience and ultimately breaks down logically (Charlesworth, 1990). Yet a similar reductionism plays a secondary role in many religious distinctions between homologous methods and third-party donation: "It does not feel 'natural' because it is not 'mine.'" Usually, however, the religious acceptability of third-party donation has more to do with religious ideals of family structure and of the connection between marriage and procreation than with simple material, genetic identity.

The more common religious objections to third-party gamete donation for married couples interestingly have little to do with biological provenance.[31] This is due in part to wide acceptance of adoption (a discussion of the Muslim and Hindu exceptions follows below): a child whose planning, conception, and birth were carried out by one set of parents is then taken by unrelated parents to raise as their own—presumably with the hope of bettering the child's situation.[32] As a result, a devotion to the idea that biological relatedness is essential to "real" parenthood cannot be the whole source of discomfort. Instead, religious objections are due in part to the belief that the use of donor gametes differs from adoption in an important way: it introduces a third (and perhaps even fourth) party into the *intentional process of fertilization and conception*, constituting a form of adultery. Even within this view there are variations. Some Roman Catholics, Greek Orthodox Christians, and Anglicans see adultery as a breach of marital fidelity: an emotional and spiritual wedge between husband and wife both symbolized by and enacted in sexual infidelity. They argue that introducing a third party into the process, as well as replacing loving intercourse with masturbation and surgical procedures, will necessarily or likely (depending on the strength of the objection) erode marital unity in the same way that an affair would (Brody, 1990, 54–56). Some of these thinkers accept some homologous procedures, arguing that they give the same result as naturally occurring sexual intercourse without interfering with the "sacred union through

---

[31] Clearly gay and lesbian couples must accept third party donation.

[32] Paul Lauritzen points out that contemporary adoption is increasingly (among other things) a consumer industry driven by the demand for healthy, white newborns and dominated by private agencies. As a result the distinction between surrogacy as a commercial, consumer transaction and adoption as an act of benevolence toward the child is eroding. See Lauritzen (1993, 119–134). Maura Ryan adds that unhappily infertile couples may make poor adoptive parents and that suggesting adoption as the "cure" for infertility not only shifts the burden for the care of needy children to the involuntarily childless but distracts our attention from the social and economic problems that cause parents to give their children up for adoption (Ryan, 2001, 56–60). Gay Becker notes that adoptive parents of infants often have initial feelings of strangeness and alienation that evaporate on acquaintance with their adoptive child or even, in some cases, with its gestational mother before its birth (Becker, 2000, 197).

which God Himself joins the two together into 'one flesh'" in marriage. Children resulting from AIH are still products of this "one flesh" (Breck, 1998, 182, see also 177). Dissenters within these traditions tend to argue a similar point, but in favor of homologous and even donor procedures: rather than being vilified by a chosen fertility procedure, the larger context of a sacred, grace-filled, loving companionate marriage sanctifies the procedure, for its higher purpose is perfectly in line with the relationship even if its aesthetics leave something to be desired.[33] For Jehovah's Witnesses the official problem is simply a transgression of Hebrew Bible dicta against transferring one man's semen to another's wife (Lev. 18:20, 29; see Brody, 1990, 54–56).

For many Muslims and Jews, as well as Confucian Asians, the adulterous implications of third-party donation arise not primarily from jeopardizing companionate marriage but from muddying blood lines. In China, where patriarchal descent is of utmost importance, women seem more willing to accept donor eggs than donor sperm; however, the latter hesitation is sometimes mitigated by a belief that donor sperm are of the highest quality and will produce the healthiest, most intelligent children—a consideration that looms large when a couple is limited to one or two offspring (Handwerker, 2002, 306–310).

Particularly for observant Jews, a convergence of several factors both strictly limits and provides fascinating openings for third-party donation. First, unlike Roman Catholic moral theology, Jewish law does not regard itself as a universal moral code but applies in particular to the Jewish people as a consequence of their divine covenant. One might assume that a community placing such a high value on belonging would shun use of outside donor gametes, and indeed the halakhic problems caused by donor ova are complex (see, e.g., Bleich, 1997b, c; Bick, 1997). But Susan Kahn's work on Orthodox Jewish attitudes to ARTs shows that the opposite is sometimes the case. For example, in cases of severe male-factor infertility, when there are no other options, some rabbis permit the use of sperm donated by a third party. This solution has been the topic of contentious debate in the rabbinic world, but all rabbis agree that if a third-party donor is used, he should be a non-Jew. The reason for this preference is that the child conceived in this way will still be Jewish, as Jewishness is transmitted through the matriline (Kahn, 2000, 85), and—in addition—using donor sperm from a non-Jewish donor circumvents three otherwise severe Halakhic obstacles to sperm donation. The first is adultery. According to Jewish law, adultery is a sexual relationship between a married Jewish woman and a Jewish man other than her husband. Thus, a child conceived with non-Jewish donor sperm would not be considered the product of an adulterous union.[34] The

---

[33] Richard A. McCormick is one representative of this style of thought, with respect both to ARTs and to contraception. For examples of such reasoning in his work, see McCormick (1987, 1989). Lisa Sowle Cahill has embraced a variation of this approach in her work; see for example Cahill (1996). As we will see below, a concession like this is not always a *carte blanche*. The elasticity it creates might be nearly infinite or might simply extend to homologous procedures not involving IVF.

[34] Most Israeli ultraorthodox rabbis agree that non-Jewish donor sperm are an acceptable solution to male infertility (Kahn, 2002, 290).

second is masturbation. Jewish law generally prohibits masturbation, but masturbation by non-Jews is not an explicit halakhic concern (Kahn, 2000, 104).[35] The third is incest. Given the relatively small size of the Jewish population, the use of anonymous Jewish donor sperm creates the potential for incestuous unions between unknowingly halakhically-related men and women; the use of non-Jewish donor sperm avoids this obstacle, as the children of two unrelated Jewish women conceived with the same Gentile sperm are considered unrelated for purposes of marriage (Kahn, 2000, 105).[36]

These laws contain loopholes that in theory permit formerly unimaginable possibilities: use of donor sperm to accomplish halakhically pure, intentionally single motherhood.[37] Likewise, an unmarried Jewish woman may provide an ovum to a married Jewish couple without involving the donor or the husband in adultery or jeopardizing the Jewish origin of the child to be born, no matter whether rabbis eventually rule that ovum source or gestation is the criterion of Jewish heritage (Kahn, 2000, 129).[38] Kahn cautions that these judgments, vigorously debated by Orthodox rabbis (Kahn, 2000, 106–107; see also Rosenfeld, 1997; Bleich, 1997b, c; Bick, 1997), are by no means universally accepted, and that "among religiously observant Jews...the nuclear family is still largely understood as the only appropriate framework for reproduction" (Kahn, 2000, 59). However, the existence of some lenient rabbis—not to mention the indifference of some Jews to rabbinic rulings—provides space for limited experimentation at these frontiers.

In Sunni Islamic cultures, as we have seen, what is at stake is not simply Muslim identity but biological inheritance. Islam can be said to privilege, even mandate, both procreation and biological inheritance, which are expressed not in the Western medical language of "genes" and "heredity," but rather in kinship idioms of "lineage" and "relations." The first paragraph of the influential Al-Azhar *fatwa*[39] on assisted reproduction begins, "Lineage and relationship[s] of marriage are graces of Allah to mankind" and ends, "Therefore, origin preservation is a most essential

---

[35] Opinions on masturbation are not unanimous; it is sometimes considered acceptable if the broad purpose is reproductive. See for example Jakobovits (1994, 58–66; 1997, 115–138).

[36] Kahn notes that secular, single Jewish women in particular are concerned about genetic incest and sometimes plan to ensure that their children do not mistakenly marry biological half-siblings (2000, 78–80). See also Kahn (2006).

[37] Kahn notes that "unmarried religiously observant women are availing themselves of the opportunity to conceive children via reproductive technology...in significantly smaller numbers than their secular counterparts and with greater attention to Halakhic concerns" (Kahn, 2000, 59). One analogy that opens the door to this possibility is bathhouse insemination (a woman's impregnation by the sperm of a man who ejaculated in the waters of a public bath), proposed in halakhic debates as a means by which a woman might become pregnant while remaining a virgin. See Reichman (1997).

[38] Kahn notes that married, infertile Jewish couples' demand for donor ova is one force driving a growing acceptance of use of ARTs by Jewish straight single women and lesbians, whose ova—by Israeli law—cannot be donated to other women unless the donors themselves are undergoing fertility treatment (Kahn, 2000, 132–133). This is one clear example of pragmatism altering common ideas of acceptability and "naturalness" in unexpected ways.

[39] See the appendix to Marcia C. Inhorn (2003, 275–279).

objective of Islamic law." The tie by *nasab* (lineage, or relations by blood) is considered to be one of God's great gifts to his worshippers; thus, knowledge and strict preservation of *nasab* is morally imperative for Muslims. The problem with third-party donation is that it destroys a child's *nasab*, which is not only immoral but also psychologically devastating to the child. In addition, Muslims believe that parents who see their child as an *ibn haram* (literally, son of sin)—as "illegal," unnatural, and stigmatized—will never treat him or her with the love and concern parents feel for their "real," natural children. Consequently only artificial insemination or IVF *with the husband's semen* is allowed in the Sunni Muslim world. Beyond the knowledge that third-party donation is forbidden, fear of lab mix-ups and anxiety over gamete origins also discourage Muslim women from using heterologous (or donor) ART procedures (Inhorn, 1994, 338–339; 2003a).

In addition, although third-party donation involves neither the sexual body contact of adulterous relations nor presumably the desire to engage in an extramarital affair, Sunni religious scholars consider it to be a form of adultery. It is the fact that another man's sperm or another woman's eggs enter the sacred dyad that makes donation of any kind inherently wrong and threatens the marital bond. Finally, sperm donation raises the specter of unintentional incest among a man's unsuspecting offspring.

These objections to donor procedures cannot be rescued by analogy with adoption. Muslims, for example, are so wary of third-party involvement and uncertain genetic heritage that western-style adoption is unknown in Sunni countries, practiced only—and rarely—in Shi'a Iran and covertly in other Middle Eastern countries (Sonbol, 1995, 60). Yet Muslim belief and practice are no more monolithic than those of other traditions. As we will show below, leaders of Shi'a communities in Iran are working within traditional Muslim notions of inheritance, adultery, and family to create new mechanisms for third-party gamete donation on Islamic terms. On the other hand, Heather Paxson notes that Greek women see even IVF as more "natural" and acceptable than adoption (Paxson, 2006, 5–6), perhaps because they retain gestational connection to their children.

In some cases the cultural and religious demand for marital fertility is so high that even where the public standard for married couples is homologous fertilization, donor gametes are widely but quietly used. In India, cultural taboos on adoption and third-party gamete donation are so strong that many couples secretly using donor gametes pull their doctors into a conspiratorial public claim that their "miracle baby" has been born of their own gametes (Bharadwaj, 2003).

1.2.3.3 Surrogacy

In most cases surrogacy overcomes some of the objections to third-party gametes but fails to solve the difficulties posed by separating sex and procreation, and it introduces a new difficulty: a nine-month rather than nine-minute commitment on the part of the donor. In surrogacy, a hopeful parent or parents contract with a woman to conceive and bear a child, if possible using gametes from one or both parents-to-be, who then pursue whatever civil legal process their nation requires in order to be recognized as the child's parents. Religious people—both hopeful couples and surrogates—who accept surrogacy in fact tend to justify it by arguing *against* the hypothesis that we have been developing: that genetic relationship is inessential to parent-child kinship. For instance, when there is no female partner, or she has no viable uterus, parents-to-be often pursue surrogacy rather than adoption because they want a genetically related child. Surrogates often take the same view. Helena Ragoné argues that American surrogates typically are morally and emotionally offended by the idea of contracting to give up their "own" children to others; consequently they prefer not to contribute their own ova when they serve as surrogates (Ragoné, 1998, 120–123).[40]

Israeli surrogacy law and practice preserve the parental genetic link as closely as possible (Kahn, 2000, 152–158). The law insists that sperm be taken from the father-to-be and that ova not be taken from the surrogate. In addition—contravening[41] traditional Jewish law—once born, the child is not considered its birth mother's offspring, and within a week of the child's birth formal proceedings to declare the contracting couple the child's parents must begin, even if the contracting parents neglect to do so themselves. Despite this irregularity, every effort is made to preserve halakhic purity: if the contracting couple is Jewish, so must be the surrogate—and except in unusual circumstances, she must also be single (Kahn, 2000, 140–148).

Members of other traditions disapprove of surrogacy for much the same reasons they reject the use of third-party gametes: Roman Catholics, most Muslims, and Orthodox Christians fall into this class. The most influential Sunni fatwa on assisted conception argues that any intrusion by third parties into "the marital functions of sex and procreation," whether through "providing sperm, eggs, embryos, or a uterus…is tantamount to *zina*, or adultery." Consequently "all forms of surrogacy are forbidden."[42] Surrogacy is frowned upon in Greek Orthodox circles as well, but not always or only because of third-party gamete or uterine intrusion; in Greece, for instance, where cultural understandings of motherhood tie mother-child kinship to full-term gestation and birth, both adoption and surrogacy (rejecting motherhood by "giving one's child away") are rare (Paxson, 2004, 221–222)—even though at least half of all Greek pregnancies end in abortion (Paxson, 2004, 163). A number of the same religious communities argue that commercialization and exploitation are inappropriate to procreation; even the Israeli law that permits controlled surrogacy strictly limits the compensation that may be paid to the surrogate (Kahn, 2000, 151–152).

---

[40] Ragoné notes that a surrogate for whom IVF is repeatedly unsuccessful sometimes consents to AI but must consciously alter her operative understanding of her role and of kinship in order to do so (1998, 124–125).

[41] Rabbis are still debating cases of surrogacy and egg donation, but the default position is that the birth mother is the halakhic mother. See Rosenfeld (1997); Bleich (1997b, c); Bick (1997).

[42] Inhorn (2003c, 97), commenting on the *fatwa* reproduced on pp. 275–279. See also Inhorn (2005).

Absent from nearly all the religious treatments—mainly because few traditions are prepared to consider the possibility of gay and lesbian parenting—is any discussion of gay couples' use of surrogates, but presumably the motive would be the same: the possibility of raising a genetically related newborn.

In sum, then, the willingness of any tradition to contemplate third party involvement through gamete donation, surrogacy, or adoption depends upon what dimensions of nature its practitioners believe the procreative process must guard as sacred. Genetic relation, gestation and birth, the unity of exclusive sexual relations with procreation, and ritual or communal purity all have their roles to play, at times with unexpected results. As we will see below, local cultural expectations for gender roles and for marriage and singleness also have their say.

1.2.3.4 Unused Embryos

A very different—and to many religious communions ultimately more important—objection to ARTs is that procedures beyond AI and GIFT, as practiced today, require the fertilization of more ova than they use. Depending on the laws of a nation and the practices of a particular clinic, unused embryos may be donated; destroyed; used for research; or frozen (cryopreserved) for possible future use by the same parents. The question for religious communities is whether each option is acceptable, and why. Responses to these concerns depend on several factors: whether fertilization constitutes the beginning of an ensouled human life that must be provided an opportunity for gestation; if so, whether statistical "natural" embryo wastage nonetheless permits one some latitude in ART procedures; whether there are parallel cases in which destruction of human life is morally acceptable; and, whether or not embryos are ensouled at fertilization, at what point along their developmental continuum they "naturally" possess a degree of dignity that obligates people to protect them. Finally, in some traditions the question of the embryo's status may not really be about the embryo simply but about the embryo in relation to the parent's sense of vocation.

It is not possible to explore all of these questions here, but two examples illustrate how important it is to proceed carefully when analyzing both positions and practices. The Greek Orthodox Church, which teaches that human life begins at fertilization, generally condemns the production and destruction of unused embryos. According to Nikolaos Hatzinikolaou, the natural process of fertilization is "sacred and secret" (Hatzinikolaou, 1996, 104). Part of what makes it sacred is that two cells unite to become something new and irreversible, a person with a specific identity. Thus fertilization marks not only the beginning of life but also the birth of the soul. In the sacred process of fertilization nature creates a spiritual being that in turn "functions beyond the laws of nature" (Hatzinikolaou, 1996, 105). Through divine design and assistance, biological nature transcends itself.

The corollary of this argument is that the progress of post-fertilization embryonic development research is morally irrelevant to the question of the embryo's nature and status. According to Breck, definitive embryological determinations regarding differentiation "would not alter the Orthodox [Christian] conviction that human life begins with conception, meaning fertilization"; he adds that "the undeniable evidence [is] that a unique human 'soul,' the divinely bestowed dynamic of animation, is present from the very beginning, when the pronuclei of sperm and oocyte fuse to form the zygote" (Breck, 1998, 143).[43]

For Breck, this affirmation nevertheless opens up the possibility for Orthodox Christians to consider IVF or ZIFT if no extra embryos are created, no fetal reduction occurs, and any extra embryos that might unexpectedly be produced in certain circumstances are "donated" to infertile couples who would then "adopt" the embryos (Breck, 1998, 187–188). The notion of adoption (which makes perfect sense, given that the embryos are regarded as very young human beings) avoids the prohibition against introducing third (and fourth) parties into the procreative process, naturalizing an otherwise illicit act—adultery—by transforming it into a licit one—adoption. The unexpected result—accepted by the Greek Orthodox Church—is an all or none scenario: either both parents must be genetically linked to the embryo/fetus/child, or neither can be. This is a wonderful example of the importance of the choice of central concern, the value that defines "naturalness" in a given analysis. In this case, because the concern is the fate of existing unused embryos, the focus shifts to embryo-as-person, and the worry about adultery no longer applies.[44]

Unsurprisingly, the Eastern Orthodox debate over the state of the embryo in IVF is linked to the debate over abortion. According to Heather Paxson, in Greece the Orthodox Church and the pronatalist New Democracy Party condemn abortion as

---

[43] See also Breck (2003, 22–24).

Breck discusses the debate surrounding when conception begins and whether or not fertilization and conception are coterminous. Some Christian ethicists follow British embryologists who differentiate between embryos and "pre-embryos," which would be defined as the entity which exists from the single-cell zygote stage until formation of the primitive streak and implantation in the uterine wall (Breck, 1998, 127–143). Conception becomes a process rather than a moment (fertilization), which takes approximately 12–14 days, at which time the embryo would have a soul. Under this scenario, very early pregnancy termination, IVF, and early embryo experimentation (before ensoulment) would be permissible.

Breck offers an embryological description that would undo this notion of the "pre-embryo." The key difference in all of these scenarios is the point at which the embryo becomes a differentiated body of cells rather than a mass of undifferentiated totipotent cells. "Differentiation," "singularity," and "biological stability" are all varying prerequisites for ensoulment. Here much is at stake: various forms of ARTs, abortion, embryonic experimentation, etc. In this case theologians appear not only to have to accommodate evolving information about nature (for example, by accepting AIH but forbidding AID and IVF because they violate nonnegotiable religious boundaries like adultery, or murder in the case of fetal reduction) but they also seem to base the morally crucial timing of ensoulment on unstable, evolving biological knowledge.

[44] The alternative would be to argue that no process that produces unused embryos should be used, but apparently—for Breck—the value of gestation and birth is high enough to permit the disengagement of sex from conception and even to permit the production of extra embryos that will have gestational homes. This is not surprising, given that Paxson argues that Athenian women consider motherhood to be defined by gestation and birth, not genetic relation (Paxson, 2004, 221–222).

the killing of a human soul (Paxson, 2004, 185–187), but the government and ordinary Greeks see legalized abortion as a sensible acknowledgement of the social reality of unplanned pregnancy (Paxson, 2004, 188–189) and as an improvement over infanticide (Paxson, 2004, 246). Urban Greek women's willingness to make use of abortion places them on the latter side of the debate. But we should not assume from this that they see embryos as "not persons." Rather, as Paxson and Georges suggest, these women may see abortions as applications of *ekonomia*, the principle that allows religious law to be excepted or selectively applied in imperfect situations in the service of a greater good. In the case of abortion, being a good mother either now or eventually may require aborting the ensouled fetus that has resulted from carelessness or misfortune.[45] Likewise, one can imagine Athenian Greek Orthodox women regretfully resisting Church teaching on ARTs by employing a traditional method of Orthodox reasoning: accepting the inevitable deaths of extra embryos—again, seen as persons or potential persons—as part of the imperfect process that enables them to realize virtuous motherhood.[46]

Both conservative and liberal Roman Catholics have a hard time making a similar argument, as both the Vatican and the progressive wing of the Roman Catholic Church have grown decidedly pacifist, rejecting arguments that traditionally justified killing in limited circumstances. On this logic, if the embryo is a person, one must preserve it; intentional embryo wastage would make moral sense only if embryos were not ensouled.[47]

Buddhist teaching and practice diverge in some of the same ways as Greek teaching and practice, and more starkly. According to Buddhist tradition animation or ensoulment can occur as early as fertilization; but even if it does not, a particular embryo may simply be on the verge of receiving a soul. Thus, on the grounds of Buddhism's first precept, which forbids killing another living being, it is difficult to argue for the acceptability of embryo research, cryopreservation of embryos, and the intentional destruction of embryos in IVF.[48] In addition, if one embodied soul is good, more are not necessarily better. Western Buddhist author Rita Gross cautions that compassion for souls awaiting rebirth cannot legitimately be turned to pronatalist purposes: an overcrowded world will work against their enlightenment, not for it, as deprivation is a barrier to spiritual growth.[49] As we have seen, however, fertility practice in Buddhist-influenced Asian nations is also formed by population policy,

---

[45] On abortion and *ekonomia*, see Paxson (2004, 61) and Georges (1996).

[46] See Georges (1996, 515). This is not to say that all Greeks see embryos and fetuses as fully invested with personhood—only to say that Greek Orthodox theology has a method by which it can accommodate loss of embryos in IVF and in abortion even if they are so invested. On the indeterminate status of the embryo and fetus in the Greek civil debate see Paxson (2004, ch. 5).

[47] Another argument that might be marshaled is the distinction between killing and letting die, which is often made in end-of-life issues. The argument that wasting embryos imitates "natural" processes would be problematic for a similar reason: it involves willfully choosing death for some embryos rather than simply accepting the results of a biological process.

[48] Keown (1995, 121–122, 135–138).

[49] Gross (1997, 292–301).

local family structure, and other devotional traditions, which can lead both to abortion and to combining PGD with ARTs, in both cases often for sex selection.[50]

The combination of two questions—the status of the embryo, and conditions under which killing an embryo is morally acceptable, whether or not it is seen as a person—has implications for the question of what to do with stockpiled embryos. According to Charis Thompson, a 2003 national survey of frozen embryos showed that

> almost 90% of frozen embryos are intended to be used by the couples from whose gametes they were derived—to use if the fresh cycle of treatment does not work or if they want to have another child at a later date. Of the approximately 4% of frozen embryos available for donation, it is estimated that 11,000 are destined for research and 9,000 for other couples. Approximately 9,000 are currently destined to be thawed without use.
> "Abandoned" embryos—which are unclaimed after a period of years and after diligent efforts have been made to contact the couple involved—are not used for research or donation.[51]

For many Greek Orthodox, Roman Catholic, and Buddhist ethicists, these statistics are simply a further argument against extra-uterine methods of fertilization. Religious women and couples in fertility treatment, however, do not always pause to consider the implications of their choices of method as their involvement and investment in the process deepen.

Finally, distaste for a strong vision of embryonic human nature actually drives some infertile couples away from ARTs. Charis Thompson recounts the story of one couple who, on learning that evangelical Christians would consider the use of a donated embryo an "adoption," refused the procedure in order to avoid appearing to support this religious view of "natural" embryo rights.[52]

Thus for many communions the question of the embryo's status—or nature—and the demands that follow from it are fundamental to the evaluation of particular ARTs. Even a communion like the Greek Orthodox Church, which has an elastic view of the connection between sex and procreation within a marriage, draws the line at embryonic life.

1.2.3.5 Gender and Gender Roles

As we have noted, infertility and fertility treatment most often reinforce traditional family patterns and gender roles. For example, ARTs can be used in support of the status quo to continue an existing identification of femininity and motherhood (when used quietly by couples who are infertile, and either ignored or tacitly permitted by religious authority) or to create or strengthen such an identification where

---

[50] Normally when East and South Asian parents select for gender, they select for males; however, in Japan, where many women anticipate that daughters will care better for them in old age, there is some indication of preference for girls (Mori and Watanabe, 2002, 422).

[51] Thompson (2005, 4).

[52] Thompson (2005, 220–221).

none was previously ascendant (when recommended publicly as a way of fulfilling a gender-linked obligation). Now that male infertility is diagnosable, ARTs might also create or reinforce an identification of maleness with virility and fatherhood. Finally, ARTs might create or magnify the idea of marriage as an institution intended chiefly for begetting and raising children in communities where this norm formerly was absent or was only a sub-theme.

Gender may be the place where a sense of personal identity intersects most powerfully with legal, social, and economic forces. Hence it is important to realize that no discussion of men's and women's private struggles with identity and worth can be undertaken in isolation from a discussion of the larger patterns and stakes that create the standards to which they hold themselves. Distinctions are possible, but the two dimensions of the question are interdependent.

As we noted above, the experience of infertility and of fertility treatment often leads to a crisis of gender identity. Gay Becker argues that infertile men and women tend to fall back on quite traditional, even stereotypical and essentialist, descriptions of masculine virility and feminine fertility and maternity in order to explain their sense of frustration and suffering. Those who remain infertile must consciously remake their gender definitions in order to be able to accept their childlessness (Becker, 2000, 28–29).[53] It would not be surprising if, conversely, those who conceived subsequently retained these newly rigidified understandings and passed them on to their offspring. Clearly, when one hears religious language about masculine and feminine "nature," one must approach it critically, with an eye to what may be at stake. Most of the work being produced on this topic focuses on women, but more recently researchers have begun to address men as well (see, e.g., Inhorn, 2002, 2003b, 2004).

Although in Euro-American cultures infertility chiefly damages people's senses of themselves as gendered beings, in other cultures larger issues are at stake. We will give three examples of infertility's implications for femininity: moral, socioeconomic, and political. Orthodox Christianity in Greece provides a moral and vocational argument for ARTs. The belief in the nature of women as bearers of children is particularly evident in a passage by Greek Orthodox author Hatzinikolaou: "During pregnancy, women experience the peak of their humanness, for the basic function of the female body, towards which the whole female nature and life is directed, is the reproductive one. Women exist as they are anatomically, physiologically and psychologically for the embryo and pregnancy" (Hatzinikolaou, 1996, 104). It could be argued then that ARTs that do not violate Orthodox principles can enable women to reach their full humanity. In other words, "women must conceive and reproduce in order to be 'real' women" (Lock and Kaufert, 1998, 20), and up to a point, they may use ARTs in order to execute this sacred project. Thus many Greek women regard ARTs as working *with* nature—enabling women to become themselves—rather than after or against nature (Paxson, 2004, 213–219; 2006).

---

[53] On adoption of exaggerated stereotypical gender identities during treatment see also Thompson (2002, 52, 65).

As Heather Paxson points out, however, modern Greek Orthodox women paint themselves a picture of womanhood that entails much more than maternity, or—perhaps more accurately—they create a vision of maternity that includes much more than childbearing. In Greek Orthodoxy, as in Roman Catholicism, the project of life is to become what one is: to develop the virtues appropriate to and even dormant in one's nature. For the Greek Orthodox, this implies the belief that there are specific masculine and feminine virtues. As Paxson notes, "Motherhood completes a woman in Greece not because it actualizes some essentially female biological capacity but because, by demonstrating they can be good mothers, women assert their proficiency at being good *at being* women and at being *good* women—and they fulfill their part of a [*sic*] unwritten contract of social reproduction" (Paxson, 2004, 18). One of the defining virtues for women is maternity, which includes service, self-control, generosity, foresight, and responsibility. Because a virtuous mother is one who plans carefully and well, she refuses to bear children whom she cannot raise. Thus "good mothering" can include the choice to have an abortion or (especially if one is poor or unmarried) not to bear children at all, instead demonstrating motherliness toward one's existing children or toward others in care and service (Paxson, 2004, 248–249). The popularity of this vision of virtue—which by embracing voluntary childlessness contradicts Greek Orthodox teaching on abortion and marriage—is evident in the low Greek birth rate and high abortion rate, both of which are surprising also in the light of Greek policies that reward parents of three or more children (Georges, 1996).[54]

In other cases not just women's religious virtue, but their safety and livelihood are at stake (Inhorn, 1996, esp. ch. 6). Marcia Inhorn's early work reveals that in patriarchal, patrilineal, patrilocal Middle Eastern Islamic societies, one realizes one's womanhood through maternity as well (Inhorn, 1996, 6–10). The identification between womanhood and motherhood is even more profound in modern Egypt than in Greece because, as a result of recent Islamicization, motherhood, charitable work, and housewifery are now considered by many Islamists to be the only appropriate work for married women of all economic brackets, and there are few real alternatives for either independent income or self-realization (Inhorn, 1996, 67–70). Women's acquisition of social and familial power can depend entirely on marrying well and then rising from daughter-in-law to mother-in-law. Consequently it is essential not just to bear children, but to bear male children (Inhorn, 1996, 4–6). Infertility or failure to produce boys threatens not just husbands' sense of virility and wives' sense of maternity but the future security of the family (Inhorn, 1996, 12–13): there is no one to inherit and no son to marry and remain at home to care for his parents. Potential ill consequences for women include mistreatment, polygyny, or divorce (Inhorn, 1996, 111).

---

[54] This does not imply that Greek Orthodox women see abortion as an unmitigated good; they see abortion as the responsible way to respond to an irresponsible pregnancy when one is not in the position to mother a new child (Paxson, 2004, 34, 61–62; Georges, 1996).

Yet her later work reveals that these consequences result less frequently than might be predicted. On one hand, the emergence of ICSI in the Middle East could have been expected to worsen the condition of women married to men with fertility problems by increasing the potential for divorce and therefore the potential for women's infertility-related financial and social suffering. ICSI allows infertile men with very poor sperm profiles—even azoospermia, or lack of sperm in the ejaculate—to produce genetic offspring. Their wives, many of whom have stood by them for years or even decades, may have grown too old to produce viable ova for the ICSI procedure. In the absence of approved donor egg technologies, such husbands occasionally forsake reproductively "elderly" wives in order to divorce and remarry or to take a second wife, believing that their own reproductive destinies lie with younger, more fertile women. The shame and stigma of divorce and supposed infertility can cause such women to be shunned by their families and prevent them from ever achieving economic security (Inhorn, 1996). However, in fact most infertile men choose conjugal connectivity over procreation, refusing to divorce wives with whom they cannot procreate (Inhorn, 1996, 2006a, b).[55] In addition, married men's personal distress over infertility and their endurance of embarrassing and even quite painful treatment loom large in their marital experience (Inhorn, 2002, 2003b, 2004, 2006a, 2007). Thus one notable consequence of failed fertility is the *de facto* emergence of a model of childless, companionate marriage built in part on the partners' shared history of suffering.

Finally, in many cultures maternity is a foundation for political authority. Where large numbers of adult men are marginalized, working in other countries, or dead as a result of war or disease, women wield palpable social, political, and economic power. In cultures with longstanding traditions of patriarchy, women's power comes from their claim to represent and protect the children of the society, from arguing that they are entering the political arena in order to be good mothers: to gain the healthcare, education, peace, and security that their (and others') children need. If they do not have children themselves, they may take on the persona of the mother of a particular community or a whole nation. Or they may produce children for the sake of the society and its survival.[56] Here maternity is not a matter of personal virtue or personal security only, but a matter of political power. Since these also tend to be cultures in which advanced medicine is available to very few people, it is unlikely that ARTs will produce a wave of children conceived under the ideal of political motherhood. But it is worth noting that the ideal of motherhood as fulfilling womanhood can take a decidedly political, empowering form.

Infertility affects men's gender identity as well. Virility is almost universally associated with masculine strength, ambition, and power, even where (as in monasticism)

---

[55] Infertile marriages in which the husband has been proven sterile survive in part for a reason having to do with preservation of traditional male gender identity: the couple allows others to assume that the wife is infertile; the husband is grateful to the wife for protecting his image of virility and making him appear benevolent; and the wife is grateful to the husband for not attempting ICSI with a younger woman (Inhorn, 2003b).

[56] On maternity as a political strategy in Palestine, see Kanaanah (2002).

it remains only potential. Infertility implies weakness, sexual impotence, and by association femininity and inferiority (Inhorn, 2003b, 238; see also Paxson, 2004, 238–239). In this context, quiet use of ARTs rescues men's gender identity as much as it does women's. In the case of Arab Muslims, for instance, successful treatment through ICSI restores men's masculinity in two ways. Most obviously, it makes them successful progenitors. But secondarily, it reinforces their authority in marriage: Although it is the husband who is infertile, the preponderance of the surgeries, all the hormone treatments, and the (often) twin pregnancies fall upon the wife, who accepts most of the discomfort and danger required to redeem her husband's masculinity (Inhorn, 2003b, 245–249).[57] Wives of impotent men often lie about their husbands' impotence even to medical practitioners, and as a consequence blame for childlessness continues to fall on the wives, with little comment from their husbands (Inhorn, 2002). However, the negative consequences of male infertility—in urban settings, at least—tend to be social and familial rather than, as for their wives, economic.

1.2.3.6 Sexual Orientation, Singleness, and Reproduction

As Helena Ragoné has observed,

> Although "the family" as it has traditionally been understood in Euro-American culture continues to be shaped by a number of factors such as race, class, ethnicity, and sexual orientation, Americans are nonetheless regularly subjected to popular depictions of "the family" as a monolithic, timeless, universal institution. In spite of well-documented historical particularities and visibly diverse current practices, definitions of the family continue to follow fairly predictable trajectories, depicting families as nuclear, heterosexual, middle-class, white, and in a state of decline (Ragoné, 1998, 118).

If we remove "white" and add "perhaps within an extended household" to "nuclear," this description fits dominant religious ideals of the "natural" family the world over.

The first question about this "natural" family is whether it is expansive enough to include ARTs as means. As we have seen, some traditions are wary of some or nearly all fertility interventions. For those that are more tolerant of interventions that preserve the "natural" family, the additional question is whether this umbrella is large enough to include not just new means to the usual end but also new ends: parenthood for gay and lesbian couples, single women, and even single men. It bears noting that many people who do not fit the description of "natural family" can already reproduce without submitting their plans, like building permits, for approval by public authorities, even though they and their offspring may suffer opprobrium for their "unnaturalness." Thus the public cannot interrupt the reproductive plans of fertile unmarried heterosexual couples or of fertile lesbian women who make informal arrangements with friends for sperm donation. The policy question is whether a

---

[57] In cases in which men's semen contains no viable sperm, sperm must be removed directly from the testicles through the painful process of needle aspiration; see Inhorn (2007).

single woman or homosexual couple is "close enough" to the accepted definition of the natural family to be permitted access to ARTs. Widespread expansion of services to these groups could have divergent consequences. If widely used by committed gay and lesbian couples, for instance, ARTs could reinforce the idea of marriage as a natural institution for begetting and raising children, reinforcing the "nuclear family" as "natural norm" (Weston, 1991). If widely used by single women, they could have the opposite effect.[58]

Probably because no groups discussed (except a few liberal Protestant communions and some Reform and Reconstructionist Jews) recognize same-sex unions, use of ARTs by same-sex couples does not even reach the table in the literature analyzed here. Official teachings of Orthodox Christianity, Conservative and Orthodox Judaism, Roman Catholicism, Islam, Hinduism and most Protestant denominations forbid same-sex unions and vowed partnerships, usually on the logic that men and women were created as sexual complements so that they could join with each other in marital, sexual procreation and that other uses of sexuality are hedonistic and disrespectful of the divine plan. Parenthood for gay and lesbian couples is typically condemned not because their desire to be parents is unnatural but because their means of achieving it is; in addition, regarded as unrepentant sinners, they are thought to provide a harmful environment for children.

As we have seen, in Israel Jewish kinship laws make it possible for unmarried women to produce Jewish children, which amounts to tacit if unintended permission for partnered lesbians and single women to use ARTs, even though social stigma remains (particularly against lesbian reproduction). The path to surrogacy is closed to gay fatherhood in Israel, where couples contracting with surrogates must be married. Yet, although it is not a widespread phenomenon, gay orthodox Jewish men occasionally contract with unmarried Jewish women or lesbian women to have children, possibly in part to fulfill their religious obligation to procreate. Similarly, many gay Israeli men contract surrogates abroad. And Israeli lesbians are taking full advantage of state-sponsored artificial insemination, creating a lesbian baby boom. Fascinatingly, in both Greece and Israel it is clear that one of the primary objections to single motherhood—that it is the result of loose, careless pleasure-seeking—is moot.[59] In these nations single women who make use of ARTs desire motherhood, which is considered virtuous, without sex, which is considered selfish; and they plan carefully for their children's births and welfare, echoing an Israeli woman's succinct comment, "the time arrived, but the father didn't" (Kahn, 2000, 11).

---

[58] Second-order factors can call both predictions in question: gay or lesbian couples who do not conform to the norm of permanently partnered, monogamous pairs may present themselves as doing so to simplify access to ARTs or adoption or to simplify social relationships (thanks to Joseph Moser for this observation). Similarly, it is not clear that intentionally single mothers will alter the nuclear norm more radically than the enormous population of (often unintentionally) single mothers already has.

[59] In China, by contrast, women's *infertility* is regarded as a consequence of "loose living." See Lisa Handwerker (2002, 185–186).

In any case, even where use of ARTs by religious singles, gay couples, and lesbian couples is on the increase, it is comparatively rare. Dominant voices in religious traditions and cultures that consider marriage truly or nearly mandatory and consider childbearing essential to a legitimate, successful union condemn use of ARTs by these groups; religious traditions and cultures that consider the "true family" to be the one that best approximates the advantages of a healthy nuclear family will allow or even welcome such exceptions when they are carried out carefully; traditions that fall toward the middle of the spectrum will be uncomfortable with the exceptions but—as in Israel and Greece—may ignore it or look the other way, especially if they can turn a blind eye to some women's lesbianism. A possible explanation for this stance is that, no matter how uncomfortable men may feel about being nearly dispensable to reproduction, it is hard to object to the willingness to become a "virgin mother." Because gay male parenting requires use of a surrogate, it will probably not be considered "natural" until and unless surrogacy for unmarried clients becomes "natural."[60]

It is worth noting that Protestant and Catholic Augustinian visions of wounded nature can have unexpected relevance here. A strong doctrine of human sinfulness or of the imperfection of the world can support religious arguments for single, gay, and lesbian use of ARTs. Put simply, whether one approves of homosexual unions and single parenthood or not, in an imperfect world, one should spend less energy arguing about inaccessible perfection and more energy identifying family settings that are "relatively excellent" or "good enough" for children and that permit adults adequate sexual, emotional, and spiritual intimacy for parenthood and spiritual growth.[61]

1.2.3.7 ARTs, Intercourse, and the Meaning of the Body

As we have already implied, for religious people the dependence of the "naturalness" of procreation on heterosexual intercourse is a basic question; this question in turn depends on the complex of meanings attached both to intercourse and to the processes that lead to conception. Traditions that assign rich, highly symbolic meanings to physical acts can be more reluctant to tamper with those acts, especially if the value of preserving their integrity is considered to be similar to or higher than the value of procreation. Yet paradoxically their theological articulations of these beliefs are still transformed by the phenomena of ARTs.

Roman Catholicism is instructive here. Contemporary Roman Catholic teaching combines a weighty understanding of sexual acts with a vision of marriage to which children are not essential. According to contemporary Roman Catholic teaching,

---

[60] IVF also overcomes distaste in some cultures over women having sex "after a certain age." See Paxson (2004, 216).

[61] See for example the development in positions on the normative family that occurs in Lisa Sowle Cahill's writing between her 1996 and 2000 works.

intercourse is nearly sacramental: more than any other act, it concretizes the divine grace that supports the couple in matrimony, strengthening them in love for each other and in mutual vulnerability, openness, and generosity. This marital act is the entire purpose for which genital sexuality was created. To use it otherwise—unlovingly, outside marriage, or intentionally obstructing any possibility of conception—is not just to disrespect and subvert the divine purpose but to do so in a way that desecrates the body as a site of grace and a holy conduit for cooperation with God's creative power. The reverse is also true: to procreate other than through this complementary, sacramental act of union and love is a desecration of the body in its union with mind and soul: "it reduces procreation to a mechanical manipulation of sperm and egg and thus neglects the embodied character of human love" (Lauritzen, 1993, 7, see also 8). Therefore,

> any type of assisted reproduction that conforms to the procreative norm just articulated, i.e., any procreative attempt that includes sexual intercourse between partners in a loving monogamous marriage, helps facilitate the natural process of procreation and is therefore acceptable. Any intervention that fails to conform to the norm is a departure from the natural law with respect to human sexuality and is therefore morally problematic (Lauritzen, 1993, 6).

Even though the Roman Catholic Church forbids all procreative interventions (with the exception of treatments that increase fertility, and possibly of GIFT after intercourse with a perforated condom, which is interpreted as being "with" or "alongside" nature) and thus would seem to have a stable view of sex and procreation, the sheer process of evaluating ARTs actually alters Roman Catholic understandings of nature. Before the discoveries of modern reproductive medicine, for example, the definition of natural reproduction would not have included sperm ascending the fallopian tubes to fertilize the ovum, nor was the distinction between women's fertile and infertile periods—now essential to descriptions of Church-sanctioned methods of natural family planning—an element of the rhythm of natural marital sexuality. Likewise, discoveries about the female fertility cycle reinforce a relatively recent shift in Roman Catholic thinking on childbearing and on the emotional importance of marital sexual relations: family planning, never actually forbidden, is now morally enjoined if not morally required, and natural family planning (a method that does not involve long periods of complete abstinence, developed with thanks to contemporary reproductive medicine and now used by it to optimize the chances of conception for infertile couples) is now seen as natural and healthy. Thus the scientific account of fertilization is deployed to reinforce traditional norms, but with details that demand ever more specific rationales for limits on ART use.

A similar pattern can be seen in Greek Orthodox theologian John Breck's opposition to ICSI (intracytoplasmic sperm injection), a procedure that would seem to solve the problem of excess embryos created through IVF. In his objection we see an unbreakable association of "the natural" with "the divine." Because ICSI involves the injection of a single sperm into an egg, it eliminates the randomness (which Breck terms "divine intentionality") of sperm selection, which allows the "strongest" and "most fit" sperm to fertilize the egg. In his opinion, AIH preserves this divine intentionality. Breck states, "Orthodox Christians hold tenaciously to the notion that there is no such thing as 'chance'....Sperm selection, then, is understood not as random, but as an example of divine-human synergy or cooperation. To put it in the simplest terms, selection of the gametes in any human conception can and should be made by God himself" (Breck, 1998, 183). "Naturally" occurring marital sexual intercourse is no longer the prerequisite for a religiously acceptable conception; the divinely intended, natural (Darwinian, if you will) process of selection of gametes has replaced it as the threshold that faithful assisted reproduction may not cross. Yet, without contemporary reproductive research one would not have known that sperm "compete" to fertilize an ovum and so could not have used this argument to reject ICSI as unnatural.

On the other hand, in the case of Orthodox Judaism, contemporary knowledge of the biology of reproduction seems to be making a new thing possible: redefining the relationship between reproductive values that formerly could not have been separated. The balance between the values of *relatedness arising from embodied parental relationship* on one hand and *reproduction of new bodies* on the other is shifting toward reproduction simply. Susan Kahn points out the irony: the main criterion for relatedness in rabbinic thought, physical relationship between Jewish bodies, is precisely the criterion that provides the metaphor for reproduction *without* body contact but *with* pure Jewish lineage nonetheless. Therefore "rabbinic enthusiasm for the new reproductive technologies has thus created an intriguing paradox. For while it reinforces the imperative to reproduce more Jewish bodies—a value of undisputed importance in a pronatalist system—it has unwittingly contributed to the conceptual disintegration of the significance of Jewish bodily experience, once a precondition for the creation of Jews" (Kahn, 2000, 171).

Thus, by asking questions that have not been thought of before, new developments in reproductive medicine force traditions to redescribe the "natural" even when their positions on the acceptability of particular acts change little or not at all, or else it forces them to privilege one of an array of criteria that formerly functioned as an integral whole.

1.2.3.8 Gamete Collection and Donation

Three kinds of issues arise around gamete collection and donation: for men, sin or impurity in the act of masturbation (clearly not an issue when, as in some cases of ICSI, sperm are removed surgically from the testicles); commercialization; and implications for improper relations among donors, parents, and children. The latter point has been mentioned above and will be discussed at greater length below, but the other two bear some mention here. There seems to be little stigma for women in any tradition against either drugs that regulate ovulation or induce superovulation or (as long as extrauterine procedures are not forbidden) against surgical removal of eggs. These processes, nearly universally described as unpleasant, produce no illicit pleasure for women.

Many traditions, however (especially Orthodox Jewish, Roman Catholic, and Islamic), look askance at male masturbation because it deposits gametes in an infertile

1 Compatible Contradictions: Religion and the Naturalization 55

place, occurs outside of marital relations, is a product of self-pleasuring, and (in the Islamic case) is polluting (Inhorn, 2007). Confucians also believe that because sperm contain vital forces one should avoid donating them—and presumably should avoid masturbation altogether (Qiu, 2002, 78). Sri Lankan Buddhists believe that celibacy increases procreative power; periods of asceticism (for instance, a short-lived monastic vocation early in adulthood) increase a man's virility, and monastic asceticism ensures human virility and agricultural fertility generally (Harris, 2001, 150–153). Thus in this case too one would want to avoid dissipating one's procreative power, and couples might also be skeptical about receiving sperm from men whose sexual habits they do not know. Compounding these moral objections are practical ones: in many countries the lack of privacy in fertility clinics is an enormous psychological obstacle to men of whom semen for testing or IVF is demanded, especially when these requests are made publicly and on short notice (Inhorn, 2007).

Commercialization is also a concern. Significant numbers (though by no means all) of leaders from all religious groups we have discussed are averse to the idea that gametes or organs could be sold. Thus even where donation is acceptable as an act of good will, the sale of gametes—and particularly the catalogue approach to choosing a sperm donor—is seen as not only exploitive but also improperly manipulative, as well as dangerously vulnerable to eugenicism. Benevolence is a more palatable rationale than financial gain if one is looking for an adequate excusing condition for the pleasure of masturbation, which many traditions view as illicit or, especially for married men, at least distasteful and excessive (Jakobovits, 1994, 1997).[62]

This discomfort with commercialization is one inspiration for the Israeli insistence that all donated ova come from women who are themselves undergoing fertility treatment. In addition, this rule prohibits abuse of the bodies of women who would not otherwise undergo the uncomfortable, complex process of hyperovulation and extraction. The image is then mutual compassion—similarly disadvantaged people assisting each other—rather than business transaction or exploitation of labor. There is not room to pursue this instinct here, but our impression is that people who do not find the high costs of fertility treatment "unnatural" or improper nevertheless find paying for gametes or embryos "unnatural." For Arab Muslims and Athenians, for example, the cost of fertility treatment (both financial and physical) is part of the sacrifice that demonstrates the seriousness of one's pursuit of motherhood,[63] whereas—perhaps—paying for gametes or embryos seems dangerously close to an act which is not only unnatural but unjust: purchasing a human

---

[62] See also Kahn (2000, 204–205, n. 9–11) and Inhorn (2007). Liberal, and some evangelical, Protestants tend to see masturbation as developmentally normal and as acceptable if not done to excess; for a representative argument, see Grenz (1990, 214–215).

[63] In 1993 the Greek parliament mandated partial insurance coverage for infertility treatments for couples (Paxson, 2004, 216); the Greek conversation thus tends to run toward the extended physical discomfort of fertility treatments (ibid., 223). See also Inhorn (2003c, ch. 2).

being. Paying for a surrogate's services is uncomfortable but can be justified as proper compensation to a woman for the extra care she must take of herself and as cheap insurance for the health of the infant she carries.

1.2.3.9 Sex Selection, Disease Prevention, and Eugenics

Although each of these subjects could fill its own chapter, a word needs to be said about uses of assisted reproductive technologies that go beyond providing a gestationally and/or genetically related child to a single woman or infertile couple. These uses raise other important questions about nature. When careful selection of donated sperm, sperm separation, PGD, or selective abortion is used in combination with IVF or GIFT to select against disease or x- or y-chromosomal abnormalities, or for a desired gender or other valued trait, is "nature" "what would otherwise occur" but for the intervention? Is intervention (especially when one selects against genetic disease) thus benevolently repairing or avoiding an imperfection in "nature"? Particularly when these methods are used to select for gender (normally, for males), do people justify the intervention on a biological basis, or on the grounds of creating the socially and culturally proper "natural" family that physical "nature" does not always deliver?

Western thinkers tend to support the use of these technologies to select against diseases that are fatal or cause great suffering, but such uses also raise the specter of eugenics, worries that widespread interventions could in the future lead either to genetically homogeneous "superkids" or, conversely, to the mandatory formation of a separate slave class. The fact that such interventions are already being widely used to select for gender, even where this practice is illegal, leads to a further question: In what sense is gender selection "natural"? In the early 1990s Alison Renteln argued that making sex selection available exacerbates rather than softens the gender preferences and privileges enforced by the culture that employs the technology (Renteln, 1992).[64] This thesis seems to be holding true. In this case intervention in "nature" as "what would otherwise occur" is justified on the basis of "nature" as the cultural/religious "ideal shape of the family."

Predictably, then, judgments about the limits of acceptability for gender selection depend on a society's view of what is natural and acceptable. A group of ethicists in Spain, where the social ideal is egalitarian, have argued in favor of gender selection not only for gender-linked diseases but also for family balancing: parents who already have a male child and who want a small family, for instance, might select for a girl in their second pregnancy (Meseguer et al., 2002).[65] Observant Jewish families, for whom the father's obligation to reproduce

---

[64] Renteln (1992) also argues that to forbid use of sex-selection technology would, nevertheless, be improperly to limit women's reproductive freedom. For a similar argument, see Schenker (2002) and Sills and Palermo (2002).

[65] The authors add a number of important conditions: parents fund the treatment, each clinic must produce roughly equal numbers of boys and girls, unused embryos must be donated, etc.

1 Compatible Contradictions: Religion and the Naturalization   57

includes at least one child of each gender, might make use these technologies for a similar reason (Schenker, 2002, 405–406). In Japan, a fundamental attitude of non-intervention may be giving way in a generation more accustomed to putting its private goals ahead of communal norms. There is some indication that Japanese parents are selecting for gender to avoid genetic disease and to balance families; in expectation of increased longevity, some Japanese parents desire girls, who are thought more likely to care for parents in their old age (Mori and Watanabe, 2002). As in the Orthodox Jewish case, this preference reinforces existing cultural stereotypes and gendered divisions of labor *without* necessarily preferring boys.

On the other hand, in some patriarchal cultures the results of gender selection are less balanced. According to Gautam Allahbadia, in India a combination of abortion, infanticide, neglect, and maternal death have already produced a deficit of 50 million women; use of ARTs for sex-selection would simply replace these existing less palatable methods and perhaps widen the gender gap by making sex-selection even more acceptable. The lucrative gender-selection business might simply shift from clinics that determine fetal gender and perform abortions (now outlawed, though not eradicated) to clinics that perform PGD and microsorting (Allahbadia, 2002).[66] In Asian societies where pressure to control family size is strong, parents are even more likely to practice sex-selective fertilization, abortion, and infanticide—as well as daughter-neglect—than they were when the possibility of another pregnancy (and another son) remained open indefinitely (Croll, 2000, 16–17). In China in particular, the strictly-enforced one-child policy, combined with strong rural preferences for boys and histories of men divorcing wives who do not produce male heirs, has led to widespread abandonment of girls (nearly one million currently) and to the use of gender selection techniques that skew the proportions of male to female births. Chinese policy analysts are hoping that the spread of urban trends (which show no preference for males) and some exceptions to the one-child rule (which could encourage parents to accept one daughter) will eventually reverse the tendency to select for boys (Chan et al., 2002). In the meantime, the current generations' gender imbalance could force unintended consequences: selective increases in women's status or greater acceptance of male homosexuality.

Thus people who accept ARTs agree that, with respect to disease, "nature" as "what would otherwise occur" can be subverted in favor of nature as healthy, perfected, and ideally realized (see, e.g., Hanson et al., 2002, 431–432). With respect to gender, though, "nature" as "what would otherwise occur" is often *reified* by western-influenced policymakers (given that uninterrupted reproduction produces fairly balanced gender on its own) but willingly *subverted* by individual women struggling to maintain stable positions in patriarchal families. In a sort of compromise between the two, some analysts argue in favor of limited use of gender selection

---

[66] As we noted above, other factors, including widespread hepatitis infections among mothers, may account for part of this difference.

to create "natural" gender balance within small families, as long as numbers of girls and boys born remain approximately equal overall.[67]

1.3 Pressures and Processes

As will by now be obvious, an enormous number of factors can enter the equation that determines both what forms of reproduction are considered "natural" and how and when it is important that this sort of "nature" be preserved. Our hope in the remainder of this essay is to highlight the some of the less obvious underlying variables, partly to suggest lines of possible further questioning. The two most important, most complex factors that we have observed are pronatalism and religious strategies for naturalizing technology.

1.3.1 Pronatalism

In some cases—for example, among some Palestinians and orthodox Jews, or in Greek government policy—pronatalism simply encourages a high birth rate in order to expand a population. But pronatalism can and normally does run far deeper than a plan for population growth and takes unlikely forms; for example, in urban Greece, sometimes a policy of population expansion goes hand in hand with a popular, religiously inspired pronatalism that paradoxically results in a low birth rate. Speaking very broadly, our research shows that the degree of value placed by women themselves on giving birth to children dictates the degree to which they are willing to depart from the "natural" standard of heterosexual intercourse in favor of ART interventions, and the degree to which their religious community or culture also values their giving birth dictates the degree to which it will support their choices officially or at least look the other way (keeping in mind that a particular religious culture—e.g., Irish American Catholicism—might be more pronatal than the larger institution of which it is a part—and vice versa).[68] In pronatal cultures childbearing is typically considered an element of women's identity or of their essential nature; childless women are considered stunted because they cannot (as in the Athenian formulation) become what they are—cannot accomplish the end for which they were created.

---

[67] See for example Meseguer et al. (2002); Hanson et al. (2002, 432) support gender selection in the case of monogenetic diseases only, because they are skeptical that people will preserve "Nature's own sex ratio" by free choice.

[68] The degree to which explicit or tacit pronatalism dictates or at least suggests the practical (as opposed to official) outcome of ART debates in a given religious setting varies. For example, conservative Roman Catholics are likely to agree that separating sex from reproduction is illicit, but traditional arguments that only reproductive intent, in marriage, excuses the disruption of reason caused by sex are likely to motivate them to employ ARTs.

Neither can their husbands demonstrate their maleness, in order to realize their own natures. In patriarchal cultures, especially for women of the middle and lower classes, procreation is not only a prominent cultural value but also women's most reliable hedge against abuse and social, religious, and economic instability, a situation that gives women an extra degree of desire and determination to bear children. And, in many cultures, if children are not the whole purpose of marriage, they are at least a significant part of the justification for married sexuality. In all these ways, then, a pronatalist culture affirms married childbearing as "natural" and as essential to the fulfillment of universal or nearly universal divinely given human ends.

The theological arguments that back this judgment are often complex, layered, and multivocal, so that several conflicting or at least different interpretations can operate simultaneously. One good example is Christian belief in the virginity of Mary, the mother of Jesus, which supports celibacy, marriage, and motherhood, all in one. Hindu beliefs too are complex and multi-voiced. Goddesses reflect, transcend, transgress, and provide models for earthly female life. Normally, in Hindu scriptures, powers and modes of being have constructive and destructive sides. So it is with women: married women are often pictured as docile, nurturing, benevolent, transforming unjust structures through clever use of their limited roles, and unmarried women are portrayed as transgressive, volatile, and vengeful. Celibacy among gods and goddesses is generally portrayed as dangerous, but also as powerful. Hindu scriptures read through this lens imply that women's path to virtue is marriage and, presumably, motherhood (Marglin, 1985). But lurking behind or within each woman/goddess, whether domestic or dangerous, is her alter-ego.

In some cases, existing "natural" pronatalist arguments have been magnified by a religious group's judgment that it needs to increase its size in the face of external threats. In the Middle East, for example, both Arab Muslim and orthodox Jewish communities have combined a renewed emphasis on religious distinctiveness with promotion of high rates of childbirth, driven in the Jewish case by a desire to overcome the decimation of the Holocaust (Kahn, 2000, 3–4; Kanaanah, 2002; Barris and Comet, 1994, 31). In nearly all situations in which women are strongly pressured to reproduce, religious leaders find ways to "naturalize" at least some forms of ARTs so that the goals of true womanhood, true manhood, and the "natural" family can be fulfilled. Means of religious naturalization will be discussed below; for the moment, the point is that pronatalism predicts which "natural" values will be sacrificed, and which supported.

The Athenian variation is worth noting, as Greek pronatalist policies (which reward families with three or more children) are not the motivating factors for well-off Greek single and partnered women who desire to bear children. These women nevertheless qualify as pronatalists by virtue of their belief that motherhood completes womanhood. Like some Israeli ART clients, they are professional and sometimes even self-supporting. They do not need to bear children in order to secure their economic futures; they only want to realize themselves, and they can afford the steep price of this privilege (Paxson, 2004).

The market is also a factor. Here pronatalism becomes a consumer value. In India, as Aditya Bharadwaj argues, pronatalism receives a push from the market through sophisticated info-mercials: "the assisted conception industry is engaged in a process of ingratiation with the journalists to ensure favorable depiction in the media" (Bharadwaj, 2000, 65) with the result that "the need for these technologies and their experts is being generated—at least in part—through media narratives" (Bharadwaj, 2000, 76). American scholar Jennifer Stone concurs, noting that although American infertility has actually declined, the media have successfully generated the impression of an "infertility epidemic" in collusion with the fertility industry and other cultural factors (Stone, 1991, 313)[69]: "today the meaning of 'infertility' has been shaped largely by a commercial mass communication system that includes interdependent mass media industries, public relations, and advertising agencies" (Stone, 1991, 314). A quick perusal of American fertility websites and newspaper articles demonstrates how easily advertising the possibility of assisted reproduction generates a need that is justified by pronatalist arguments or sentiments.

Clearly pronatalism's influence is much more complex and varied than offhanded references imply. The highly complex Greek situation—which combines Greek pronatal policies; visions of womanhood-as-motherhood; and an anti-abortion, ART-wary Church that has hinted that the childless may divorce to find fertile spouses—could be expected to produce a high birthrate and nearly mandatory, covert use of ARTs, but in fact ART use (though growing) is thoughtful and selective, and the birthrate is below replacement levels (Paxson, 2004, 262). On the other hand, as we have seen, in officially-anti-natal China strictly enforced family limits, widespread diseases that compromise fertility, and continuing cultural preference for large families and male heirs may make childbearing, and therefore ARTs, nearly mandatory where they are available and affordable.

1.3.2 Naturalizing ARTs: Holy Paths, Religious Authority, and Strategies for Change

As we have been arguing, ARTs cause religious anxiety by demonstrating that understandings of "natural" family structure, gender, marriage, sexuality, procreation, and inheritance that have been seen as indivisible wholes are actually composed

---

[69] According to Stone, the apparent epidemic is result of the following factors: (1) contraception gives people the impression that they can control fertility, turning it "off" when it is unwanted and turning it "on" when it is desired; (2) the numbers of working women are rising, leading to more concern about baby timing; and (3) the proliferation of fertility clinics leads to the need to generate demand through "educational seminars" and the like. Stone also charges that for commercial reasons the media create the impression that classic infertile woman is white, married, middle-to-upper class, professional. Actually she is African-American, black, older, with no previous children, and with less than high school education. The media also grossly downplay the link between environment and infertility (Stone cites government estimates "that 15 to 20 million jobs in the United States expose workers to chemicals that might cause reproductive injury" and a Centers for Disease Control study that "called human reproductive failure 'a widespread and serious' problem, and one of the ten most prevalent work-related diseases" (1991, 324–325).

of parts that are *de facto* distinguishable, even separable.[70] Religious people, communities and individuals can respond to this development in one of two ways. Both involve naturalizing a vision of reproduction by divinizing it: redescribing reproduction in terms of their vision of divine creation, divine law, or the shape of the holy life. First, they may renaturalize by redivinizing: acknowledging scientific discoveries; taking them into account by revising and reinforcing the theological arguments behind their original, unified visions or law; and preserving these traditional visions or laws in newly-credible, finely-textured scientific language, thereby subtly transforming them.[71] Official Roman Catholic teaching on ARTs is a good example: the Church rejects ARTs but employs knowledge gained through reproductive research to sharpen its own description of a holistic, procreation-driven, divinely created, gendered human nature and reproductive process.[72] Or, they may naturalize by selectively divinizing: accepting the *de facto* dissection of the formerly indivisible vision; anointing particular elements *de jure* natural-because-divinely-intended; and willingly sacrificing others. Single Athenian women's valuation of motherhood over conception through marital intercourse illustrates this approach.

The strategy chosen depends heavily on the tradition's sources for knowledge of the divine will or holy path, on the traditional location of interpretive authority, and on the relationship of the individual to that authority—not to mention the degree and kind of local pronatal pressure. In what follows we will—without attempting to be comprehensive—reflect on the different strategies of response produced by varying understandings of authority. We will focus on the divinization of ARTs in a few traditions where religious authority is centralized and in a few in which it is not, with emphasis on the latter, because decentralized religious authority is a widespread phenomenon less familiar to American readers.[73] Nearly universally, however, religious people regard infertility as an obstacle that has religious significance and may in fact have divine origins; seek knowledge of God's will generally and in their specific situations; view the success or failure of their treatment as a consequence of God's will for and intervention in their lives;

---

[70] Mette Bryld notes the same anxiety in the political community. "With their capability of radically eroding the apparently 'natural' order of life ('natural' human beings, 'natural' motherhood, 'natural' nuclear families, 'natural' kinship, etc.), reproductive technologies and reprogenetics evoke so much anxiety and such strong fears in the political mind that, as if in defiance against all this monstrosity, its cultural imaginary clings to what is perceived as either 'natural' or, for lack of anything better, as 'close to the natural as possible'" (Bryld, 2001, 301–302).

[71] On the transformation of language see Franklin (1997, 104–105).

[72] On this transformation of "the natural" generally see Franklin (1997, 104–105). This is perhaps the most conservative possible example of what she calls the "substitutability" of nature and technology. See Franklin (1997, 209).

[73] In our experience, professional organizations and the press tend to ask scholars to outline *the* position of a religious communion on a moral issue. The question bespeaks the assumption that all religious communions possess clear bureaucratic structures and centralized authorities and value unanimity on all moral issues highly. This is rarely the case.

and believe that God acts through the clinicians who attempt to help them conceive. In non-theistic traditions, the idea of karma or cosmic balance may stand in for the divine.

Orthodox Judaism is a good example of decentralized authority. Orthodox Jews tend to understand new reproductive technologies as a set of tools and strategies that can be readily appropriated and harnessed to achieve individual and collective goals. The threat they pose to traditional notions of the natural family, or the role of God in conception, can be negotiated and resolved through textual interpretation aimed at strengthening and benefiting the community. The interpretation of these technologies takes place on both the formal and informal levels.

On the formal level, the interpretation is embodied by the multivocal rabbinic tradition of debating and of mining traditional texts for beliefs about the mechanics and meanings of conception. In Jewish cultural life the divine texts of the Torah are a privileged conceptual space that is believed to contain all knowledge. There is no "new" knowledge, for human innovation cannot invent knowledge that is absent from the Torah; human innovation can simply present challenges to those invested with the authority to interpret the Torah by revealing hidden knowledge through textual excavation. Rather than weakening this authority, the rabbinic response to novel reproductive technologies—to search traditional texts for precedents that determine the nature and location of relatedness—actually reinforces the belief that the Torah embodies all knowledge and that the answer to every question can be found through interpretation of traditional texts. As a consequence the perhaps irreversible social processes that ARTs set in motion do not displace orthodox Jews' foundational assumptions about authority and kinship. On the contrary, orthodox rabbis have full confidence that their interpretive strategies can create social uses for these technologies that are coherent with traditional assumptions about kinship. If so, the social uses of reproductive technology in Israel will enhance the authority of rabbinic conceptions of kinship, a result that may also both entrench the importance of maternity as the ultimate determinant of Jewishness and reinforce the cultural imperative to reproduce.

Policymakers and others who may be in search of *the* orthodox Jewish position on ARTs must be aware, however, that rabbinic authority is not centralized. A rabbi gains authority through a reputation for wisdom in debating within and against a complex, multi-voiced tradition of rabbinic casuistic commentary. When a rabbi puts forward a novel interpretation of the tradition, it may over time be embraced, rejected, or simply ignored, with corresponding implications for its power in the orthodox Jewish community. Not surprisingly, there is no single, authoritative rabbinic position on any ART therapy. Rather, formal rabbinic responses have been complex and wide-ranging.

Since the late 1940s, when artificial insemination among humans first became a medical alternative, orthodox Jewish rabbis have hotly debated every nuance concerning the possible use of these technologies, resulting in a plethora of often-conflicting rabbinic opinions. From a halakhic point of view, as we saw above, concerns include introduction of third-party donor material into marital conception, appropriate extraction of sperm and ova, appropriate protocol surrounding the artificial conception of embryos, surrogacy, and terms for reckoning kinship established through reproductive technology.[74]

Rabbinic flexibility regarding the use of these technologies typically rests on four basic principles, all of which connect to a theological vision of covenantal law[75]:

- The technologies aid in the fulfillment of the divine commandment to be fruitful and multiply (Genesis 1:28), considered an obligation of central importance in Judaism and therefore a marker of "natural" family life.
- The suffering caused by involuntary childlessness (i.e., being thwarted in fulfilling the command to bear children) is evil.[76] The mitzvah to practice loving kindness obligates the rabbis to do everything they can to alleviate that suffering, including the prescription of assisted reproductive treatments.
- Family integrity—the natural family as "whole" family—is a nearly absolute halakhic value. Permission to use ARTs can prevent the kind of serious marital difficulties that can arise for a childless couple, including the traditional (though seldom applied) obligation to divorce after ten years of childless marriage.
- Since the medieval period halakhic tradition has exhibited extraordinary receptivity to advances in medical science and to reconciling philosophical concepts of rationality with the Torah.

On the informal level, individual Jews struggle to make sense of these technologies in their own ways, depending both on the role religion plays in their lives and on their metaphorical spheres of reference. For orthodox Jews, the principle of *hishtadlus* is instructive here. *Hishtadlus* requires the individual to exert maximum effort within halakhic guidelines in order to fulfill halakhically imposed obligations. In the case of marital infertility, "making hishtadlus" might mean undergoing fertility treatments in order to fulfill the obligation to be fruitful.[77] However, success in that effort is ultimately determined not by human effort but by God, who through *hashgacha* (divine providence) is a partner in all human efforts and may grant or deny fruitfulness. When queried as to whether it is technology or God that accomplishes assisted conception, one observant doctor explained that "truly religious people understand that at the end of the day it is up to God to smile on the process in order to make it work" (Kahn, 2006, 472). Orthodox couples who use fertility services believe that God chooses their doctors; they develop special prayers and rituals to accompany the treatments; they employ numerology

---

[74] For a fuller discussion of the halakhic issues involved in artificial insemination, egg donation, and surrogacy, see Kahn (2000); see also Bleich (1997b, c) and Bick (1997).

[75] For items 1–3, see Kahn (2006, 471).

[76] It is hard to exaggerate the profundity of this suffering among infertile Orthodox couples. See Barris and Comet (1994).

[77] According to Kahn, recent rabbinic opinions limit this obligation by noting that people cannot be required to do something of which they are physically incapable; that is, an infertile person who is physically incapable of contributing to conception is not required to undergo fertility therapy. These opinions emerged as a way to end the suffering of patients who exhausted themselves physically, emotionally, and financially in order to achieve conception, without success.

and other traditional theological and exegetical methods to develop explanations and descriptions that divinize and therefore naturalize the treatment process and its components (Kahn, 2006).

Finally, regardless of the tenor of the larger debate, local rabbinic authority often dictates practices within the local community. Despite the enormous variations in acceptable rabbinic views, infertile orthodox couples may not simply shop for a sympathetic opinion. Rather, they are bound to obey the opinion of the same rabbi concerning infertility treatment to whom they turn for guidance in other aspects of their life. Consequently, in keeping with the fourth principle above, each orthodox couple's fertility problem is evaluated by both their rabbi and a doctor, who work together to balance and integrate halakhic considerations with available and appropriate medical protocol.

At the opposite end of the spectrum, less observant Jews, whose concern to adhere to halakhah is less intense, may either ignore rabbinic guidelines entirely or establish practices rooted in new interpretations of halakha. Reconstructionist rabbi Renée Bauer, for instance, reasons that a child conceived by an observant but non-Jewish lesbian with her Jewish partner is also Jewish—not because of biological origin of the sperm, or through adoption or conversion, but because its parents intend to coparent the child they have conceived together in a Jewish household.[78]

Clearly orthodox Jews have embraced the potential of new reproductive technologies both formally and informally. This embrace has not challenged traditional notions of kinship, nor has it destabilized traditional understandings of the role of God in conception. Rather, extraordinary efforts have been made, both conceptually and practically, to integrate these technologies into existing systems of religious meaning and existing social and professional networks. Rabbinic flexibility with regard to the appropriate uses of these technologies is perhaps extraordinary when juxtaposed with other religious traditions, but when understood in context of the historical processes of rabbinic decision-making regarding medical and technological developments more broadly, it is in fact commensurate with traditional practice. It should also be noted that the freedom this flexibility bestows on infertile orthodox Jewish couples is actually, rather than only potentially, a two-edged sword: it explicitly requires them to make greater efforts than ever before to fulfill their obligation to reproduce, further solidifying the heterosexual, married, Jewish, gestating couple as the halakhically normative family.

In Sunni communities, moral deliberation on important issues is centralized and scripturally based, and authoritative *fatwas* are issued by clerics who in some cases are even appointed and paid by national governments (Inhorn, 2003c, 104). As we have seen, these *fatwas* permit use of ARTs in support of homologous, marital reproduction, but concerns about adultery and inheritance have led the Sunni clerics to forbid the use of third-party gametes. Notably, these limits are very

---

[78] Bauer (2006, 33). Reconstructionist Jews practice egalitarian descent: a child is a Jew if either of its parents is a Jew.

rarely transgressed in practice, thanks to Sunni Muslims' concerns to procreate in Islamically correct ways.[79]

Iranian Shi'ite Muslims are no less concerned than their Sunni counterparts to conform to religious tradition.[80] But two factors lead to greater potential flexibility in the uses of ARTs among Shi'ites. First, they employ an individualized form of moral reflection, known as *ijtihad*, which makes use of ᶜ*aql*, or intellectual reasoning, rather than relying primarily on scriptural exegesis. This method has opened a door among some adventuresome Shi'ite religious leaders toward third-party donation to resolve infertility in married couples through a potential subtle transformation in concepts of kinship. Rather than breaking with traditional visions of kinship entirely, these new approaches employ them innovatively. The process of transformation began with a tentative step away from absolute prohibition. In the late 1990s, the Supreme Jurisprudent of the Shi'a branch of Islam, Ayatollah Ali Hussein Khamanei (successor to Iran's Ayatollah Khomeini) issued a *fatwa* declaring that egg donation "is not in and of itself legally forbidden," but that *both* the egg donor and the infertile mother must abide by the religious codes regarding parenting. Thus, the child of the egg donor has the right to inherit from her, as the infertile woman who received the eggs is considered an adoptive mother. Sperm donation is a slightly different case (Inhorn, 2006b). The baby born of sperm donation follows the name of the infertile father rather than of the sperm donor and, as with egg donation, the donor child can inherit only from the biological father, since the infertile father is considered to be an adoptive father.

Second, Shi'ite clerics overcome these problems of potential adultery by justifying gamete donation for their followers under a form of Shi'ite temporary marriage called *mutᶜa* (also called *sigheh* in Iran) (Zuhur, 1992; Inhorn, 2006b), which allows couples to avoid the complications of adultery. *Mutᶜa* is a union between an unmarried Muslim woman and a married or unmarried Muslim man, which is contracted for a fixed period in return for a set amount of money. In the past, divorced or widowed women often engaged in *mutᶜa* marriages for financial support, and Shi'ite men contracted *mutᶜa* marriages while traveling, or as a means of achieving marital variety and sexual pleasure. Since the arrival of donor technologies, *mutᶜa* has been invoked to make egg donation legal within the parameters of marriage. In Iran, unmarried women who agree to participate as egg donors may enter one- or two-day *mutᶜa* marriages for a fee. Such marriages require no witness and are not officially registered; thus, they can take place in confidence in the back rooms of IVF clinics. Indeed, donors who wish to remain anonymous enter these *mutᶜa* marriages by written agreement, without ever meeting the recipients of their eggs or their

---

[79] A global survey of sperm donation among assisted reproductive technology centers in 62 countries provides some indication of the degree of convergence between official discourse and actual practice. In all of the Muslim countries surveyed, sperm donation in IVF and all other forms of gamete donation were strictly prohibited. Meirow and Schenker (1997), cited in Inhorn (2003c).

[80] Shi'ites, the minority branch of Islam, can also be found in parts of Iraq, Lebanon, Bahrain, Saudi Arabia, Afghanistan, Pakistan, and India.

temporary husbands or divulging their own personal information to them, receiving their money following egg harvesting (usually around US$550).[81] Thus oddly, despite meticulous concern for adherence to religious law, egg donation is largely a financial transaction.

But donor sperm are more problematic than donor eggs because—as a Muslim wife cannot have more than one husband, and single women cannot legitimately bear children—they inevitably produce an out-of-wedlock child, or *laqit*. The rarely-exercised option in this case is for a woman to divorce her husband; wait three and a half months to ensure that she is not pregnant; enter a temporary marriage with a fertile man; and later divorce him and remarry her first husband (Tremayne, 2005).

In Iran, religious rulings regarding gamete donation have continued to evolve quickly. A law on gamete donation passed in 2003 in the Iranian parliament (*majlis*) and approved by the Shi'ite Guardian Council restricted gamete donation to married persons. Egg donation is allowed, as long as the husband marries the egg donor temporarily, thereby ensuring that all three parties are married. Sperm donation, on the other hand, is forbidden, because polyandry is illegal: an already-married woman whose husband is infertile cannot marry her sperm donor. However, quite interestingly, embryo donation—which involves both sperm and egg from another couple—*is* allowed in order to overcome both male and female infertility. Because an embryo comes from a married couple and is given to another married couple, it is considered *hallal*, or religiously permissible (Tremayne, 2005; Inhorn, 2006b). This allows Iranian couples affected by male infertility to bypass the problem of the husband's weak (or absent) sperm. Embryo donation, like egg donation, is primarily a financial transaction, but it is much more heavily regulated.[82] According to Tremayne, the recent law in Iran specifies clearly that infertile couples requiring an embryo apply in writing to a court for permission for embryo transfer. The law specifies that, like adoptive couples, the recipient couple must be morally sound and suitable as parents and must be Iranian citizens. Yet the donor embryo law is so new that most IVF clinics in Iran do not yet own a copy of the legislation and do not necessarily abide by it. If the husband is infertile, the couple simply receives another couple's embryos, with most donor couples choosing to remain anonymous. One author of this paper, Marcia Inhorn, observes that absent a national law, similar patterns of reasoning seem to exist among clinics that serve Shi'ites in Lebanon.

---

[81] Tremayne (2005). In Iran, women commonly bring their sisters as potential egg donors. But this is not allowed, as Islam is explicitly against the marriage of one man to two living sisters. Apparently, men also bring their brothers as potential sperm donors, despite the prohibitions outlined below. Tremayne observed one case where the husband did so without his wife's knowledge; the wife believed that she was receiving her husband's sperm instead of that of her brother-in-law. She also reports that non-anonymous donations sometimes lead later to tension between the birth family and the female donors who wish to exercise parental rights.

[82] This is very similar to organ donation, which is being done through financial transactions between donors and non-related recipients in Iran; see Tober (2004).

Consequently, some Shi'ite Iranian and Lebanese couples are beginning to receive both donor eggs and donor embryos[83] as well as to donate their gametes to other infertile couples. Their reasons hint at the path future clerical justification may take. Infertile Shi'ite couples who accept the idea of third-party donation—as well as Ayatollah Khamanei—describe donor technologies as "marriage saviors" that deflect the "marital and psychological disputes" that may result from otherwise untreatable infertility (Inhorn, 2006a). They fit this treatment into a view of life in which God not only prescribes marriage and childrearing for all but is present and intimately active in all events. Like orthodox Jews, Muslims of all stripes believe that all creation, including the creation of human beings, is ultimately in God's hands; or, as Egyptians are fond of saying, "Human beings are incapable of creating even a fingernail." Each individual's life is "written" by God; events occur at predestined times according to a plan the purpose and meaning of which can be known only by God. Most important to this discussion, human procreation is beyond human control. God not only decides who will be fertile and who will be infertile but also imbues each fetus with a soul and a gender (Quran 42:49–50; Inhorn, 2003c, 102–103). Thus, infertility is ultimately explained by most practicing Muslims as an example of God's will: if God does not allow a woman to become pregnant, it is because God has reasons for waiting—for example, to prevent the birth of a child who will be abnormal or grow up to lead a bad life. Thus, women who fail to become pregnant often remark, "My time hasn't come yet." Yet, as in orthodox Judaism, the belief that life is "written" does not imply that human beings are passive creatures, devoid of volition and will. Science and technology are part of the divinely scripted human story. God expects human beings to exercise their minds, to make use of available means, and to make choices, including decisions about seeking treatment for conditions such as infertility. Infertile Middle Eastern Muslims argue that since God expects those who are sick to seek treatment, they are obligated to do so (see Inhorn, 1994, 212–218; 1996, 65, 76–81).

Thus in both the orthodox Jewish and Muslim cases, the divine will is made known through debates among religious leaders who work within the developed notions of divinely-ordained kinship to promote a heterosexual, gestational vision of marriage and family. Both see medicine as a divinely-ordained means for pursuing this divinely-ordained, ideal (even obligatory) vision of family life. In both traditions determinations about means are made "on the ground," often for particular couples, with significant pastoral concern, in full awareness of the spirited debates being conducted among religious authorities. And in both traditions, infertility, its treatment, and its unknown outcome are all seen as part of God's wise but impenetrably mysterious plan for individual lives.

---

[83] Furthermore, in Lebanon, anonymous sperm donation—using frozen sperm from overseas sperm banks or fresh sperm samples from mostly medical and graduate students—is also "quietly" practiced in IVF clinics. One of Marcia Inhorn's azoospermic Lebanese male informants produced a donor child this way, and several others, both Muslim and Christian, had also made the decision to use donor sperm.

A similar pattern can be found in Hinduism. Westerners especially must be cautious about making blanket statements about Hindu moral thought, for Hinduism "encompasses a broad array of traditions, sects, and religious-philosophical schools" (Marglin, 1985, 39), and its beliefs, customs, and emphases take on highly particular local forms. The moral tradition, though unified by very general principles of life-promotion, is also multivalent and multi-voiced, without any sort of central authority or system of doctrinal dissemination. This multivocity is also due in part to the importance of narrative in Hindu moral reflection, which in turn feeds on the proliferation of gods and goddesses and the scriptural accounts of their personalities and interactions.

Like Arab Muslims and Israeli orthodox Jews, Indian Hindus have traditionally seen childbearing—especially the production of male children—as essential to marriage and to female identity. Begetting a son is not merely a this-worldly duty but fulfills ritual obligations: it delivers male ancestors from *putra*, a hell (Fagley, 1965, 333; Bharadwaj, 2003, 1870–1871), and ensures that rituals necessary to the health and successful passage of a father's soul can be performed for him after his death (Mohapatra et al., 2001, 34; Bharadwaj, 2003, 1870; Crawford, 2003, 116–119). Infertility and adoption are stigmatized, elevating the pressure to conceive (Malpani and Malpani, 1992, 49–50; Bharadwaj, 2003, 1870–1871). In addition, "Hindu law is very flexible and accepts anything which is good for mankind provided that it does not inflict any injury to the moral values and sentiments of others" (Chakravarty, 2001, 11–14). Here clearly "good for mankind" is the operative value; enormous numbers of Hindus welcome ARTs as means to fulfill the social and religious obligation to procreate. In fact, they are so welcome that Malpani and Malpani argue forcefully against use of expensive, inefficient techniques like ICSI and for the development of lower-technology, lower-cost methods that, while they might be less effective per treatment cycle, are cheap enough to be universally accessible (Malpani and Malpani, 1992, 49–50).

In addition, Hindus too see the divine—unified in one God, but manifest to people in the local actions of multiple gods and goddesses—as involved in the minute details of daily life. Hindu ART patients often make temple pilgrimages to petition for successful treatment, and patients and clinicians tend to place each case in the hands of God, attributing success or failure to God's will (Bharadwaj, 2006; "Medical Ethics", 2002). But the particular character of Hindu relationships to divinity tends to differ slightly from some monotheistic traditions. In Hinduism the disciple is to the god roughly as child is to parent, and it is understood that the god expects and tolerates childlike behavior—the religious equivalent of tantrums—and will eventually reward or at least comfort this persistent pestering. In addition, the wall between divinity and humanity is a bit lower than in many monotheistic traditions, allowing doctors and medical therapy generally to be treated as quasi-divine (Bharadwaj, 2006).

We cannot speak with utter confidence on the popularity of these ideas, but there is some evidence that the concerns that ARTs raise about divinely intended nature are slightly different in Hinduism than in either Islam or Judaism. Since the goal of Hindu life is to be reborn as a higher being, the spiritual auspiciousness of any event has moral overtones. An author in *Hinduism Today* notes,

1 Compatible Contradictions: Religion and the Naturalization

> From the Hindu point of view, conception connects a soul from the next world to this world, and the state of mind at the moment of conception including the purity and spiritual intent of both partners is a major factor in determining who is born into the family. Prospective parents often offer prayers at the temples, perform spiritual disciplines and visit saints for their advice and blessings in their effort to conceive a worthy child. In Western thinking, no emphasis is placed on the state of mind of the parents at conception, and there is little understanding of the ways parents can affect the "quality" of the souls born to them ("Medical Ethics", 2002).

The same article hints that IVF, third party donation, and other procedures, while not absolutely forbidden, must be used cautiously because they may introduce *spiritually* problematic unknowns into the conception of a child; the implication is that homologous procedures are less spiritually and psychologically risky than heterologous ones, which not only weaken the child's relationship with the biologically unrelated parent but may bestow the bad *karma* of adultery upon the child. The "unnaturalness" in this case is explained as being a matter of virtue and spiritual auspiciousness, not a just a matter of legal right-relationship.

Another evidence of this difference can be found in references to *niyoga*, an ancient custom that has been suggested as a justification for third-party sperm donation ("Medical Ethics", 2002; Schenker, 1992, 4; Crawford, 2003, 116–119), but also as worthy of revival in its old form, because it avoids the possibly emotionally and spiritually inauspicious intervention of "unnatural" medical instruments in conception and typically maintains some genetic connection between the social and donor fathers (Mohapatra et al., 2001, 33–38). In this case achieving the "natural" family—specifically, the production of a son—*and* the ideal of "natural" heterosexual intercourse seem to share primacy of place. This now-defunct custom allowed a childless widow, a woman betrothed to a man who died before their marriage, or a woman with a sterile husband to contract with one of her husband's male relatives to impregnate her. The relationship was to last only until impregnation occurred, permitted intercourse only once per month during her fertile period, and involved a number of regulations in dress and conduct designed to reduce the possibility of "extra" visits and of full enjoyment of sexual contact. It could also be renewed later if the child conceived turned out to be a girl, or if the family desired a second son. Any resulting children were considered the offspring of the woman's husband (Mohapatra et al., 2001, 33–35). But it was clearly an asymmetrical solution, providing only a means for fertile women to conceive sons for their infertile or dead husbands and (because it was available to widows) valuing production of a son above creation of a mother-father-child unit.

Mohapatra, Dash, and Padhy find precedent for *niyoga* in the *Manusmrti*, or Laws of Manu, dating from the second and third centuries AD. But they are also quick to point out that if their solution is orthodox, this does not mean that it is obligatory, as—if fulfillment of ritual obligations is really the only concern—there are a number of other traditional ways for a man to "acquire" a ritually acceptable son. As alternatives to *niyoga*, a husband might traditionally have pursued various kinds of adoption; recognized a child possibly or definitely conceived by his wife with another man; purchased a son; or recognized a son he had conceived before marriage. Failing all of these, a woman could appoint a son conceived through her second marriage to perform rites for her first husband, or (if a man had a daughter but no sons) he could, at his daughter's marriage, claim rights to her future first son (Mohapatra et al., 2001, 35–37). Seen from this angle, the criteria of the divinely-ordained natural family seem to be a man and his son, as a woman's role is really to bear a son for the father.

The case of *niyoga* illustrates how, in Hindu thought, although divine will can be discerned in scripture and tradition, and local custom or theology can make a convincing case for a particular approach or set of guidelines, there is no theological means of universalizing any single conclusion. The discussion remains inherently multi-vocal. In contrast to Western monotheistic traditions, few Hindus consider unanimity on concrete moral questions even to be a religious ideal. The result is a flexible system that has both the advantages and disadvantages of responsiveness to individual cases and local customs.[84] In addition, *niyoga* and other variant means of acquiring sons demonstrate that Hinduism's practice of accomplishing "orthodox" family formation flexibly to accommodate infertility is backed by a tradition of nearly two millennia. This suggests that ARTs, like the myriad solutions that predate them, will never *by themselves* dislodge traditional ideals of divinely prescribed natural family form. These may crumble eventually, but only if novel internal or external forces collude.

In all three cases mentioned above, only *the list of acceptable, natural ways of creating the religiously orthodox family* has expanded; *the structure of this traditional natural family* has been reinforced. ARTs shore up the ideals of heterosexual, procreative marriage and of women as mothers by nearly guaranteeing that they can be realized. Promoters of ARTs need merely convince religious people that ARTs support regnant understandings of the divinely-ordained nature of things, whether that is a nature that mainly sets limits (by outlining the boundaries of purity) or a nature that points toward some form of ideal realization that is practical, spiritual, or both. The argument's success absolves religious people from the responsibility to develop alternative ideals for holy childlessness for couples (and especially for women) while they paradoxically and simultaneously remind women and couples who remain infertile even after approved treatment that their childlessness is somehow also a product of God's will for them.[85] As a result acceptance of ARTs does not merely ratify existing beliefs about the clarity and continuity of the divine will but also narrows the criteria of adherence to this will in comparison with earlier periods. Several authors have noted this effect in communities where ARTs and newly-energized identity movements coincide (Kahn, 2000, 172–175; Inhorn, 1996, ch. 2; 2003c, 90).

---

[84] For more on local moral worlds, see Arthur Kleinman's work on the "commitments of social participants in a local world about what is at stake in everyday experience" (1995, 45).

[85] Inhorn notes that faithful adherence to restrictive religious guidelines for ARTs actually thwarts women's accomplishment of their religious ends. We argue that wherever this is true, the religion's pronatalism must not be *de jure* absolutely monolithic, even though it may be so *de facto* in local practice.

1 Compatible Contradictions: Religion and the Naturalization

In Euro-American cultures fertility patients are less commonly explicitly concerned to adhere to religious law. Charis Thompson, who studied American infertility and the principle of privacy, noted that her subjects virtually never made any appeals to religious law that were directly relevant to their treatment decisions. On the other hand, she found that many subjects explicitly invoked religion during their infertility treatments (Thompson, 2005, 217–219; 2006).[86] Thompson sees this phenomenon as a product of the way America society organizes privacy, which encourages private acts of "meaning-giving" not only for all discrete experiences and events—in this case, for infertility generally—but for each decision and stage in treatment as well.[87] Religiously, private meaning-giving reflects the Protestant tradition of privileging the conclusions of a conscience formed in prayer, study, and consultation, as well as the tendency of American adherents of traditions with central authority (e.g., Roman Catholicism) to lean toward the Protestant end of their tradition's approach to moral law. It also reflects the tendency of religious Americans to give personal religious meaning to momentous events and decisions independently of the formal theological meaning that religious institutions may supply. Finally, in the absence of extreme economic and social pressure to procreate, some religious couples who reject the conscience model and whose traditions strictly forbid use of most ARTs may simply not investigate fertility treatment and so would not show up in a study like Thompson's.

Importantly, claims that God's will and power are at work in a person's body, in a clinician's work, or in a Petri dish do not have the same significance in all circumstances. In the orthodox Jewish, Muslim, and (perhaps to a lesser degree) Hindu cases, religious people and their religious authorities agree that God promotes procreation, limits the licit means and circumstances for achieving it, and is directly and intentionally responsible for each success or failure within these limits. Local communities within the traditions may disagree—and be aware that they disagree—over the exact content of these guidelines, but they do not disagree with the fundamental claim that God's hand is directly at work in each attempt at conception.

Similar statements about divine activity by Protestants who are members of so-called conservative communions would have comparable meanings. Conservative evangelical Protestants (e.g., adherents of the Southern Baptist Convention and the Assemblies of God), who tend to see God as active in the minute details of individual lives, also give the highest level of assent to the inerrancy of Scripture and the greatest weight to biblical norms in moral decision-making. But even for the many Protestant communions that have a more "hands-off" vision of God, Scripture is the primary touchstone for individual and communal moral discernment. We should

---

[86] See also Thompson (2006). Compare Greil (1989, 13): "My interviews with infertile couples indicate that such indifference to religious objections is widely shared. For the couples I have interviewed, practical concerns overshadow ethical concerns. These couples have one overriding goal: to become parents. They judge treatment options primarily on whether they are efficient and practical."

[87] Thompson (2005, 219).

expect to see ethical questions related to assisted reproduction framed in terms of the biblical vision for sexuality and reproduction.

The Lutheran Church—Missouri Synod, a conservative Protestant denomination, adheres to this model: "God's word establishes as the appropriate context for procreation the loving relationship of husband and wife conceiving a child from their bodily lives as father and mother" (Missouri Synod, 2002, 18). This statement from a report on the ethics of cloning captures several typical features of Protestant approaches to reproductive technologies, both liberal and conservative. First, the starting point for weighing the ethical or theological significance of medical interventions into reproduction is God's intention for procreation as revealed in Scripture. Drawing most often from the first part of the book of Genesis, Protestant analyses generally affirm marriage as the divinely sanctioned locus for procreation. Further, they frequently invoke a normative connection between sexual differentiation (as intentionally established by God's creative activity), procreation, and marriage. Using language borrowed from Anglican theologian Oliver O'Donovan, Lutheran theologian Gilbert Meilaender explains the first axis of connection this way: "The creation story in the first chapter of Genesis depicts the creation of humankind as male and female, sexually differentiated and enjoined by God's grace to sustain human life through procreation. Hence, there is given in creation a connection between the differentiation of the sexes and the begetting of a child" (Meilaender, 1997, 41–42; see also O'Donovan, 1984). The Missouri Synod report on cloning describes the second axis: "God has designed the procreation of a child to be a complex uniting of two similar but significantly different individuals whose union bears a new flesh that is rooted in but different from the flesh of the parents" (Missouri Synod, 2002, 16). In the context of a committed, enduring marriage sex achieves its dual function of uniting the partners and at the same time turning their love outward toward the world in the form of a new, unique life. Thus, writes Meilaender, "by God's grace the child is a gift who springs from the giving and receiving of love. Marriage and parenthood are connected—held together in a basic form of humanity" (Meilaender, 1997, 42).

As we have seen already, this does not mean that Protestants uniformly refuse to separate sex and procreation, or even marriage and procreation. It is simply that Protestants take biblical models of family procreation as their starting point and draw analogies to them in order to justify deviations, cognizant that even their reading of the Bible is selective (for instance, few Protestants embrace polygamy, which is certainly a biblical model of family, or press St. Paul's enthusiasm for celibacy). In addition, moderate and liberal Protestants leave most decisions about sex and procreation to the individual conscience, informed by scripture and communal moral reflection. Thus within these communions the faith that "God is powerfully at work" in a particular person or procedure is a statement of devotional confidence and personal conscience rather than a universalizable dogmatic assertion; it is likely to be met with polite respect but makes no normative claim on fellow churchgoers' own fertility decisions.

There is one important further circumstance we have not explored. For singles and couples whose use of ARTs contradicts the teachings of a highly centralized religious communion, the claim that "God is at work" in their treatment is both an act of resistance that constitutes a partial or even wholesale rejection of their tradition's view of "natural" procreation and an appropriation of personal religious authority against the claims of the tradition to know and interpret God's will and laws. It is, in this sense, an act of rebellion. Athenian women who freely employ abortion and donor IVF, Arab Sunni Muslims who travel to Shi'ite countries for treatment, and American Roman Catholics who use contraception and fertility treatment freely (not to mention single straight and single or partnered lesbian women and gay men of almost any tradition who conceive through ARTs) are in this sense claiming not only parenthood but also the ability to reinterpret and reform their own religious traditions.[88]

This phenomenon has been noted among Roman Catholics in Latin America. Elizabeth Roberts reports that Ecuadorian fertility patients and clinicians are absolutely devoted to scientific and clinical rigor and are also aware that the Roman Catholic Church opposes ARTs. Yet—much like the Indian Hindus, Arab Muslims, and Israeli orthodox Jews mentioned earlier—they insist that God is nonetheless present and active in all events. They consistently invoke God's blessing at all stages of reproductive interventions, believing that God empowers their work and the work of science itself. Roberts sees this behavior "partially as a means to challenge Church condemnation of their practice" (Roberts, 2003; also 2006). Although Gwynne Jenkins does not draw any conclusions about resistance, she reports that Costa Rican fertility patients also interpret the particular ups and downs of infertility, fertility treatment, adoption, and conception as direct products of the divine will (Jenkins et al., 2002).[89] It is worth asking whether use of ARTs in conscious defiance of religious authority may not shift the understanding of the "natural" family—or of the source of its authority—slightly. To flaunt orthodox religious authority in order to create what would otherwise be a "natural" orthodox family is to erode the authority that originally reified that family model.

Liberal Roman Catholic responses to *The Instruction on Respect for Human Life in Its Origin and on the Dignity of Procreation* (*Donum Vitae*) hint at the beginning of such a shift—and give evidence that even in a highly centralized, apparently monolithic institution, authority structures are more complex than official documents might lead one to believe.[90] As we have seen, in its more or less blanket condemnation of assisted reproduction, the encyclical *Donum Vitae* invokes a "natural language of the body," which sets the parameters for legitimate reproduction and gives a rationale for the Catholic Church's insistence in its sexual ethic on the

---

[88] Of these traditions, worldwide Orthodox Christianity is for theological reasons decentralized, although all communions look to the Patriarch of Constantinople for leadership. However, particular communions within Orthodoxy may be highly centralized, especially—as in the Greek case—if they are national churches.

[89] See also Paxson (2004) and Kahn (2000) on the intimate, spiritually-charged relationships that develop between patients and clinicians in Israel and Greece.

[90] For a full discussion of the theology of *Donum Vitae*, see pp. 33–34 above.

inseparability of sex, reproduction and marriage. Critics of this document who question the inherent, created moral authority of the structure of the sexual body therefore also question the authority of the Roman Catholic Church to impose it as a norm. They have argued that *Donum Vitae*'s conclusions about what sort of practices are consistent with the character and dignity of human reproduction are drawn deductively from a narrow, "physicalist," deterministic reading of natural human sexuality and reproduction and uncritically enshrine as "nature" what may in fact be time-bound cultural customs. As theologian Lisa Sowle Cahill put it,

> appeals to 'natural moral norms' [such as those made in *Donum Vitae*] seem to rigidify into moral absolutes human realities that either are the merely biological preconditions of human decisions (like the genetic code or the reproductive system) or are but culturally enshrined though nonnecessary patterns of decision and action (like feudal socioeconomic organization or patriarchal marriage) (Cahill, 1988, 11).

This development is particularly vexing for feminist theologians (Post and Andolsen, 1989; Traina, 1999). The difficulty of identifying "objective" human goods leads some Catholic moralists to reject the Church's official method of natural law reasoning altogether in favor of values such as "justice" and "responsibility" for judging what is fitting in reproductive interventions (Farley, 1983). Others, like Cahill, hold that it is possible to draw conclusions about what is normatively human in reproduction, but only through a process that is inductive, empirical, progressive, and provisional, rather than deductive and conclusive (Cahill, 1988, 13). In any case, even many Catholic moral theologians who defend the normative relationship between marriage and reproduction emphasize the marital relationship as a sexual-procreative whole rather than evaluate (as in the Vatican position) the morality of each potentially procreative act individually. Thus, they accept homologous AI and IVF as legitimate alternatives for realizing the human drive toward married generativity. A minority of Catholic theologians accept donor-assisted reproduction, arguing that the willingness and capacity to care for children, rather than biological relation, are the preconditions for truly good parent-child relationships (Lauritzen, 1993). Significantly, these same authors also depart from Roman Catholic teachings on gender and other issues on the basis of their claim that where the Church's moral reasoning is logically faulty, it carries no moral authority.

1.3.3 The Divine in and Against Nature

Clearly, it matters to what degree—and in which events—God is understood to be present, manipulating developments, and to what degree this intervention is seen as being "with nature" or "against nature." Even when nature is identified with physical process, the divine will may or may not determine its course in a particular moment. For example, the entrance of a particular sperm into a given ovum can be seen as divine anointing of a "winner," a sperm that will result in the specific child these particular parents were meant to raise (and the implantation of that ovum would be read as another signal; the miscarriage of the resulting fetus as another; etc.); as the product of a divinely-designed process that usually, but not always, works in favor of fertility and health and therefore is susceptible to limited human assistance; or as a random event, and therefore fully open to human control for constructive purposes. These differences can lead to quite separate understandings of what occurs in fertilization and implantation; to separate assessments of the means and motives for "intervention" or "assistance" in nature; and therefore to very different explanations and evaluations of particular ARTs. In addition, it matters where the divine will is written (in scripture, and if so, in which passages; in human reason; in the structure and function of the human body; in answers received to petitionary prayer; in developed extra-scriptural tradition, etc.), as well as who is the authoritative interpreter (a large body of leaders who debate over many years; a single religious leader; the infertile couple; or the woman who seeks to conceive). Global cultural relations play unexpected roles. For example, in Japan ARTs, like all Western technologies, are greeted warily because of their association with "unnatural" Western individualism and self-seeking greed; the implication is that they must be naturalized and divinized by being turned to community-building ends (Lock, 1998, 215–216, 232–233). In this case the West is an unnatural anti-authority that throws the continuing power of traditional Japanese cultural values into relief.

In any case, clearly infertility inspires many religious people to make choices about which strands of their locally experienced religious tradition to choose as authoritative. These choices are usually (especially outside Euro-American culture) but not always in line with the forms of family structure that the traditions have promoted or called "natural." But these moments of decision are also (especially in Euro-American culture) often moments for the critical reevaluation of the religious tradition itself, and therefore for its transformation. Generally, however, the tradition is criticized and transformed on its own terms. For instance, liberal Catholics judge Roman Catholicism's current version of natural moral reason insufficiently rational. Finally, as many commentators have noted, religion provides people ways to find meaning, hope, and consolation in circumstances whose resolution is complex, outside human control, and to a great extent even beyond human understanding. Infertility calls for theodicy.

Finally, in nations in which there is either an official religion or a tradition with an overwhelming majority, the distinctions between religious and political debates often are either nonexistent or at least highly fluid, suggesting that "naturalization" is driven partly by political concerns. The identity of civil and religious law in Iran is the clearest example. The results of such influence are complex and sometimes surprising. Nia Georges notes that the Greek Orthodox Church in Greece opposed legislation that would have made artificial insemination legal, a proposal that would seem to have promoted the church's own emphasis on marital procreation. However, the church apparently objected not to the "unnaturalness" of artificial insemination but to a perceived connection between the legislation and interests that supported AID, IVF, and use of ARTs by unmarried women. The influence of legal thought on moral deliberation in Japan is another example. Yoshinao Katsumata points out that because few Japanese medical schools have departments of medical ethics, the task of medical ethics education falls to departments of legal medicine (Katsumata,

2000, 491). This state of affairs has profound influence upon the kinds of moral questions people raise. For example, in a democratic culture the basic legal question tends to be "What can we permit?" while the basic religious question is "For what shall we strive?"

1.3.4 Eagerness or Reluctance of Traditions to Use "Nature" as an Interpretive Category

It seems worthwhile to point out that the willingness of religious people and institutions to talk about the "naturalness" of ARTS has a great deal to do with whether, and how, the evaluation of human behavior is related to "nature." If "nature" is seen as a Western category with a content that is antithetical to religious values (Japan, e.g.) or that is insufficient for religious moral reflection (Western traditions of the "book"), or if it is seen as unredeemed/unredeemable (some Protestants), then "nature" will be used not at all or with qualifications. This strategy tempers the authority of science and medicine. On the other hand, sometimes the divine will and biological nature are seen as synonymous, in which case they may be friendly to some sciences but cautious toward technology, which is seen as potentially unnatural.

This conundrum might also be expressed as the question, at what level is the argument from "nature" authoritative when accounts seem to conflict? For instance, there is the microcosmic question about nature—by what biological process is it natural and acceptable to conceive a child?—and the macrocosmic one—how does reproduction fit in to the natural social order, and what is the latter? "Against nature" might also simply refer to anything that is empirically observed to harm flourishing—for instance, the increasing incidence of risky ART-related multiple births (Bryan and Denton, 2001). Strathern's observations about Western views of nature and kinship also explain reluctance to martial "nature" in arguments: not everyone regards scientific "nature" as the foundation of reality.

1.4 Conclusion

1.4.1 Nature Contested

As we have seen, the answer to the question, "how do religious people use 'nature' in their moral descriptions of infertility and ARTs?" has more answers than there are religious traditions, and even more answers than the number of religious traditions multiplied by the number of national cultural variations of each. Gender, class, education, position within the religious community, and a number of other factors also enter the equation. More revealing—and ultimately more useful to policymakers—is the process of tentative naturalization, divinization, and/or rejection of these technologies within religious institutions and among religious laypeople. This process reveals religious people's finely shaded understandings of gender and kinship, the rank and order of their authoritative sources, the channels through which religious authority flows, and many other nuances that do much more to help us understand their likely responses to new technologies or to technological proposals than would the simple, material answer to the question, "are ARTs natural?"

However, if we were asked to draw a tentative conclusion from our work, we would say that in a competition between achieving the "natural" or acceptable family and using "natural" or non-technological means, in most cases the latter value gives way to the former. This suggests that religious people will eventually tend to support the use of ARTs for married, heterosexual couples. Our strong reservation about this trend is that in societies with uneven economic development and few opportunities for women, wide use of ARTs will not only entrench gender roles and gender inequality but likely even reduce the proportion of female to male births. Approval of ARTs in other cases may be less universal, although the American and Athenian cases suggest that an educated, cosmopolitan society will be willing to use ARTs to create other kinds of families. In all cases, however—whether ARTs are embraced or rejected—the scientific knowledge that ART research generates and ART debates disseminate alters religious descriptions of "natural" reproductive processes and the "natural" family.

1.4.2 *Policy Implications*

The foregoing study raises a number of important points for policy and for policymaking. First, a study of the moral statements of major traditions is not an adequate way to predict religious people's responses to ART policy proposals. To begin with, because most traditions do not create bureaucracies, not all have centralized discussions and publications of moral positions, and their members' moral opinions of ARTs are therefore varied. Second, not all traditions with bureaucracies are centrally concerned with bioethics or exhibit real unanimity on bioethical questions. Third, not all religious people are members of the "big traditions." Fourth, in some cases official religious teaching is local—a rabbi, or imam, or head of an autocephalous Eastern Orthodox communion may speak to and for a community of adherents. Yet—fifth— even the local mores generated by this local moral world are rarely applied literally and inflexibly. Religious advisers, and often religious people themselves, often adjust them pastorally in order to relieve suffering and preserve relationships in quite particular personal circumstances formed by dynamic relationships among factors like the following: official religious teaching; cultural views of gender; cultural views of marriage; economic relationships; political relationships; social relationships; class and race; expectations of levels and kinds of available health care. Finally, in these processes of interpretation and adaptation, we see distinct but related processes: *communal* (*de jure* and official, either local or global), *individual*, and *collective* (*de facto*, but acknowledged and widespread, common patterns of decisions by individuals in a tradition) transformation of ideas of what is "natural" with regard to reproduction.

The implication for policymakers is that although religious organizations sometimes lobby, adherents of religions do not follow their leaders docilely on this issue (see Thompson, 2006, 2005; Paxson, 2004; Greil, 1989). Examination of one tradition's official teachings or consultation with another's leader will not provide a reliable measure of individual adherents' support for particular policies. These investigations are of very limited usefulness. They can only alert policymakers to the kinds of goals people from particular religious backgrounds may be especially interested in pursuing or provide a context for the arguments and objections they may be raising.

A second lesson is that in both ethics and policy mention of "nature" or the "natural" is not an argument in itself. References to "nature" raise more questions than they answer, as the meaning of the word varies according to the religious and cultural world of the speaker and sometimes varies for the same person depending on the precise question asked. "It's natural" marks the beginning, not the end, of the search for moral understanding.

A more substantively helpful consequence of this investigation is that it provides a clearer understanding of the process by which ART procedures that apparently contradict some religious commitments nevertheless win the support of religious people. We have argued that reproduction is so tightly bound with religious identity and religious ideas of the natural that religious people's decisions about ARTs are rarely revolutionary or discontinuous with tradition. Rather, their decisions exhibit *subversive continuity* with accepted traditions. That is, religious people are willing to transform some of their deeply-held assumptions about nature significantly, but they do so by preferring or even absolutizing other equally traditional assumptions about nature. We argue further that when push comes to shove these decisions most often permit alteration of "natural" biological processes in favor of preserving "natural" family and social relations. Thus policy innovations phrased to support religious visions of the "natural family" are likely to be better received than those that openly embrace its erasure. As the chart below shows, not even the "great traditions" embrace identical reductive definitions of the natural family, but secondary concerns about proper procreation create significant practical overlap in a practical benchmark: heterosexual, two-parent families with biologically related children.

Religious tradition	Reductive core of the "natural family"[91]
Hinduism, Confucianism	Father and son
Judaism	Mother and child
Roman Catholicism, Eastern Orthodoxy	Husband and wife, bonded by God (children anticipated and desired; childlessness does not invalidate marriage)
Islam, Protestantism	Husband and wife, bonded by God through children
Buddhism	Depending on local traditions, husband and wife (children optional but anticipated and desired)

---

[91] We did not explore, but think it worth asking, whether traditions that strongly support vocational vowed celibacy as an alternative to marriage view ARTs differently than those that do not.

Finally, as we have already indicated, intent is not result: even policies designed to shore up the "natural" family will transform it subtly. ARTs will often produce unintended consequences when applied within local moral and religious frameworks. We have already alluded to probability that in more affluent settings liberal ART policies will have paradoxical results: policies originally intended primarily to help married women to realize motherhood will lead to an expansion of intentional single motherhood, and policies that invite openly gay and lesbian couples to become parents through ARTs will end up supporting the traditional ideal of the two-parent nuclear family. These secondary results in turn may encourage a tertiary consequence: in both cases ARTs encourage a subtle privileging of biological or gestational relatedness over social parenting, implicitly jeopardizing cultural acceptance of adoption. In less affluent settings, these unexpected consequences may be a bit more insidious; for example, we noted above the (in this case unrealized) fear that in Egypt the advent of ICSI would create an epidemic of divorce, depriving women of social standing and livelihood (Inhorn, 2003b, 251–252). In all cultural settings, widespread availability of ARTs puts pressure on couples—and in particular on women—to attempt increasingly expensive treatments with decreasing chances of success. As Arthur Greil warned nearly two decades ago, acceptance of one's own infertility used to be a virtuous, humble act of resignation; now it is in danger of becoming evidence of laziness or selfish willfulness (Greil, 1989, 13). Assisted reproduction could become morally mandatory.

Thus despite cautions that first-wave feminists may have overreacted when they condemned ARTs for reinforcing patriarchy, we believe that in most cases religious conversations about ARTs fall short by ignoring just this sort of critique.[92] These conversations tend either to focus on the "natural" biological reproductive process and the "natural" family abstractly and universally or to address individual situations in their personal, concrete detail. This double focus on the universal and the particular obscures crucial moral questions that arise at the intermediate level of analysis: social justice concerns about gender, racial, and economic stratification of demand and of availability and quality of service; reservations about the commercialization of fertility services; questions about women's reproductive freedom and autonomy in health care; and questions about unjust uses of technology for (for example) purposes of eugenics or sex selection. Policy makers have a responsibility to keep larger social patterns and justice considerations, both local and international, before the public and not to permit the debate to circulate only around the foci of abstract "nature" on one hand and wrenching intimate accounts of unwanted childlessness on the other.[93]

---

[92] This is not to imply that all religious conversations about ARTs ignore justice. See for example Ryan (2001).

[93] Class, above all, is important here (see, e.g., Jenkins, 2002). For example, if infertility is less often interpreted through patriarchal lenses in the United States and Europe than in the developing world, that is due not to their predominantly Christian religious heritage but to the existence of a proportionately larger upper-middle and middle-class population, in which both women and men can afford to be freer in their vocational and reproductive choices. Indeed, if recent predictions about the spread of Christianity in the poorer cultures of the southern hemisphere and Asia bear out, Christianity will again provide religious justifications for patriarchalism (Jenkins, 2002).

References

Allahbadia, Gautam N. (2002). "The 50 Million Missing Women," *Journal of Assisted Reproduction and Genetics* 19(9), 411–416.
Barris, Sara, and Joel Comet (1994). "Infertility: Issues from the Heart," in Richard Grazi (ed.), *Be Fruitful and Multiply*. Spring Valley, NY: Genesis Jerusalem Press, 19–37.
Bauer, Renée (2006). "Patrilineal Descent' and Same-Sex Parents: New Definitions of Identity," *The Reconstructionist* 70.02(Spring), 26–33.
Becker, Gay (2000). *The Elusive Embryo: How Women and Men Approach New Reproductive Technologies*. Ewing, NJ: University of California Press.
Bharadwaj, Aditya (2000). "How Some Indian Baby Makers Are Made: Media Narratives and Assisted Conception in India," *Anthropology and Medicine* 7(1), 63–78.
Bharadwaj, Aditya (2003). "Why Adoption Is Not an Option in India: The Visibility of Infertility, the Secrecy of Donor Insemination, and Other Cultural Complexities," *Social Science and Medicine* 56, 1867–1880.
Bharadwaj, Aditya (2006). "Sacred Conceptions: Clinical Theodicies, Uncertain Science, and Technologies of Procreation in India," *Culture, Medicine, and Society* 30, 451–465.
Bick, Ezra (1997). "Ovum Donations: A Rabbinic Conceptual Model of Maternity," in Emanuel Feldman and Joel B. Wolowelsky (eds.), *Jewish Law and the New Reproductive Technologies*. Hoboken, NJ: Ktav Publishing House, 83–105.
Bleich, J. David (1997a). "Sperm Banking in Anticipation of Infertility," in Emanuel Feldman and Joel B. Wolowelsky (eds.), *Jewish Law and the New Reproductive Technologies*. Hoboken, NJ: Ktav Publishing House, 139–154.
Bleich, J. David (1997b). "In Vitro Fertilization: Questions of Maternal Identity and Conversion," in Emanuel Feldman and Joel B. Wolowelsky (eds.), *Jewish Law and the New Reproductive Technologies*. Hoboken, NJ: Ktav Publishing House, 46–82.
Bleich, J. David (1997c). "Maternal Identity Revisited," in Emanuel Feldman and Joel B. Wolowelsky (eds.), *Jewish Law and the New Reproductive Technologies*. Hoboken, NJ: Ktav Publishing House, 106–114.
Breck, John (1998). *The Sacred Gift of Life: Orthodox Christianity and Bioethics*. Crestwood, NY: St. Vladimir's Seminary Press.
Breck, John (2003). *God with Us: Critical Issues in Life and Faith*. Crestwood, NY: St. Vladimir's Seminary Press.
Brody, Baruch (1990). "Current Religious Perspectives on the New Reproductive Technologies," in Dianne M. Bartels, Reinhard Priester, Dorothy E. Vawter, and Arthur L. Caplan (eds.), *Beyond Baby M: Ethical Issues in New Reproductive Technologies*. Clifton, NJ: Humana Press, 45–63.
Bryan, Elizabeth, and Jane Denton (2001). "Reproductive Health Care Policies Around the World: The Work of the Multiple Births Foundation," *Journal of Assisted Reproduction and Genetics* 18(1), 8–10.
Bryld, Mette (2001). "The Infertility Clinic and the Birth of the Lesbian: The Political Debate on Assisted Reproduction in Denmark," *The European Journal of Women's Studies* 8(3), 299–312.
Cahill, Lisa Sowle (1988). "Women, Marriage, Parenthood: What Are Their 'Natures?'," *Logos* 9, 11–35.
Cahill, Lisa Sowle (1996). *Sex, Gender, and Christian Ethics*. Cambridge: Cambridge University Press.
Cahill, Lisa Sowle (2000). *Family: A Christian Social Perspective*. Minneapolis, MN: Fortress.
Campbell, Courtney (1997). "Cloning Human Beings: Religious Perspectives on Human Cloning," in *Cloning Human Beings*, Vol. 2. Rockville, MD: National Bioethics Advisory Commission, D1-D66. http://www.georgetown.edu/research/nrcbl/nbac/pubs/cloning2/cc4.pdf (accessed July 8, 2006).
Chakravarty, B.N. (2001). "Legislation and Regulations Regarding the Practice of Assisted Reproduction in India," *JARG* 18(1), 11–14.

Chan, Cecelia L.W., Paul S.F. Yip, Ernest H.Y. Ng, P.C. Ho, Celia H.Y. Chan, and Jade S.K. Au (2002). "Gender Selection in China: Its Meanings and Implications," *Journal of Assisted Reproduction and Genetics* 19(9), 426–430.

Charlesworth, Max (1990). "Human Genome Analysis and the Concept of Human Nature," *Human Genetic Information: Science, Law and Ethics*, Ciba Foundation Symposium 149 (article 180–189, discussion 189–198). New York: Wiley, 180–198.

Chrysostom, St. John (1997). "Sermon on Marriage," in Catharine P. Roth and David Anderson (trans.), *On Marriage and Family Life*. Crestwood, NY: St. Vladimir's Seminary Press.

Congregation for the Doctrine of the Faith (1987). *Instruction on Respect for Human Life in Its Origin and on the Dignity of Procreation: Replies to Certain Questions of the Day* [*Donum Vitae*]. http://www.vatican.va/roman_curia/congragtions/cfaith/documents/rc_con_cfaith_doc_198.

Crawford, S. Cromwell (2003). *Hindu Bioethics for the Twenty-First Century*. Albany, NY: SUNY.

Croll, Elizabeth J. (2000). *Endangered Daughters: Discrimination and Development in Asia*. New York: Routledge.

Dubner, Stephen J., and Steven D. Leavitt (2005). "The Search for 100 Million Missing Women: A Detective Story," *Slate*, May 24. http://www.slate.com/id/2119402/

Fagley, Richard M. (1965). "Doctrines and Attitudes of Major Religions in Regard to Fertility." *The Ecumenical Review* 17, 332–344.

Farley, Margaret A. (1983). "An Ethic for Same-Sex Relations" in Robert Nugent (ed.), *A Challenge to Love*. New York: Crossroad Publishing, 93–106.

Franklin, Sarah (1991). "Fetal Fascinations: New Dimensions to the Medical-Scientific Construction of Fetal Personhood," in Sarah Franklin, Celia Lury, and Jack Stacy (eds.), *Off-Centre: Feminism and Cultural Studies*. New York: Routledge, 190–205.

Franklin, Sarah (1997). *Embodied Progress: A Cultural Account of Assisted Reproduction*. London: Routledge.

Franklin, Sarah (1998). "Making Miracles: Scientific Progress and the Facts of Life," in Sarah Franklin and Helena Ragoné (eds.), *Reproducing Reproduction: Kinship, Power, and Technological Innovation*. Philadelphia, PA: University of Pennsylvania Press, 102–117.

Franklin, Sarah (2006). "Origin Stories Revisited: IVF as an Anthropological Project," *Culture, Medicine and Psychiatry* 30, 547–555.

Georges, Eugenia (1996). "Abortion Policy and Practice in Greece," *Social Science and Medicine* 42(2), 509–519.

Greil, Arthur (1989). "The Religious Response to Reproductive Technology," *Christian Century* January 4–11, 11–14.

Grenz, Stanley J. (1990). *Sexual Ethics: An Evangelical Perspective*. Louisville, KY: Westminster/John Knox.

Gross, Rita M. (1997). "Buddhist Resources for Issues of Population, Consumption, and the Environment," in Mary Evelyn Tucker and Duncan Ryuken Williams (eds.), *Buddhism and Ecology*. Cambridge, MA: Harvard University Press, 291–311.

Handwerker, Lisa (1998). "The Consequences of Modernity for Childless Women in China: Medicalization and Resistance," in Margaret Lock and Patricia A. Kaufert (eds.), *Pragmatic Women and Body Politics*. New York: Cambridge University Press, 178–205.

Handwerker, Lisa (2002). "The Politics of Making Modern Babies in China: Reproductive Technologies and the 'New' Eugenics," in Marcia C. Inhorn and Frank van Balen (eds.), *Infertility Around the Globe: New Thinking on Childlessness, Gender, and Reproductive Technologies*. Berkeley, CA: University of California Press, 298–314.

Hanson, Charles, Lars Hamberger, and Per Olaf Janson (2002). "Is Any Form of Gender Selection Ethical?," *Journal of Assisted Reproduction and Genetics* 19(9), 431–432.

Harakas, Stanley (1980). "For the Health of Body and Soul: An Eastern Orthodox Introduction to Bioethics." http://www.goarch.org/en/ourfaith/articles/article8076.asp (accessed July 8, 2008).

Harris, Ian (2001). "Magician as Environmentalist: Fertility Elements in South and Southeast Asian Buddhism," *Eastern Buddhist* 32(2), 128–156.

Hatzinikolaou, Nikolaos (1996). "The Orthodox Christian Approach" in *Part Seven: Nature and Status of the Embryo: Scientific, Philosophical and Legal Aspect*. 3rd Symposium on Bioethics: Medically Assisted Procreation and Protection of Human Embryo and Fetuses, 103–108.
Inhorn, Marcia C. (1994). *Quest for Conception: Gender, Infertility, and Egyptian Medical Traditions*. Philadelphia, PA: University of Pennsylvania Press.
Inhorn, Marcia C. (1996). *Infertility and Patriarchy: The Cultural Politics of Gender and Family Life in Egypt*. Philadelphia, PA: University of Pennsylvania Press.
Inhorn, Marcia C. (2002). "Sexuality, Masculinity, and Infertility in Egypt: Potent Troubles in the Marital and Medical Encounters," *The Journal of Men's Studies* 10(3), 343–359.
Inhorn, Marcia C. (2003a). "Global Infertility and the Globalization of New Reproductive Technologies: Illustrations from Egypt," *Social Science and Medicine* 56, 1837–1851.
Inhorn, Marcia C. (2003b). "'The Worms Are Weak': Male Infertility and Patriarchal Paradoxes in Egypt," *Men and Masculinities* 5(3), 236–256.
Inhorn, Marcia C. (2003c). *Local Babies, Global Science: Gender, Religion, and In Vitro Fertilization in Egypt*. New York: Routledge.
Inhorn, Marcia C. (2004). "Middle Eastern Masculinities in the Age of New Reproductive Technologies: Male Infertility and Stigma in Egypt and Lebanon," *Medical Anthropological Quarterly* 18(2), 162–182.
Inhorn, Marcia C. (2005). "*Fatwas* and ARTs: IVF and Gamete Donation in Sunni v. Shi'a Islam," *The Journal of Gender, Race and Justice* 9(2), 291–317.
Inhorn, Marcia C. (2006a). "'He Won't Be My Son': Middle Eastern Muslim Men's Discourses of Adoption and Gamete Donation," *Medical Anthropology Quarterly* 20(1), 94–120.
Inhorn, Marcia C. (2006b). "Making Muslim Babies: IVF and Gamete Donation in Sunni Versus Shi'a Islam," *Culture, Medicine and Psychiatry* 30, 427–450.
Inhorn, Marcia C. (2007). "Masturbation, Semen Collection and Men's IVF Experiences: Anxieties in the Muslim World." *Body & Society* 13(3), 37–53.
Inhorn, Marcia C., and Carolyn F. Sargent (2006). "Medical Anthropology in the Muslim World: Ethnographic Reflections on Reproductive and Child Health," *Medical Anthropology Quarterly* 20(1), 1–11.
Jakobovits, Yoel (1994)."Male Infertility: Halakhic Issues," in Richard V. Grazi (ed.), *Be Fruitful and Multiply: Fertility Therapy and the Jewish Tradition*. Spring Valley, NY: Genesis Jerusalem Press, 55–78.
Jakobovits, Yoel (1997). "Male Infertility: Halakic Issues in Investigation and Management," in Emanuel Feldman and Joel B. Wolowelsky (eds.), *Jewish Law and the New Reproductive Technologies*. Hoboken, NJ: KTAV, 115–138.
Jenkins, Gwynne L. with Silvia Vargas Obando and José Badilla Navas (2002). "Childlessness, Adoption, and *Milagros de Dios* in Costa Rica," in Marcia C. Inhorn and Frank van Balen (eds.), *Infertility Around the Globe: New Thinking on Childlessness, Gender, and Reproductive Technologies*. Berkeley, CA: University of California Press, 171–189.
Jenkins, Philip (2002). "The Next Christianity." *The Atlantic* 290(3), 3–68.
Kahn, Susan Martha (2000). *Reproducing Jews: A Cultural Account of Assisted Conception in Israel*. Durham, NC: Duke University Press.
Kahn, Susan Martha (2002). "Rabbis and Reproduction: The Uses of New Reproductive Technologies Among Ultraorthdox Jews in Israel," in Marcia C. Inhorn and Frank van Balen (eds.), *Infertility Around the Globe: New Thinking on Childlessness, Gender, and Reproductive Technologies*. Berkeley, CA: University of California Press, 283–297.
Kahn, Susan Martha (2006). "Making Technology Familiar: Orthodox Jews and Infertility Support, Advice, and Inspiration," *Culture, Medicine and Psychiatry* 30, 467–480.
Kanaanah, Rhoda Ann (2002). *Birthing the Nation: Strategies of Palestinian Women in Israel*. Berkeley, CA: University of California Press.
Kass, Leon (2002). *Life, Liberty and the Defense of Dignity: The Challenge for Bioethics*. San Francisco, CA: Encounter Books.
Katsumata, Yoshinao (2000). "Organ Transplantation, In-Vitro Fertilization, and Euthanasia in Japan," *Forensic Science International* 113, 491–493.

Keown, Damien (1995). *Buddhism and Bioethics*. New York: St. Martin's Press.
Keown, Damien (2005). *Buddhist Ethics: A Very Short Introduction*. Oxford: Oxford University Press.
Khanna, Sunil K. (1999). "New Reproductive Technology in India: Social Context, Legal Implications, and Health Outcomes," *Anthropologist* 1(1), 61–71.
Kleinman, Arthur (1995). *Writing at the Margin: Discourse Between Anthropology and Medicine*. Berkeley, CA: University of California Press.
Lauritzen, Paul (1993). *Pursuing Parenthood: Ethical Issues in Assisted Reproduction*. Bloomington, IN: Indiana University Press.
Layne, Linda L. (1992). "Of Fetuses and Angels: Fragmentation and Integration in Narratives of Pregnancy Loss," in Linda L. Layne and David J. Hess (eds.), *Knowledge and Society: The Anthropology of Science and Technology*. Greenwich, CT: JAI, 29–58.
Lewin, Ellen (1998). "Wives, Mothers, and Lesbians: Rethinking Resistance in the US," in Margaret Lock and Patricia A. Kaufert (eds.), *Pragmatic Women and Body Politics*. New York: Cambridge University Press.
Lock, Margaret (1998). "Perfecting Society: Reproductive Technologies, Genetic Testing, and the Planned Family in Japan," in Margaret Lock and Patricia A. Kaufert (eds.), *Pragmatic Women and Body Politics*. New York: Cambridge University Press, 206–239.
Lock, Margaret, and Patricia A. Kaufert (1998). "Introduction," in Margaret Lock and Patricia A. Kaufert (eds.), *Pragmatic Women and Body Politics*. New York: Cambridge University Press, 1–27.
Malin, Maili (2002). "Made in Finland: Infertility Doctors' Representations of Children," *Critical Public Health* 1(24), 291–308.
Malpani, A., and A. Malpani (1992). "Simplifying Assisted Conception Techniques to Make Them Universally Available—A View from India," *Human Reproduction* 7(1), 49–50.
Marglin, Frédérique Apffel (1985). "Female Sexuality in the Hindu World," in Clarissa W. Atkinson, Constance H. Buchanan, and Margaret R. Miles (eds.), *Immaculate and Powerful: The Female in Sacred Image and Social Reality*. The Harvard Women's Studies in Religion Series. Boston, MA: Beacon, 39–60.
May, William E. (2006). "Cloning Humans vs. Begetting Children." http://www.culture-of-life.org/?Control=ArticleMaster&aid=1415&c=4&p=2 (accessed July 6, 2006).
McCormick, Richard A. (1987). "Surrogate Motherhood: A Stillborn Idea," *Second Opinion* 5, 128–132.
McCormick, Richard A. (1989). *The Critical Calling: Reflections on Moral Dilemmas Since Vatican Council II*. Washington, DC: Georgetown University Press.
"Medical Ethics" (2002). *Hinduism Today* July–September. http://www.hinduismtoday.com/archives/2002/7-9/40-47_ayurveda.shtml (last accessed April 19, 2006).
Meilaender, Gilbert (1997). "Begetting and Cloning," *First Things* 74 (June/July), 41–43.
Meirow, D., and J.G. Schenker (1997). "The Current Status of Sperm Donation in Assisted Reproductive Technology: Ethical and Legal Considerations," *Journal of Assisted Reproduction and Genetics* 14, 133–138.
Meseguer, Marcos, Nicolás Garrido, José Remohí, Carlos Simón, and Antonio Pellicer (2002). "Gender Selection: Ethical, Scientific, Legal, and Practical Issues," *Journal of Assisted Reproduction and Genetics* 19(9), 443–446.
Missouri, Synod (2002). *What Child Is This? Marriage, Family and Human Cloning*. A Report of the Commission on Theology and Church Relations of the Lutheran Church—Missouri Synod April. http://www.lcms.org/graphics/assets/media/CTCR/45061CloningCTCRfinal.pdf (accessed July 8, 2006).
Mohapatra, Ratnaprava, S.K. Dash, and S.N. Padhy (2001). "Ethnobiological Studies from *Manusmruti*: IX. *Niyoga Prathaa*, A Natural Process of Artificial Insemination," *Journal of Human Ecology* 12(1), 33–38.
Mori, Takahide, and Hirohiko Watanabe (2002). "Ethical Considerations on Indications for Gender Selection in Japan," *Journal of Assisted Reproduction and Genetics* 19(9), 420–425.
Mulkay, Michael (1994). "Science and Family in the Great Embryo Debate," *Sociology* 28(3), 699–715.

O'Donovan, Oliver (1984). *Begotten or Made?*. Oxford: Clarendon.
Orthodox Church in America (1992). Tenth All-American Council. "The Procreation of Children." http://www.antiochian.org/moral_issues_family (accessed May 31, 2007).
Paxson, Heather (2004). *Making Modern Mothers: Ethics and Family Planning in Urban Greece*. Berkeley, CA: University of California Press.
Paxson, Heather (2006). "Reproduction as Spiritual Kin Work: Orthodoxy, IVF and the Moral Economy of Motherhood in Greece," *Culture, Medicine and Psychiatry* 30, 481–505.
Pope John XXIII (1961). *Encyclical Mater et Magistra*, III, AAS 53(1961), 447.
Pope Paul VI (1968). *Humanae Vitae*, AAS 60(1968), 489.
Post, Stephen G., and Barbara Andolsen (1989). "Recent Works on Reproductive Technology," *Religious Studies Review* 15(3), 210–218.
Qiu, Ren-zong (2002). "Sociocultural Dimensions of Infertility and Assisted Reproduction in the Far East," in E. Vayena, P. Rowe, and D. Griffin (eds.), Report of a meeting on Medical, Ethical, and Social Aspects of Assisted Reproduction, 17–21 September 2001. Geneva, Switzerland: WHO, 75–80. http://www.who.int/reproductive-health/infertility/12.pdf (accessed July 8, 2007).
Ragoné, Helena (1998). "Incontestable Motivations," in Sarah Franklin and Helena Ragoné (eds.), *Reproducing Reproduction: Kinship, Power, and Technological Innovation*. Philadelphia, PA: University of Pennsylvania Press, 118–131.
Reichman, Edward (1997). "The Rabbinic Conception of Conception: An Exercise in Fertility," in Emanuel Feldman and Joel B. Wolowelsky (eds.), *Jewish Law and the New Reproductive Technologies*. Jersey City, NJ: KTAV Publishing, 1–35.
Renteln, Alison Dundes (1992). "Sex Selection and Reproductive Freedom," *Women's Studies International Forum* 15(3), 405–426.
Roberts, Elizabeth F.S. (2006). "God's Laboratory: Religious Rationalities and Modernity in Ecuadorian In Vitro Fertilization," *Culture, Medicine and Psychiatry* 30, 507–536.
Rosenfeld, Azriel (1997). "Generation, Gestation and Judaism," in Emanuel Feldman and Joel B. Wolowelsky (eds.), *Jewish Law and the New Reproductive Technologies*. Hoboken, NJ: KTAV Publishing, 36–45.
Ryan, Maura A. (2001). *The Ethics and Economics of Assisted Reproduction: The Cost of Longing*. Washington, DC: Georgetown University Press.
Ryan, Maura A. (2004). "Beyond a Western Bioethics," *Theological Studies* 65, 158–177.
Satha-Anand, Suwanna (2001). "Buddhism on Sexuality and Enlightenment," in Patricia Beattie Jung, Mary E. Hunt, and Radhika Balakrishnan (eds.), *Good Sex: Feminist Perspectives from the World's Religions*. New Brunswick, NJ: Rutgers University Press, 113–124.
Schenker, Joseph G. (1992). "Reproductive Health Care Policies Around the World: Religious Views Regarding Treatment of Infertility by Assisted Reproductive Technologies," *Journal of Assisted Reproduction and Genetics* 9(1), 3–8.
Schenker, Joseph G. (2002). "Gender Selection: Cultural and Religious Perspectives." *Journal of Assisted Reproduction and Genetics* 19(9), 400–410.
Sen, Amartya (1990). "More Than 100 Million Women Are Missing." *New York Times Review of Books* 37 (December 20). http://www.nybooks.com/turing.library.northwestern.edu/articls/3408 (accessed March 13, 2008).
Sheean, Leon A. (2004). "Protecting the Unborn." http://www.oca.org/PDFS/christianwitness/2004-PMConf-LScheean.pdf (accessed July 12, 2006).
"Should Surrogate Motherhood Be Banned?" Forum (2001). *Beijing Review* (May 10, 2001), 28–29.
Sills, E. Scott, and Gianpero D. Palermo (2002). "Preimplantation Genetic Diagnosis for Elective Sex Selection, the IVF Market Economy, and the Child—Another Long Day's Journey Into Night?," *Journal of Assisted Reproduction and Genetics* 19(9), 433–437.
Sonbol, Amira el Azhary (1995). "Adoption in Islamic Society: A Historical Survey," in Elizabeth Warnock Fernea (ed.), *Children in the Muslim Middle East*. Austin, TX: University of Texas Press, 45–67.

Stone, Jennifer L. (1991). "Contextualizing Biogenetic and Reproductive Technologies," *Critical Studies in Mass Communication* 8, 309–332.
Strathern, Marilyn (1992a). *After Nature: English Kinship in the Late Twentieth Century*. Cambridge: Cambridge University Press.
Strathern, Marilyn (1992b). *Reproducing the Future: Anthropology, Kinship, and the New Reproductive Technologies*. New York: Routledge.
Thompson, Charis (1998). "Producing Reproduction: Techniques of Normalization and Naturalization in Fertility Clinics," in Sarah Franklin and Helena Ragoné (eds.), *Reproducing Reproduction: Kinship, Power, and Technological Innovation*. Philadelphia, PA: University of Pennsylvania Press, 66–101.
Thompson, Charis M. (2002). "Fertile Ground: Feminists Theorize Infertility," in Marcia C. Inhorn and Frank van Balen (eds.), *Infertility Around the Globe: New Thinking on Childlessness, Gender, and Reproductive Technologies*. Berkeley, CA: University of California Press, 52–78.
Thompson, Charis M. (2005). *Making Parents: The Ontological Choreography of Reproductive Technologies*. Cambridge, MA: MIT.
Thompson, Charis M. (2006). "God Is in the Details: Comparative Perspectives on the Intertwining of Religion and Assisted Reproductive Technologies," *Culture, Medicine and Psychiatry* 30, 557–561.
Tober, Diane (2004). "Shi'ism, Pragmatism, and Modernity: Islamic Bioethics and Health Policy in the Islamic Republic of Iran." Presentation at the University of Michigan, April 14.
Traina, Cristina L.H. (1999). *Feminist Ethics and Natural Law: The End of the Anathemas*. Washington, DC: Georgetown University Press.
Tremayne, Soraya (2005). "The Moral, Ethical, and Legal Implications of Egg, Sperm, and Embryo Donation in Iran." Unpublished manuscript. Presented at Reproductive Disruptions: Childlessness, Adoption, and Other Reproductive Complexities, University of Michigan, May 19–22, 2005.
Waters, Brent (2001). *Reproductive Technology: Towards a Theology of Procreative Stewardship*. Ethics and Theology Series. Cleveland, OH: Pilgrim Press.
Weston, Kath (1991). *Families We Choose: Lesbians, Gays, Kinship*. New York: Columbia University Press.
Williams, Paul (1991). "Buddhism and Sexuality–Some Notes," *Middle Way* 66(August), 101–106.
Zuhur, Sherifa (1992). *Revealing Reveiling: Islamist Gender Ideology in Contemporary Egypt*. Albany, NY: SUNY.

Chapter 2
Religion, Conceptions of Nature, and Assisted Reproductive Technology Policy

John H. Evans

In Chapter 1 of this volume, Traina and her colleagues examine conceptions of the natural regarding assisted reproductive technologies (ARTs) in religious traditions across the world. They focus on the strategies by which average religious people cognitively transform the "un-natural" technology to the "natural," as long as the technology is in service of larger religious ends, typically having to do with families. The strategies are many, as are the starting points for each tradition's considerations of what is "natural." Since the actual beliefs of the religions about nature and ARTs are so divergent, and since the religions discussed are primarily not in the U.S., I will not focus on the details of beliefs about nature and ARTs since they are unlikely to be introduced in the policy process. Rather, I will discuss the challenges that will be faced by those who want to introduce religious conceptions of nature into policy-making processes concerning ARTs. To do so I will incorporate a broader point that the authors of the earlier chapter make about religions in general: that actual lived religion is more important for policy in liberal democratic societies than official religion.

I begin with a discussion of contemporary U.S. policy regarding ART. I then discuss the types of moral discourse that can be accepted in different policy-making venues and the challenges this will create for those who want to incorporate religious conceptions of nature into policy regarding ARTs. I then turn to an additional challenge that specifically religiously-based conceptions of nature will face to be included in policy deliberations: the problem of translating theological language to secular language. To provide at least some broader context beyond the U.S., next I apply my theoretical typology of policy application to Europe and discuss the extent to which similar challenges exist there in integrating religious conceptions of nature into ART policy.

2.1 Existing Policy and Regulation of ART

I follow the earlier chapter by Traina and her colleagues in defining ART as including artificial insemination, GIFT (gamete intra-fallopian transfer), IVF (in-vitro fertilization), ICSI (intra cytoplasmic sperm injection), ooplasm donation, sperm sorting for sex, and pre-implantation genetic diagnosis. Existing policy on ART is very much a patchwork, with policy being set by various state governments, federal agencies and professional societies.[1]

On the federal level, there are various consumer protection and embryo laboratory standards. There is one policy specifically directed at ART, which aims to provide consumers with reliable information about the efficacy of ART services offered by fertility clinics and to create a model certification program for states to use in regulating embryo laboratories. There are also more general medical or scientific federal regulations that indirectly provide some oversight to ART applications. Federal executive branch agencies such as the Food and Drug Administration (FDA) indirectly regulate ART practice through regulations concerning topics such as the spread of communicable diseases and the use of medical devices and drugs. Sperm, ova and embryos are defined as medical products derived from the human body, and thus fall under this regulatory authority. For example, the FDA has determined that a cloned embryo inserted into a woman's uterus is the equivalent of administering an unapproved new drug. Quality assurance of laboratory tests touching on ARTs are regulated by the Center for Medicare and Medicaid Services, and the truthfulness of advertised claims by ART clinics is regulated by the Federal Trade Commission.

There are also state policies regarding ARTs. Most of these policies concern access to ARTs, such as mandating that insurance companies cover certain procedures. Others try to protect consumers from unethical practitioners. A number of states dictate parental rights and obligations in ART contexts. A few states, such as Pennsylvania, have more comprehensive regulation, where policies focus upon record-keeping and maintenance of clinical facilities. There are a few states where laws forbid embryo experimentation, although this is not thought to influence ART practice itself. In one state, Louisiana, it is unlawful to intentionally destroy a viable embryo.

States also engage in indirect regulation of ARTs, primarily through the states' traditional role as the regulator of medical practice. ART procedures are considered medical, and therefore are subject to all of the state regulations that regulate medicine. These generally include the requirement of informed consent, that physicians be licensed, that they obtain hospital credentials, board certification, have malpractice insurance and so on. States also engage in indirect regulation through being the location for litigation. For example, medical malpractice litigation has the result of controlling the behavior of ART practitioners.

The courts have also been a location of policy-making, typically when the legislature has provided no guidance. For example, who should be the parent in a custody dispute between a surrogate mother and a biological mother? The person who gestated and bore the child, or the person who contributed the genetic material? Lacking laws from legislatures, courts have had to step in and become de-facto policy-setting venues.

---

[1] The information in the next four paragraphs comes from President's Council on Bioethics (2004).

Finally, non-governmental regulation by professional societies such as the American Society of Reproductive Medicine proscribe certain practices concerning safety, efficacy and privacy, but also ethical concerns. Similarly, more general medical societies such as the American Medical Association and the American Board of Obstetrics and Gynecology have their own standards. Compliance is voluntary, but presumably there is a normative sanction for being in a profession and acting outside of its standards.

For the purposes of this chapter, the central issue is the moral logic of existing ART policy, and whether religious conceptions of nature could be included as motivation for additional policies. According to an extensive study by the President's Council on Bioethics, existing regulation and policy is "motivated by concerns for consumer protection, quality assurance in laboratory procedures, safety and efficacy of products according to their intended use, and the delivery of medical care according to accepted standards of practice" (President's Council on Bioethics, 2004, 167). I would say that these policies are being driven by a limited set of values such as "do no harm" and "tell the truth" which are the consensual values of the scientific and medical professions. While all religions of which I am aware would share these values, most religions have additional values that go beyond these limited few, and which would clash with the interests of scientists and medicine. So, while religious traditions would be supportive of promoting safety and efficacy, they would tend to want to bring additional values to the fore, such as concerns about "nature."

The review of ART policy by the President's Council agrees, concluding that while even regulation based on safety is a bit lacking, "there do not seem to be significant oversight activities or effective guidelines that address larger ethical concerns relating to the enhanced control over human procreation," and that "the system of regulation currently in place does not reflect the concerns many people have about the use and destruction of human embryos attendant on the practice of assisted reproduction" (President's Council on Bioethics, 2004, 168). That is, there is no policy at present motivated by the type of concerns articulated by religious groups, be it about "nature" or any other aspect of religious belief. So, while it seems clear that regulation concerning safety and efficacy exists but could be improved, regulations based upon "deeper" ethical concerns that have not been agreed upon by the scientific profession are a clean slate.

2.2 Types of Moral Discourse and Policy Venues

Policy is made in different venues, and these venues will be differentially receptive to religiously-based claims about "nature." Before continuing, a distinction must be made between two types of moral discourse used in our society, which I have elsewhere called "thick" and "thin" (Evans, 2002). "Thick" discourse is a debate about the ends we as a society should pursue. Should be protect nature? Should we maximize wealth? Should we serve God? Should we create a society with income equality?

These are all debates about the ends we as a society should pursue, and to perhaps get ahead of myself, this is where debates about "nature" lie. That is, are we as a society to follow conception of nature X (e.g., that particular end) when it comes to setting policy about ARTs?

With "thin" discourse, there is no debate about ends to pursue, but these ends are assumed and unexamined. The debate using "thin" discourse concerns the most efficacious means for achieving these assumed ends. Should we have a policy of tax cuts or government spending? That is, which of these is the most efficacious means to advance the assumed and unquestioned end of economic growth? Or, more to the point of this essay: should facilities that use ARTs be required to report on success rates? That is, would the means of required reporting maximize the unspoken and assumed end of autonomous decision-making of the consumers of ART services?

To incorporate religious conceptions of the "natural" into ART policy is to engage in "thick" discourse, because what is "natural" is then the end that we should be pursuing. Contrary to enlightenment-based reason—embodied in contemporary philosophy and bioethics—which holds that you cannot derive the "ought" of proper action from the "is" of nature, for the public "nature often continues to serve, as it did prior to the Enlightenment, as a moral touchstone."[2] To debate what is "natural," or, more strongly, to set policy on ART based on what is "natural," is to debate the ends that we should pursue, since "nature" is thought to be "that which is to be followed." For example, if Orthodox Jews think that God, who created nature, wants them to have large families, then the only question is: what is the most efficacious means of achieving this end that does not violate other ends created by God? That means is, apparently and according to Traina and her colleagues, the use of IVF.

Different policy venues have different appetites for these two types of moral discourse. Why? First it is important to note that "thin" discourse has a calculable quality to it, in that one means can *more* efficiently maximize the end than some other policy. We could in principle calculate whether building dams or windmills maximizes the end of providing inexpensive energy. There is then an official reason that the policy-maker can point to justify their policy selection: it was better. On the other hand, "thick" discourse about ends is not calculable and indeed, the ends are most likely incommensurable. Shall we pursue conception of nature X or Y? As Weber said long ago about value spheres: there are no objective grounds with which to pick one over another. There are no objective grounds for picking worldview X over worldview Y, no objective grounds for picking Protestantism over Catholicism over Hinduism. As an end, there are no objective grounds for the policy maker in picking conception of nature X or conception of nature Y. It is a matter of personal preference or belief.

The further a policy-maker is from the legitimating power of being elected, the more they prefer "thin" discourse to "thick." The assistant under-secretary of the National Institutes of Health is very distant from democratic legitimation in that he or she is not even a political appointee of the president, but rather is a more or less

---

[2] Lock and Kaufert, cited in Traina et al., this volume page 18.

permanent bureaucrat, only distantly accountable to the citizens. This person would tend to prefer "thin" discourse. The president of the U.S. is fairly directly accountable to the citizens, as is your member of the House of Representatives. They *can* use "thick" discourse as well as "thin."

The reason for this distinction is as follows. Government authority in Western liberal democracies such as the U.S. needs to be transparent to the citizens. Unlike decrees coming from the subjective perspective of the European sovereign, the U.S. political system was partly founded "on the idea that politics is transparent, that political agents, political actions, and political power can be viewed" (Ezrahi, 1990, 69). This would not be a problem in the city-state described by Plato where all decision-makers knew each other, but in representative democracies this becomes a critical issue. The surface manifestation of this impulse is that government decision-making proceedings have open meetings, allowing the public to view the decision-making process undertaken on its behalf by its elected officials (and appointed officials, like Commissioners). The more subtle manifestation of the need for transparency is that government decision-making that is distant from public accountability tends toward methods that purport to be objectively transparent.

Historian Theodore Porter explains the popularity of commensurable scales in government decision-making, such as cost-benefit analysis, to be the result of government officials needing to appear to not really be exercising judgment, but following transparent and objective laws or rules. As he puts it, in other countries government officials are "trusted to exercise judgment wisely and fairly. In the United States, they are expected to follow rules" (Porter, 1995, 195). This is because, put simply, it is part of the U.S. political culture to not trust authority, especially government authority, and particularly the authority of bureaucrats. A complex decision means that we would have to trust the judgement of the government functionary, because they cannot readily explain how they reached their judgement. "In a country where mistrust of government is rife, the temptation to substitute supposedly impersonal calculation for personal, responsible decisions… cannot but be exceedingly strong" (Richard Hammond, in Porter, 1995, 195).

If Americans do not trust government officials to exercise their discretion over where to build a dam (one of Porter's examples), they certainly do not trust the unelected government official deciding what nature demands of us. Therefore, a panel at the NIH devoted to reviewing ART policy cannot simply approve of some ART because they think that it is ethical, but must rather "show" or, better yet, "prove" with objective, transparent methods, that the research is ethical (Ezrahi, 1990, ch. 3). They must show their reasoning in a manner that the public can judge. For the same reason that cost-benefit analysis is popular with bureaucrats—because it purports to be a value-neutral calculation—creating an ART policy based upon maximizing assumed ends purports to be value free and provides the same "rule-following" quality.[3] The ends that can safely be maximized are ends that few would disagree with, such as it being better to not harm people than harm them.

---

[3] The preceding three paragraphs follow an argument made in Evans (2000).

The result of this is that any policy that is based upon "morality" is most likely to be set by people in policy-making venues who are more directly accountable to the citizens. Elections are essentially picking people who will promote certain ends. If your preferred candidate with your preferred ends loses, their policy-making is still legitimate because they were elected by a majority of the citizens. This preference for different types of moral discourse has strong implications for explaining what policy about ARTs currently exists, and what policy may exist in the future—particularly policy that would include conceptions of nature in their justifications.

2.3 Policy Venues and Conceptions of Nature

There are many venues that set policy, as the previous discussion of existing ART policy shows us. For simplicity, I will discuss five venues: executive branch agencies at the state or federal level; professional societies; government commissions; legislative bodies; and the courts. In each of these venues there will be different abilities to invoke conceptions of the natural in ART policy.

2.3.1 Executive Branch Agencies

While employees of the executive branch act on the authority of an elected official, the president on the federal level or the governor on a state level, they are themselves quite distant from democratic accountability. The number of decisions that these people make on any given day is enormous, and the elected official will only be aware of a tiny fraction of them. Unless explicitly commanded to create a policy by the elected official—such as when president Bush created a very specific policy on stem cell lines—the employee of the executive branch will generally not want to be perceived as determining the ends that policies should pursue. That would be setting values. Rather, they generally want to be seen as following rules, and in the context of policies having to do with ethics, the closest one can get to following rules are following institutionalized unquestioned ends.

This is also an explanation for why executive branch agencies prefer pursuing ends that are portrayed as universally held, as in what are called "common morality" theories of ethics such as principlism (Evans, 2006). If the ends that you are pursuing are held by all rational beings (as is claimed by some common morality theories), then you cannot be criticized for "imposing" your ethics on everyone else through your bureaucratic authority.

Consider, for example, the fury over the case of the Food and Drug Administration's policy on over-the-counter emergency contraception called "Plan B." The drug had passed the normal bureaucratic review at the FDA, but approval was still being held up by the director for various reasons. When an assistant FDA commissioner resigned because of the way the Plan B application was being handled, she said that she "can no longer serve as staff when scientific and clinical evidence, fully evaluated and recommended for approval by the professional staff here, has been overruled." She thought that the decision to withhold approval was "political" (Kaufman, 2005).

That is, in her model of bureaucratic authority, the ends that are to be followed are "scientific" and "clinical evidence," which implicitly means the maximization of health of patients. What is not allowed is "politics." In this case this means embryonic politics—debates about the status of embryos.

Current federal and state policy on ARTs implicitly invokes common morality. As described above, the current policies are maximizing the ends of safety and patient autonomy. Policies determine basic safety standards of IVF procedures, others require information be sent to potential patients so that they can make accurate decisions. In all of these instances these policies are fairly non-controversial because to the extent there actually is an end that is commonly held among people, one would be that it is better to not harm people than harm them. While there may be a "thick" debate about which of the limited number of acceptable ends to focus upon in any given case, the executive branch agencies are generally not going to be open to discussions of possible new ends.

Executive branch agencies—such as the Food and Drug Administration—are then the least amenable to making claims based on "the natural." Again, this is because in the public's mind determining something as "natural" is like setting an end, since the "natural" is to be followed. It is simply hard to imagine the head of the Food and Drug Administration holding a press conference to announce that he had determined what was "natural" regarding ART, and that policies would flow from this understanding.

That said, particular religiously based notions of "the natural" could be used as ends by bureaucratic agencies if "the natural" could be construed as "natural fact." As the above complaint by the assistant FDA commissioner notes, bureaucratic agencies mostly prefer to be following "facts," and "facts" are best determined by science. If there is a conception of "the natural" held by a religious group that can be said to be "true" by the standards of contemporary Western science, then it could be used. However, to the extent the conception of nature is seen as "religious" this would work against its adoption because "religious" is, in our pluralistic society, akin to saying "non-universal" and thus not necessarily "true."

Unfortunately, from the perspective of those wanting to incorporate religious conceptions of nature into ART policy, the chapters in this book show that "nature" is hardly an uncontested concept with universally agreed upon meaning. Moreover, and more specifically, something like the Orthodox Jewish claim that there is a "natural" family form (Traina et al., this volume p. 29). that ART policy should maximize is shown to not be "fact" because of the other religious groups that do not think of this family form as a "fact." The overall assessment would be that executive branch agencies are generally not a conducive location to incorporation of religious conceptions of the natural into ART policy.

2.3.2 Professional Societies

Professional societies regulate not based upon law, but upon a form of moral suasion. To have your ART clinic seen as a legitimate member of the community, and for the employees to be seen as legitimate professionals, they need to follow the standards of the profession. Additionally, these standards can take on the force of law in that they are considered "good practice" and can be admitted into legal proceedings that question the behavior of professionals. We are familiar with this idea in more established areas. Physicians have professional standards established by professional societies, and patients would think twice about seeing a physician who has foresworn the professional standards of the American Medical Association. Even my own profession, academic sociology, has professional standards having to do with submitting articles for publication in professional journals.

In the area of ARTs the professional societies are not as well known as the American Medical Association, but they serve a similar function. For example, the American Society of Reproductive Medicine (ASRM) and the Society for Assisted Reproductive Technology (SART) have their own standards for their members.

One must ask, what we are to make of these societies? Do they exist to forward the collective good, or do they exist to advance the good of their members? While both are undoubtedly true to some extent, I would argue that they lean toward the latter. As the American Medical Association exists to forward the interests of physicians (see, e.g., Starr, 1982), the ASRM and the SART exist to forward the interests of their members. (Luckily, the interests of the members most often coincide with the interests of society.) What then are the ends that these types of groups are trying to maximize with their policies?

If we turn to their existing policies, we see, in the view of the President's Council on Bioethics, that "the chief values ASRM seeks to promote through its opinions and guidelines are safety (of ART participants), efficacy (of techniques and procedures), and privacy (of ART patients)" (President's Council on Bioethics, 2004, 72). I see two ways to interpret this list of ends to maximize. First, the members of these societies tend to think of themselves first and foremost as scientists and physicians, so they also adopt the values or ends of science and medicine, which can best be summarized by the values or ends of mainstream bioethics: autonomy, beneficence, non-maleficence and justice (Beauchamp and Childress, 2001, 5). Safety and efficacy are part of beneficence and non-maleficence, while privacy is part of autonomy. The second way to interpret this list is that it is a means of weeding out unscrupulous practitioners who will damage the reputation of the entire field. What, after all, do patients want from an ASRM affiliated physician? The want to not be harmed, to efficiently produce a baby and to have their experience remain private.

Would professional societies welcome the incorporation of ends or values about "nature" into their policy deliberations? The answer is: they would generally not. Conceptions of nature are not part of the ends of science and medicine, and it is hard to see how pursuing a particular notion of the natural would help the profession. In fact, it is in the interest of the profession to only be seen as promoting what are portrayed as the universally held ends of the medical profession.

One advantage of purporting to follow universally held ends is that they cannot be portrayed as particularistic ends that only benefit ART practitioners. Creating professional guidelines that forward safety, for example, is not only in the interest of the members of the profession, but is also a universal value to forward. This gives the societies more legitimacy than what could be perceived as the simply mercenary interests of other societies that cannot be said to be working for the collective values or the collective good. (One thinks here of the Chamber of Commerce which, while claiming to be acting on behalf of the public good, is probably also perceived as advancing policies that most effectively forward its interests.)

In sum, while executive branch agencies will generally avoid religiously based notions of the "natural" in policy making because they do not want to be debating any ends (but rather applying taken for granted ends like safety), professional societies have a somewhat different reason to limit discussions of ends. They generally want to limit discussions of ends to universal ends because of the legitimacy it provides, but also they need to adhere to the values and ends promoted by the scientific and medical professions. This too is not a good venue for incorporating discussions of nature into policy of ARTs.

2.3.3 Legislatures

Policies regarding ARTs can also be set by elected officials on the state and federal level. While "thin" arguments are preferred by unelected executive branch officials, elected officials can make either "thick" or "thin" arguments. That is, elected officials can either set policies based upon assumed or universally held ends like safety, *or* they can set policies by first setting ends of values that may not be universally held, and then crafting means (policies) that will forward those ends.

The reason that elected officials can do this—while unelected officials generally do not—is that elected officials have a mechanism for dealing with non-universal ends called majoritarianism. As of the writing of this chapter in late 2006, if your values do not coincide with those of conservative Republicans, your values are not represented in federal policy making. Similarly, the elected officials of Louisiana have decided to set an end in place that is not universally held (on the value of embryonic life), and have crafted a set of means (policies) that purport to maximize that end. Louisiana now defines in-vitro embryos as "juridical persons" "with nearly all of the attendant rights and protections of infants. It stipulates that the use of an in vitro embryo must be solely for 'the support and contribution of the complete development of human in utero implantation.' The production, culture, or use of human embryos for any other purpose is proscribed" (President's Council on Bioethics, 2004, 52–53).

Majoritarianism is how representative democracy is supposed to function, unless it can be claimed that the law violates the constitutional rights of some group in society. Constitutionally based civil rights, a constraint on majoritarianism on certain aspects of human experience—such as religious practice—could in theory be applied to ARTs. However, while some have argued that the privacy constitutional framework articulate in the *Roe* decision protects all matters of reproduction (Robertson, 1994), this seems to be an interpretation that would have little traction in contemporary federal constitutional jurisprudence.

It is then the venue of the legislature where conceptions of nature could be used in creating policy concerning ARTs. It could be argued that God has determined that the "natural" family has genetically related children, and that therefore IVF needs to remain legal as a means for pursuing that end. Of course, in the U.S. context elected officials cannot claim *only* divine reasons for policy because that would in general make the law unconstitutional. However, given how "God's will" and "nature" are taken as nearly equivalent by the public (Traina et al., this volume), one can make a religious argument secular by simply calling God's will "natural." For example, in the debate about homosexual marriage, much clearly religious reasoning is discussed using the language of "nature." While of course most official theology posits a more complicated relationship between God and nature, I agree with Traina and her colleagues that what really matters is what the general members of the religion actually do. Therefore, from the perspective of those interested in using religious arguments about nature in policy deliberation, the elected officials seem like the perfect venue.

Note some drawbacks. Majoritarianism means that it is only the religious conception of nature favored by the numerically dominant group within a jurisdiction that will be used. Specific conceptions of nature held by Shiites will, for example, be entirely ignored by any state legislature in the U.S. However, this is also the case with the previous two venues I have discussed in that Shiite values and ends are certainly not universal. The problem is that whereas in the previous two venues ends that are not universally shared are ignored, in this instance an end that is not universally shared may be applied to everyone, whether they adhere to it or not. Christian conceptions of nature will be determinative in the U.S., not Muslim conceptions of nature.

2.3.4 *Courts*

To return to high school civics class, courts are not designed to create laws, but rather to interpret laws. Is this doctor violating this particular law as set in place by the California legislature and signed by the governor? Do the practices of this government commission actually violate the state law about open meetings? Or, on a federal level, the Supreme Court typically engages in the question of whether a new law violates the constitution, and does not create new laws. As one could imagine, the line between interpretation of existing laws and setting legal precedence that has the function of being like a law is very fuzzy. Indeed, I have probably just referred to a good portion of all legal scholarship.

Structurally, the courts would prefer to have the legislatures make all of the decisions about ends. Given that legislatures do not do this—either because they do not want to or because they cannot anticipate all situations—courts are often put in the role of determining ends. They are even put in the role of determining what is "natural."

Sheila Jasanoff cites the example of *In re Marriage of Buzzanca*, a case adjudicated in the California courts. In this case a couple employed a married surrogate to carry an embryo created from the egg and sperm of unknown donors. The couple who hired the surrogate broke up, and a court decided that this child, "who could potentially have claimed kinship to any of six different adults," did not have parents at all. On appeal, another court gave custody to the parents who had contracted the woman to carry the embryo to term (Jasanoff, 2005, 166).

Jasanoff considers this a typical case where courts were forced to "make law in a place where law was otherwise silent." Moreover, the "U.S. state courts generally took the lead in shaping the legal order surrounding assisted reproduction. In the process, they also laid down the rules governing what would count as natural or unnatural uses of the new reproductive technologies" (Jasanoff, 2005, 166). While state legislatures should be doing this laying down of the rules, for various reasons, they do not.

Does this mean that the courts would be a good location for the insertion of issues of the natural into public policy? Perhaps, but there are some clear cautions. First, the avenue of influence is variable. You could create *amici curiae* briefs that outlined one's conception of nature, but ultimately an individual judge is going to make an initial decision that could ignore your perspective. Second, even if one were able to use the courts to advocate a particular notion of nature, another group could simply get the legislature to pass a law effectively overturning that decision, which would then be binding on the courts. Unless one could somehow read one's conception of nature into a federal or state constitution, which is hard to imagine, the courts are structurally following the legislature. Therefore, the courts will always be a second choice for these issues compared to the legislature. However, it is a choice advocates have often made when they knew that the legislative route would not produce results. Examples would be the early civil rights movement and the abortion legalization movement before the 1973 *Roe* decision.

2.3.5 Government Commissions

There is a final venue that is a hybrid form. The government commission typically does not have actual policy-making power, but rather has agenda-setting power. For example, federal bioethics commissions have the power to frame the debate for the actual decision-makers, shaping policy-making in a somewhat subtle yet powerful way. The current government bioethics commission was able to raise alternatives to embryonic stem cell research into public prominence through a series of hearings and a report. The current commission also proposed various laws that were introduced in Congress. The previous government bioethics commission cooled down much of the furor over cloning when they concluded that the only ethical concern they could reach consensus upon was that cloning was not currently safe (National Bioethics Advisory Commission, 1997).

These government commissions prefer either thick or thin discourse, depending upon who they perceive their audience to be. If it is the federal bureaucracy, then thin discourse will be preferred. If it is the public or elected officials, then either thick or thin discourse can be used. Most of these commissions officially state that they have two audiences—the government and the public—but in practice they focus upon one or the other. The government commission of the 1980s, titled the President's Commission for the Study of Ethical Problems in Medicine and Biomedical and Behavioral Research, focused, at least in its work on human genetic engineering, on translating "thick" public concerns into a "thin" language amenable to executive branch action (Evans, 2002). Tellingly, its report on human genetic engineering led to the formation of another commission under the aegis of the National Institutes of Health. In contrast, the current bioethics commission has thought of its audience more as the public, a public which is interested in the "thick" debate. It has explicitly engaged in conversations about ends to pursue instead of simply the most efficacious means toward assumed ends. It has produced various educational materials for the public and tellingly, instead of resulting in a new executive branch commission or communicating with executive branch officials, it took its recommendations directly to the Congress in the guise of new laws to pass that would reflect is chosen ends.

The "thin" oriented government commission would be roughly as good a venue for introducing religious conceptions of nature in ART as the executive branch venue. That is, it would be equally bad in that while it may at least listen to such perspectives, they will be translated into universal ends to make them palatable for the executive branch policy-making audience. The "thick" oriented government commission would be better, in that they want to have the discussion about ends. While not ideal from the perspective of advocating a religious conception of nature because commissions lack actual policy-setting authority, from the perspective of minority religious traditions (like those discussed in the Traina chapter), this sort of commission would be the best opportunity to have one's concerns embedded in policy. That is because their processes are more reflective than legislatures, and if Shiite concerns about nature could be incorporated into a suggested policy in a way that does not interfere with the majority Christian view, it would probably be incorporated.

2.4 The Translation Problem

The first challenge for incorporating religious conceptions of nature in public policy is that various venues of policy making do not encourage such discussions, as discussed above. The second, even more subtle problem is well known to practitioners of public theology in the U.S.: the translation problem. Put simply, "thick" debate about ends to pursue cannot be made in explicitly religious language, even in the venues that accept "thick" discourse. They must be translated into secular equivalents.

Consider Methodist theologian Paul Ramsey, one of the founders of modern debates about science, ethics and society. He was one of the first advocates in modern medical ethics of forwarding the end of autonomy in medical decision making through pursuing the consent of patients. (It may be hard for younger people to realize, but it was only in the past 40 years that it was considered important to always get a patient's consent.) For Ramsey this was a translation of his pursuit of the explicitly religious ends of agape[4] and holding the covenant with God. According to one reviewer, he made use of consent because he "sought to find a language accessible to as many people as possible despite their theological convictions and 'consent' appeared to cross over communities and traditions" (Long, 1993, 125–126).

In fact, I would argue that at least in the case of ARTs, the ethnographic observations of Traina and her colleagues about nature indicate that "nature" is a secular translation of more explicitly theological notions of God's will. Again, while this conflation would scandalize religious experts, the religious citizens seem to do this anyway. For many religious traditions, the authors point out, God's will is for us to have children. This is expressed (translated) as "it is natural for us to have children." This is a common way to articulate religious concerns in policy contexts. Taking advantage of the conflation of God's will and nature found among most religious people, people can say in the debate about homosexual marriage, for example, that it is "not natural" when they really mean it is "against God's will."

The question that concerns us here is the extent to which this translation is a problem, and it is a problem to the extent that something gets lost in translation. Put simply, if religious conceptions of nature are going to have an actual influence on public policy about ARTs, these conceptions need to be accurately represented in these discussions.

Intuitively we can imagine that any translation is not as precise as the original. We often hear of jokes that "lose something in translation" or see foreign words in English translations of academic texts that were left in place because the translator does not want to lose the precise meaning of the term in the original language.[5] When we pick a word that will stand in for what we would actually like to say, we also must accept the alternative meanings of this new word that we do not intend.

That is probably the case with people translating God's will to nature. As earlier chapters point out, at least in the official theology of the Christian majority in this country, God cannot be equated with nature. God has a more complicated relationship with nature, starting with the idea that God created nature (but is not nature).

---

[4] *Agape* means, in the Christian context, universal love, including love of one's enemies.

[5] For example, consider the following "translator's note": "another troublesome term is bürgerlich, an adjective related to the noun Bürger, which may be translated as 'bourgeois' or 'citizen.' ... Bürgerliche Öffentlichkeit thus is difficult to translate adequately. For better or worse, it is rendered here as 'bourgeois public sphere" (Habermas, 1989, xv).

If one is pressed, because of the need to use secular language in policy debates, to not say that "God wants families to have children," but instead say that "it is natural for families to have children," one is adopting the meanings of "nature" one does not intend. As pointed out in Chapter 1, Volume One of this project ("Spiritual and Religious Concepts of Nature"), there are (at least) 14 different definitions of nature. One of these is "natural," as in, not modified by humans. Many Christians will not want to endorse that version of "natural," in that God calls upon them to intervene in nature. However, using the translation "nature" to some extent weds them to that foreign concept of nature, distorting their true meaning.

2.4.1 Avoiding Severe Loss of Meaning: The Well Maintained Path

Translation is inevitable in modern pluralistic societies. However, there are better and worse ways to translate, and engaging in better translation practice will help those with a religious conception of nature accurately influence ART policy. Let's start with a metaphor. Imagine that the explicit theological tradition is a well, and that believers walk on a path away from the well to encounter others who do not share their theological language. On the path away from the well they translate their language into language that the others can understand and appreciate. This works well, as long as they regularly return to the well to replenish their ideas and get new language for new problems. However, if they lose the path, when faced with a new problem they are only left with their already translated language and the language of other wells they encounter along the way.

The loss of the path is exemplified by the debate over the end or value called "autonomy" in bioethics. Returning to the case of Paul Ramsey, this end was introduced into these debates in the 1960s as a Kantian idea, where we are to have respect for persons as ends unto themselves and not means toward ends. Ramsey claimed that he was, in my terms, translating theological discourse to secular discourse: he was just "using" Kantian language, and that as far as he was concerned, Kant's meaning of autonomy "does not supplant the grounds for care that I espouse" which remained theological (Long, 1993, 205). Critics of Ramsey's translation at the time encouraged him to "write something where it is clear that you only 'use' Kantian language," because "it is indeed a tricky business of how one can use one position without buying into its deepest commitments" (Long, 1993, 206).

The path between the religious and its secular translation was never made and, not surprisingly, when people tried to return to the well to address related issues in the public sphere, they returned to a philosophical well. A subsequent theological interpreter of the use of autonomy in these debates, Karen Lebacqz, now complains that autonomy has lost not only its religious connection, but any social connection as well. Autonomy simply means the right to do whatever one wants, derived largely from liberal political philosophy, due to the lack of a path drawn back to Kant. She wants to modify the principle of autonomy by drawing on the "tradition of covenant," a Protestant theological idea (Lebacqz, 2005, 105). In my terms, connecting autonomy to the idea of "covenant" would create the discursive path back to the well of religious discourse, a path that had been lost due to under-use.

What would a well-maintained path look like? A good example is the abortion discourse of mainline Protestant denominations (Evans, 1997). For example, the United Methodist Church has, since 1972, advocated for the continued availability of legal abortion. The dominant discourse among secular pro-choice advocates has been liberal individualism or "autonomy:" women should be free to control their own reproductive lives. While the United Methodists agree with this position, they have also traditionally grounded that freedom of choice in a more explicitly religious discourse. That is, they have created the discursive path between the secular discourse of "freedom of choice" and the explicit religious discourse of balancing the God-given sanctity of the life of the woman and the fetus. Because the Methodists have retained the path between the religious well and the secular arena, when a new issue arises and they must return to the well, they know where to find it. For example, when the human cloning issue arose, they could have claimed that cloning fell under the guise of "reproductive rights," as many have done. Instead, they argue for banning cloning, based upon their return to the well of religious discourse in Methodist theology (Hanson, 2001), not too different from the religious discourse that legitimated the more secular translation about abortion. While this path is often constructed in the public sphere, it is not always maintained, and the loss of maintenance means the religious discourse becomes simply secular yet thick discourse.

For advocates of a religious conception of nature, they will have to maintain their path, so that their discussion of "nature" remains grounded in its original theological motivations. The best way to do this is to keep up a vivid conversation at the well, so that traveling translators feel that they must at least occasionally return in order to remain a legitimate member of the community. Imagine, for example, an Orthodox Rabbi who carries translated theological ideas about nature to the public policy making arena, but who does not ever return to Rabbinical debate. Another Rabbi travels to the policy making arena but regularly returns to Rabbinical debate to keep fluent in the original language. It is the latter Rabbi who will be most able to represent the tradition. Therefore, for religions to truly represent themselves in the policy arena, they must simultaneously keep conversations on these topics alive in their own communities. It is my sense that these conversations about religion and nature, and in particular the application of these concepts to ARTs, is nearly non-existent in the dominant religious communities in the U.S. This is evident by the conflation of God's will and nature. Anything that can be done to strengthen the conversation at the well will help.

2.5 Western European Comparison

Unfortunately, space considerations preclude a study of policy-making throughout the world, but a short comparison with Western Europe will broaden our understandings of the challenges facing incorporating religious conceptions of nature into ART policy. The most obvious feature of Western Europe is that its ART policies are generally more restrictive than those in the U.S. For example, in the most comprehensive comparative study of these policies to date, Bleiklie et al. (2004) categorize the eleven European and North American countries into three categories. The first are the highly restrictive nations of Germany, Norway and Switzerland. The second are the intermediate nations of France, Netherlands, United Kingdom and Spain. The third are the most permissive nations of Italy, Belgium, Canada and the U.S.

To give a sense of the difference with the essentially unregulated U.S., the authors of the study provide a succinct summary of the policies in the restrictive countries, which

> aim at protecting patients and society at large from the potential negative effects and assumed dangers of the new techniques. In these countries many techniques are prohibited, and whatever is allowed is strictly regulated. They limit the use of and access to ART considerably, and prohibit several techniques – namely egg donation, pre-implantation diagnostics and embryo donation – and define strict conditions and rules for special licensing, reporting and controls for what is allowed. Furthermore, they strongly protect the embryo, to the extent that IVF is regulated in such a way as to exclude the production of left-over embryos. Embryo research is fully or almost prohibited...human cloning and chimera and hybrid building are prohibited in all three countries. For the techniques allowed, access is limited to stable heterosexual couples. For some techniques involving gamete donation, access is limited to married couples (Switzerland) (Bleiklie et al., 2004, 229–230).

For example, in Germany—one of the restrictive nations—the policy declares some uses of ART to be criminal, premised on the idea of protecting embryos. The law prohibits egg donation and embryo donation, and prohibits transferring more than three embryos to a woman within a cycle. All created embryos must be transferred to the woman, and all research on embryos is forbidden. ARTs can only be used by married heterosexual couples, or those with stable heterosexual relationships (European Society of Human Genetics Public and Professional Policy Committee, 2005, 94).

In France, an intermediate nation, the law allows most ARTs, but there is tight regulatory surveillance. Only heterosexual couples are eligible, and procedures must be justified on therapeutic grounds. Unlike Germany, research on surplus embryos is allowed with the consent of the parents, but research and reproductive cloning is banned, as is surrogacy (European Society of Human Genetics Public and Professional Policy Committee, 2005, 94). The U.S., which we have examined extensively above, is an example of a permissive nation.

The difference in policies between the nations in Europe and North America have been explained by structural features of the polity in each nation. For example, Varone and his colleagues explain these differences by asking, first, whether the political parties in the government are in agreement with a certain policy approach. One of the reasons Germany has such strong restrictions is that both the German right and left wanted to limit ART, albeit for different reasons (Varone et al., 2006, 325). In the U.S. case there is a split between the two dominant political parties. A second reason for differential restrictiveness in policies between countries is the unity and strength of the medical community, and more unity leads to regulation.

These scholars conclude that the U.S. does not have a unified medical community on these issues.

A third reason is whether interest groups were mobilized, and if so, the authors expect a less permissive policy regime. A fourth reason is the number of policy arenas in play. Nations that set policies in multiple locations (such as national and state levels) are less likely to enact ART policies, and thus freedom will reign. Nations that set policies in fewer locations (such as at the national level only) will more likely have policies and thus restrictions. The U.S. is an example of a nation with multiple policy-setting venues, and the UK is an example of a nation with one policy-setting location. The final reason is whether power is shared or whether majorities simply rule. In the U.S. the party in power does not need to compromise with the other, and therefore it is more likely that restrictive policies will be created (Varone et al., 2006). Therefore, one explanation for U.S. difference from Europe is in these institutional design features of the polity.

While I will focus on the structure of the polities, I must note that obviously this is not a complete explanation for the more conservative ART policies in Europe. For example, in Europe, activists have been able to effectively use what I would call an end of "human dignity" in debates about various ARTs. Its power perhaps harkening back to the Nazi experience, the idea that certain ARTs reduce human dignity has been a powerful argument against liberalization.[6] Arguments about human dignity are by and large not made in the U.S.

While I have no intention of trying to further explain differential policies across nations, I do want to try to assess the ability of religious concepts of nature to influence ART policies in Europe, in light of different structural features of their polities. Obviously, each nation will have its own particularities, but certain generalizations can be made. Let me begin by clearing some conceptual brush. First, clearly fewer individuals in Europe participate in organized religion than in the U.S. Forty-six percent of Americans report weekly religious service attendance. Eight percent of French, 3% of Danes, 16% of Germans and 14% of British—to use a few examples—report such a level of devotion (Norris and Inglehart, 2004, 74). However, this does not mean that religion has no influence in public affairs or policy formation. In her comparative study of the formation of policies regarding contraception, abortion and ARTs in Britain and France, Melanie Latham concludes that

> religious and moral opposition to assisted conception has been very much in evidence during parliamentary debates in both the UK and France...religious groups, particularly the Roman Catholic Church in France and the Anglican Church in the UK, have been accorded a privileged position within the policy community, more so indeed than on the previous two conscience issues of contraception and abortion (Latham, 2002, 181–182).

Similarly, in his analysis of the UK debate that resulted in the Human Fertilisation and Embryology Act of 1990, Mulkay focuses an entire chapter on how secular proponents of a liberal approach to ARTs ridiculed the religious opposition for

---

[6] For recent debates about "dignity" as a concept in policy debates, see Ashcroft (2005) and Caulfield and Brownsword (2006).

being akin to the medieval Catholic Church's persecution of Galileo, which suggests that religious groups were a primary source of opposition (Mulkay, 1997, ch. 7). Second, and briefly, I think that the translation issue identified above also exists in Europe, and would impose a similar impediment toward the use of religious discourse about nature.

2.5.1 European Policy Venues and Thick and Thin Arguments

There is a critical distinction between Europe and the U.S., which then allows for differential ability to introduce religious conceptions of nature into debates about ARTs. That is that, as I alluded to before, Europe and America are very distinct in the type of discourse that they will allow in different policy making venues. I will argue that "thick" discourse about ends to pursue can be considered by all policy venues in Europe—in contrast to the U.S., where this thick discourse can primarily be considered in a legislative context. Therefore, to the extent that European policy is made by officials who are distant from democratic legitimation, which is often the case, concerns of nature will be more admissible.

I need to go further into this difference in bureaucratic authority between Europe and the U.S. James Q. Wilson notes that both Sweden and the U.S. have inspections by government officials for worker safety in workplaces. In the U.S., the officials "go by the book" and follow the rules to the letter, writing up formal citations when warranted. The book itself is quite lengthy, with myriad rules and forms of evidence. In Sweden, the equivalent government officials "are expected to use their discretion and not go by the book. There is scarcely any book to go by; the procedures they are supposed to follow are outlined in general language in a six-page pamphlet." They spend their time offering friendly advice to factory managers on how to improve conditions. In a similar study of these differences, a study of regulation in America and three European nations found that "the administrative system afforded the bureaucrats more discretion than was enjoyed by their American counterparts and encouraged them to use informal procedures in formulating and enforcing rules" (Wilson, 1989, 295–297).

In sum, in contrast to the U.S. which distrusts government authority, "in some societies the right of government officials to make decisions is taken for granted. People may disagree about the substance of the decision but they do not question the authority behind it. As a consequence they defer to officials who act for the government" (Wilson, 1989, 303).

Now, instead of thinking of workplace inspectors, consider the Human Fertilisation and Embryology Authority (HFEA), a regulatory and licencing body established in the UK by the Human Fertilisation and Embryology Act of 1990. (The U.S. has no similar regulatory agency.) Unlike in the U.S. context, this agency does not follow the rules so much as it makes the rules. It is currently engaged in a public consultation to determine the limits on the number of children sperm and egg donors can produce, how donor's characteristics should be matched with patients, and how much to pay donors. Similarly, the HFEA issues licenses to engage in pre-approved applications of pre-implantation genetic diagnosis, deciding which genetic conditions are "serious." For example, they recently gave a licence to a clinic in London to use PGD to screen out embryos with Familial Adenomatous Polyposis Coli, which tends to lead to cancer in middle age. In July of 2004 the HFEA started allowing PGD to allow for the creation of children whose tissue could be used to treat a seriously ill sibling. The Chair of the HFEA—an appointed, not an elected official—wrote that "Our review of the evidence available does not indicate that the embryo biopsy procedure disadvantages resulting babies compared to other IVF babies. It also shows that the risks associated with sibling to sibling stem cell donation are low and that this treatment can benefit the whole family" (European Society of Human Genetics Public and Professional Policy Committee, 2005, 101).

The important point here is that this bureaucratic agency in the UK just approved two technologies that would not be approved by a bureaucratic agency in the U.S. but by a legislative body: PGD for late onset disease, which calls into question the distinction between disease and improvement, and using PGD for "savior siblings," essentially creating a child with particular genetic qualities so that their tissue can be more effectively used to heal another child.

This comports with Jasanoff's study of genetic technology policy across the U.S., the U.K. and Germany. She finds that whereas public accountability for decisions made by bureaucrats in the U.S. is akin to what I have described previously, the "more insulated regulatory processes of both Britain and Germany historically depended on greater trust in expertise" (Jasanoff, 2005, 262). Whereas in the U.S. objective knowledge is produced by numbers, "objective knowledge is sought in Britain through consultation among persons whose capacity to discern the truth is regarded as privileged." And, in "Germany, by contrast, expert committees are often constituted as microcosms of the potentially interested segments of society; knowledge produced in such settings is objective not only by virtue of the participants' individual qualifications, but even more so by the incorporation of all relevant viewpoints into the output that the collective produces" (Jasanoff, 2005, 266–267). In short, in these European countries, these elite experts can have a discussion about what the ends should be in ART policy, and remain legitimate in the eyes of the public.

Returning to conceptions of nature, Traina and her colleagues point out that ARTs come to be "naturalized" if they support what are thought to be other religious commandments, such as what is perceived in many religions to be a commandment to have children. Given the greater freedom of European bureaucrats to use their own judgement to set the ends that policies should forward, we can imagine that conceptions of nature such as this would at least potentially have greater influence in the UK than in the U.S.

In addition, the chance of having any discussions of ends in any policy making arena are higher in Europe than in the U.S. because of the more collectivist orientation of Europe. The more individualistic America is less likely to set ends at the national level, instead letting individuals set their own ends. As Daniel Callahan puts it, in the U.S., unlike Europe, we have a society that forswears communal goals, that has "tried to replace ultimate ends with procedural safeguards; that [has] resolutely worked to abolish the most profound questions of human meaning to the depths of hidden, private lives only; and that [has] striven to sanctify the morally autonomous agent as the cultural ideal" (Callahan, 1981, 264). In short, Europe is more likely to set any end—be it of "nature" or anything else—than is the U.S. This is I think evident in European policy in general, in that it is more substantive and less procedural than the U.S.

2.6 Conclusions

I have focused this essay on the challenges those who hold religiously based conceptions of nature will face in trying to have these views influence the policy making process. In the U.S., different policy making venues will be differentially open to a debate that tries to set the ends that we should pursue, such as conceptions of what is "natural." Moreover, religious conceptions of nature face the challenge that the explicitly religious qualities of these claims must be translated to secular terms, thus potentially losing some of their meaning. While the translation problem remains for Europe, the different orientation toward government authority in Europe means that a debate about ends such as what is "natural" could occur in any policy making venue, unlike in the U.S.

The premise of the entire project represented in the essays in this volume is that the public debate over biotechnology would be improved if we as a society were to think more deeply about the commonly made claims about "nature." I think that all involved in this project, and hopefully by this point the reader, would agree. However, if these discussions of nature cannot make it to the policy process, then the contribution to the broader debate will be limited.

References

Ashcroft, Richard E. (2005). "Making Sense of Dignity," *Journal of Medical Ethics* 31, 679–682.
Beauchamp, Tom L. and James F. Childress (2001). *Principles of Biomedical Ethics*, 5th ed. New York: Oxford University Press.
Bleiklie, Ivar, Malcolm L. Goggin, and Christine Rothmayr (2004). *Comparative Biomedical Policy: Governing Assisted Reproductive Technologies*. New York: Routledge.
Callahan, Daniel (1981). "Minimalist Ethics: On the Pacification of Morality," in A. L. Caplan and D. Callahan (eds.), *Ethics in Hard Times*. New York: Plenum, 261–281.
Caulfield, Timothy and Roger Brownsword (2006). "Human Dignity: A Guide to Policy Making in the Biotechnology Era?," *Nature Reviews Genetics* 7, 72–76.
European Society of Human Genetics Public and Professional Policy Committee (2005). *The Interface Between Medically Assisted Reproduction and Genetics: Technical, Social, Ethical and Legal Issues*. European Society of Human Genetics, 110 pp.

Evans, John H. (1997). "Multi-Organizational Fields and Social Movement Organization Frame Content: The Religious Pro-Choice Movement," *Sociological Inquiry* 67(4), 451–469.

Evans, John H. (2000). "A Sociological Account of the Growth of Principlism," *The Hastings Center Report* 30(5), 31–38.

Evans, John H. (2002). *Playing God? Human Genetic Engineering and the Rationalization of Public Bioethical Debate.* Chicago, IL: University of Chicago Press.

Evans, John H. (2006). "Between Technocracy and Democratic Legitimation: A Proposed Compromise Position for Common Morality Public Bioethics," *Journal of Medicine and Philosophy* 31, 213–234.

Ezrahi, Yaron (1990). *The Descent of Icarus: Science and the Transformation of Contemporary Democracy.* Cambridge, MA: Harvard University Press.

Habermas, Jurgen (1989). *The Structural Transformation of the Public Sphere.* Cambridge, MA: MIT.

Hanson, Jaydee (2001). *Testimony Before a Hearing of the Oversight and Investigations Subcommittee of the House Energy and Commerce Committee*, March 28. U.S. House of Representatives. Government Printing Office. Washington, DC.

Jasanoff, Sheila (2005). *Designs on Nature: Science and Democracy in Europe and the United States.* Princeton, NJ: Princeton University Press.

Kaufman, Marc (2005). "FDA Official Quits Over Delay on Plan B," *Washington Post* (Washington, DC), 1/September, A, 8.

Latham, Melanie (2002). *Regulating Reproduction: A Century of Conflict in Britain and France.* Manchester: Manchester University Press.

Lebacqz, Karen (2005). "We Sure Are Older but Are We Wiser?," in J. F. Childress, E. M. Meslin, and H. T. Shapiro (eds.), *Belmont Revisited: Ethical Principles for Research with Human Subjects.* Washington, DC: Georgetown University Press, 99–110.

Long, D. Stephen (1993). *Tragedy, Tradition, Transformism: The Ethics of Paul Ramsey.* Boulder, CO: Westview Press.

Mulkay, Michael (1997). *The Embryo Research Debate.* Cambridge: Cambridge University Press.

National Bioethics Advisory Commission (1997). *Cloning Human Beings: Report and Recommendations of the National Bioethics Advisory Commission.* Rockville, MD: NBAC.

Norris, Pippa and Ronald Inglehart (2004). *Sacred and Secular: Religion and Politics Worldwide.* New York: Cambridge University Press.

Porter, Theodore M. (1995). *Trust in Numbers: The Pursuit of Objectivity in Science and Public Life.* Princeton, NJ: Princeton University Press.

President's Council on Bioethics (2004). *Reproduction and Responsibility: The Regulation of New Biotechnologies.* Washington, DC: Government Printing Office.

Robertson, John A. (1994). *Children of Choice: Freedom and the New Reproductive Technologies.* Princeton, NJ: Princeton University Press.

Starr, Paul (1982). *The Social Transformation of American Medicine.* New York: Basic Books.

Traina, Christina, Eugenia Georges, Marcia Inhorn, Susan Kahn, and Maura Ryan (2008). "Compatible Contradictions: Religion and the Naturalization of Assisted Reproduction," in B. A. Lustig, B. Brody, and G. McKenny (eds.), *Altering Nature: Volume II. Religion, Biotechnology, and Public Policy*, Dordrecht, The Netherlands: Springer, this volume.

Varone, Frédérik, Christine Rothmayr, and Eric Montpetit (2006). "Regulating Biomedicine in Europe and North America," *European Journal of Political Research* 45(2), 317–343.

Wilson, James Q. (1989). *Bureaucracy: What Government Agencies Do and Why They Do It.* New York: Basic Books.

Chapter 3
Religious Traditions and Genetic Enhancement

Ted Peters, Estuardo Aguilar-Cordova, Cromwell Crawford, and Karen Lebacqz

Among important new biotechnologies are those that offer the promise of genetic intervention, both to correct genetic anomalies and to enhance human capacities. In this chapter, we ask: How do different religious traditions assess the possibilities for genetic enhancement? Understandings of "nature" become crucial here. Do religious traditions have a view of nature that impacts their approach to new genetic technologies? Do they see nature as a given that should not be changed? Do they see genetic enhancement as an alteration of nature, and if so, is that alteration acceptable? How does the approach to nature of a tradition impact its evaluation of the ethical acceptability of genetic enhancement? Is altering nature a shift to something 'non-natural' or 'unnatural,' or is nature itself understood as always in flux, so that alterations are simply in accord with nature? Inherent in the question of altering nature is an incipient moral charge, an ethical electricity, so to speak. If the form of altering nature to be considered is genetic enhancement, it is intuitively and immediately considered a moral matter.

Our approach to these questions takes the following form. We begin with some caveats, definitions and clarifications: what is "enhancement," how does it differ from "eugenics," what is "nature," how is the concept of nature relevant to moral norms or ideals? We then offer four case studies or scenarios involving possible genetic enhancement. Our third section reviews in turn a number of traditions: Judaism, Roman Catholicism, Protestantism, Hinduism, and Buddhism. In this section, we include a brief discussion of feminism as an important contemporary tradition that emerges in and draws from several religious traditions. As we look at each of these traditions, we will attempt to indicate how this tradition would reflect upon the four case scenarios.

The vastness and complexity of millennia of religious practice provide a formidable challenge. For sources, we have elected to analyze primarily texts ancient and modern, especially the writings of theologians or ethicists whose thinking brings us close to the issue at hand, namely, genetic enhancement. We report the classic or contemporary spokespersons at the heart of a religious tradition, or cite those who have commented on the focal ethical issue. We recognize that this dependence on the written word may not reflect actual practice in those traditions. We further recognize that no great religious tradition is monolithic; all include various perspectives and even subversive discourses. Yet, we believe that if we can report a theological commitment or spiritual thrust that authentically articulates a tradition's fundamental vision, we will have provided a service to those sorting through today's ethical challenges.

3.1 The Shadow of Eugenics

No contemporary discussion of genetic enhancement can escape the legacy of suspicion and distrust left by the earlier eugenics movement. The publication of Sir Francis Galton's *Hereditary Genius* in 1869 sparked a long and controversial tradition of interest in the genetic 'improvement' of humans. The United States became one of the centers of eugenic thought. It is important to note that many eminent scholars in the first half of the twentieth century were associated with this movement (see Kevles, 1992, 5; Peterson-Iyer, 2004, 49–51; Lombardo, 2003, 209). That association has left a legacy of distrust of genetics in some communities.[1]

The worst atrocities of the eugenics movement involved compulsory measures such as compulsory sterilization and racial 'cleansing.' In the United States, more than 60,000 people were sterilized in the twentieth century (Lombardo, 2003, 202). Legal support for sterilization laws drew from the famous declaration of Chief Justice Holmes in *Buck v. Bell* that Carrie Buck could be forcibly sterilized on grounds that she, her mother, and her child were all 'feeble-minded' and that "three generations of imbeciles are enough" (Robitscher, 1973, 10–11). In Nazi Germany, of course, eugenics took the form not only of elimination of those who were considered feeble-minded, but also, explicitly racially, of the elimination of hundreds of thousands of Jewish people. In the United States as well, the eugenics movement focused not only on mental 'defect' but on blindness and other physical attributes. Fear that any emphasis on genetic enhancement will fuel the seemingly dead embers of that compulsory movement can still be seen.

However, the eugenics movement was not all compulsory. From the beginning, the eugenics movement incorporated voluntary efforts to have children who were advantaged mentally or physically. More recently, the development of molecular genetics has allowed, in the view of an eminent scientist, a 'new eugenics' focused not on large population groups but on individual parents; and "the new eugenics would permit in principle the conversion of all the unfit to the highest genetic level" (Robert Sinsheimer, quoted in Keller, 1992, 289). Even where compulsory efforts are condemned, therefore, voluntary efforts at improving the health or characteristics of children are seen as contributing to a new emphasis on eugenics. Some see this new emphasis as dangerous precisely because of its subtlety and its distance from compulsory measures (see, e.g., Duster, 1990).

---

[1] It is significant that feminist and African American approaches to bioethics and genetics often begin with or include a discussion of eugenics. See Peterson-Iyer (2004), Tong (1997), and King (1992).

Thus, "the manner in which the eugenics movement developed cast a long shadow over the growth of sound knowledge of human genetics...."[2] In spite of the fact that the eugenics movement took many forms and that there is no single meaning to the term 'eugenics,' the term today carries a largely negative connotation (Lombardo, 2003, 202). Any association between genetic enhancement and eugenics will pick up this negative connotation.

As we approach our clarifications and descriptions, therefore, we are mindful of the need to remember and honor the painful legacy of the past. At the same time, we believe that contemporary genetic interventions need not carry a negative association with eugenics. To that end, some clarifications are in order.

3.2 Therapy vs. Enhancement

Many contemporary discussions of genetic intervention make a distinction between therapy and enhancement. For example, the President's Council on Bioethics defines *human gene therapy* as directed genetic change of human somatic cells to treat a pathological situation, a genetic disease or defect (President's Council on Bioethics, 2002). By *human genetic enhancement* they refer to the use of genetic knowledge and technology to bring about improvements in the capacities of living persons or future generations. "Genetic enhancement" generally refers to the transfer of DNA material intended to modify non-pathological human traits. The distinction between therapy and enhancement implies drawing a line between what is necessary to heal and what is desirable for reasons going beyond good health. Enhancement involves efforts to make someone not just well, but better than well, by optimizing attributes or capabilities. The goal might even be to raise an individual from standard to peak levels of performance. Eric Juengst defines enhancement this way: "The term *enhancement* is usually used in bioethics to characterize interventions designed to improve human form or functioning beyond what is necessary to sustain or restore good health" (Juengst, 1998, 29).

There are thus two ways to distinguish therapy from enhancement. One way is to identify therapy with a pathology and enhancement with the nonpathological. Another way is to identify therapy with genetic intervention that would bring an individual up to what is average and enhancement with interventions that would bring an individual beyond the average up to a level of excellence above others. We find the second way most helpful for our conversations, though we will raise questions about the distinction between therapy and enhancement.

The distinction between therapy and enhancement often carries a value connotation: 'therapy' is a good word and 'enhancement' is morally suspect if not outright objectionable. Note the assumption in a passing remark by Francis Fukuyama: "One obvious way to draw red lines is to distinguish between therapy and enhancement,

---

[2] From the 1961 Presidential address of L.C. Dunn, quoted in Lombardo (2003, 191).

directing research toward the former while putting restrictions on the latter" (Fukuyama, 2002, 208). Much ethical discussion to date has concentrated on the distinction between therapy and enhancement, asking whether a clear line can be drawn between them. The importance of this distinction will become clear when we turn to the discussion of Roman Catholic, Protestant, and Feminist ethics in particular.

In what we present in this paper, we will not concentrate on providing a precise distinction between therapy and enhancement. Indeed, we find such a distinction blurry and, hence, problematic. Nonetheless, we find that much of the literature uses such a distinction as though the line between therapy and enhancement is morally relevant.

Another distinction has been important in ethical discussions to date—the distinction between 'somatic cell' gene intervention and 'germline' intervention or Inherited Genetic Modification (IGM). When the objective is enhancement, the inserted gene may supplement the functioning of normal genes or may be superseded with genes that have been engineered to produce a desired enhancement. Such gene insertion may be intended to affect a single individual through somatic cell modification, or it may target the gametes, in which case the resulting effect could be passed on to succeeding generations. Thus, enhancement may apply to somatic cell modification or it may apply to what is sometimes called "germline" or inheritable genetic modification. Our focus will be on somatic cell interventions. There is presently considerable unanimity against germline intervention; hence, it is somatic cell interventions that raise the most pressing ethical issues.

At the level of the genome, we observe, enhancement might be accomplished in one of two ways, either through genetic selection during screening or through directed genetic change. Genetic selection may take place at the gamete stage, or more commonly take the form of embryo selection during pre-implantation genetic diagnosis (PGD) following in vitro fertilization (IVF).

Directed genetic change could be introduced into early embryos, thereby influencing a living individual, or by altering the germline, thereby influencing future generations. To attain genetic change, exogenous genes would be sent into existing cells aboard a vector, most likely a modified virus. One or two genes can be introduced in this manner.[3] This makes genetic change an effective technique for dealing with a disease precipitated by a single mutant gene—that is, it is effective for some forms of genetic therapy. It is less likely to be effective for purposes of enhancement. The most desirable human traits—the traits most likely to be chosen for enhancement—are thought to be the result of interactions of many genes and their products.[4] Successful enhancement may require the introduction or modification of

---

[3] The understanding of gene with which we work here is this: a gene is a segment of DNA that provides the source of a phenotypic trait. Some traits are due to multiple genes interacting with environmental factors.

[4] Steven Pinker, a brain and cognitive science researcher at M.I.T., sees the multiple gene factor as the "Achilles heel of genetic enhancement.... We know that tens of thousands of genes working together have a large effect on the mind." Single gene changes, which may be technologically feasible, will not suffice for enhancement (President's Council on Bioethics, 2003).

3 Religious Traditions and Genetic Enhancement

numerous genes. Such a proposal would require technical capabilities beyond what is currently available in therapy vector techniques.

Modest forms of enhancement are becoming possible. For example, introduction of the gene for IGF-1 into muscle cells results in increased muscle strength as well as health. Such a procedure would be quite valuable as a therapy, to be sure; yet, it lends itself to availability for enhancement as well. For those who daydream of so-called "designer babies," the list of traits to be enhanced would likely include increased height or intelligence as well as preferred eye or hair color. As of this moment, the science of genetics and the technology of gene transfer are not sufficiently developed to inaugurate a new industry of genetic enhancement.

Instead of depending on a distinction between therapy and enhancement, therefore, we will place this discussion within the context of asking: Can an ethic of genetic intervention for either therapy or enhancement or both be grounded in nature? Do different traditions understand nature as entailing its own ethic? Would nature forbid our altering human DNA as evolution has bequeathed it to the present generation? Can nature provide us with the ethical guidance we need at this juncture of scientific research and medical advance?

3.3 What Is Meant by 'Nature'?

What do we mean by nature? Does the word 'nature' simply describe the world as natural science describes it? Does the word 'nature' refer to a moral norm? Do the multiple meanings slip and slide in such a way that ambiguities enter into our discourse about it? Among the many ways that we speak of nature, here are six assumed meanings that flow through everyday speech (see Coates, 1998, 3).

First, nature is a *place*. Nature is the place where civilization is absent. Wilderness is natural, and we can go to the wilderness to visit it. Some of the forces of life present in the wilderness are included in parks and gardens, so that visiting parks and gardens give us a feel for nature and an appreciation for nature. Prior to the rise of human cities we assume nature was wild, untamed, and everywhere. In this sense, nature is both place and time. It appears to us as a *given*, what we have inherited, what originally existed prior to our control.

Second, nature is our *comprehensive cosmos*. Nature is the composite totality of the cosmos, from the big bang in the past to the present fifty billion stars and anticipating its future demise due to entropy. The evolutionary history of life on planet Earth is one tiny chapter in the epic history of nature. Nature in this sense is not something opposed to human civilization, but something larger that encompasses civilization as well.

Third, nature provides us with *essence*. As essence, the nature of something can be minimalist or sublime. In its minimalist form, to speak of the nature of something is to point out its actual characteristics; its nature (*natura*) or quiddity (*quidditas*) defines a thing by its primary qualities. In its sublime form, nature (*natura*) identifies the fundamental—even transcendental—characteristics of an entity or species in essence (*essentia*) and sets criteria for its existence (*esse*). In the latter case, the actual existence of something can undergo estrangement from its essential nature. This implies that what we see or perceive might not be true nature, but rather a distortion. The phenomena we daily see both hide and reveal an underlying nature, and this underlying nature constitutes the essential identity and even the value of an entity or a person. Nature is a quality or principle that informs or shapes or directs the empirical physical world, but it is not reduced to the physical reality. Nature is more real than what is merely physical. When nature is thought to be equivalent to essence, it takes on ethical valence. Nature becomes the source of value and the reference point for moral normativity.[5]

Fourth, building on what was just said above, nature can become the *guide* and even the inspiration for directing social affairs. Nature in this sense would be normative. It provides moral authority and orientation for governance. Green movements convey this sense of nature. The romantic tradition evokes a sense of mystical bond, wherein nature is at first alien but then intimate. The realization of this intimate unity with nature becomes the spiritual goal.

Fifth, nature is the *opposite of culture or artifice*. What human beings add to the history of nature is technological, and products of human technology are thought to be artificial. What is artificial is the opposite of what is natural. Whereas culture and its technology are anthropocentric—that is, oriented toward the human race—nature is that domain wherein humanity is not central but rather one expression among others of life on our planet. As the opposite of culture, this fifth definition is the flip side of the first definition of nature as place or what is given.

Sixth, nature is the result of a divine act of *creation*. Religions in the biblical tradition—Judaism, Christianity, Islam—view the origin of the natural world as the product of a divine action. Nature is not eternal; rather, its existence is contingent upon God's will and God's act of drawing the world into being from nonbeing. According to the doctrine of creation, human beings are fully natural, embedded in nature, creatures among other creatures. Nature is thought to give glory to God. It is dubbed "good" even if still in need of transformation.

Not all religious traditions proffer a doctrine of divine creation. Buddhism, for example, sees the physical world as the current state of co-dependent co-arising, *pratityasamutpada*. The ceaseless interaction of physical and mental processes provide the backdrop for spiritual striving, according to which human consciousness strives for enlightenment, for transcendence of the natural world. We human beings are born natural; our goal is to transcend our birth.

These alternative definitions of nature are not necessarily rivals. They can complement one another. When dealing with the ethical question of genetic enhancement,

---

[5] Treating nature as essence has played a significant role in theological deliberation. Roman Catholic theologian Richard P. McBrien (1995) provides a relevant definition. Nature is "that which is basic and unifying in anything existing independently of human action or chance, contrasted to what in it is superficial or transient." Note the transcendental and impassible status of nature as essence here.

our focus will be on nature as essence or guide, but it will be important to see whether concerns arise for what is 'artificial' or 'unnatural' as well.

A crucial question when we address the meaning of "nature" is how *human* nature is to be understood. Are we simply part of nature? Are we different from nature? When Descartes declared "I think, therefore I am," did he extricate human subjectivity from the objective natural world? Postmodern hermeneutical philosophers and holists hold that this extrication of the human subject from the natural world is an abstraction, maybe even a delusion. The human being is as much a part of nature as is the natural world the human being studies. Furthermore, the study and even the technological alteration of the world are as natural as the world being understood or altered. Nature is inclusive of both subject and object, of both human and non-human. Theologian Jürgen Moltmann makes this point: "Human beings no longer stand *over against* nature, as the determining subjects of knowledge and endeavour. As the determining subjects of knowledge and endeavour they are also part of a history *with* nature. They *have* nature—possess it—and yet they *are* themselves, at the same time, nature, which goes on developing in them and in their world" (Moltman, 2003, 15, italics in original). Still, there are some strands within religious traditions that place humans either outside of "nature" or as having a special place within "nature," such that "human nature" becomes a special category. Where alteration of nature might not be problematic for those strands of traditions, alteration of human nature can still be problematic.

The concept of therapy reminds us that nature can kill us, and in the end it finally will. While on the way toward our death, greater or lesser degrees of pain and suffering become options. Therapy employs science and technology as well as loving care to reduce suffering, to increase human wellbeing, to make life meaningful and even joyful. In the case of inheriting genetic predispositions to debilitating disease, we need to ask: does our genome belong to our nature, to the essence of who we are? Does our inherited genome say anything that counts as a moral guide? Does nature destine us to suffer, thereby making therapy an alteration of who we essentially are? Does therapy require altering the genome, or is it sufficient to offer loving care in order to reduce suffering?

Does what we say about therapy also apply to enhancement? The concept of enhancement in the abstract seems to entail an assumption that nature is what is given. Figuratively, nature is a place prior to alteration by human technology. What is being enhanced is a course of nature that would otherwise be merely natural, not yet modified by intentional intervention. To this extent, human enhancement could look like making a garden out of wilderness. Yet, in our experience, it is more. For many critics and skeptics, enhancement assumes that what is given in nature is what is essential and, therefore, normative. To enhance, then, becomes a deviation from the norm, an alienation from essence. Is our nature prior to intervention definitive of our essence? Does unaltered nature serve as a moral guide to proscribe altering it by technology?

Contributing to this moral valence is the high value contemporary society places on DNA. At the advent of the Human Genome Project in the late 1980s and early 1990s, scientists spoke of DNA as providing the blueprint for human beings. DNA became associated with the essence of life, and our individual genome became associated with our individual identity. Our very personhood has come to be associated with the particular set of genes we have inherited. In addition, idealized images of biological evolution were lifted up in our culture so that the human genome became thought of as a gift of nature. Nature, the benevolent giver of our genetic code, evokes amorphous religious sensibilities for many, leading some to attempt a "resacralization" of nature.[6]

With these various definitions of nature in mind, we ask: can "nature" be our guide when it comes to moral norms? Does the attempt to move from "nature" to ethics constitute a form of the "naturalistic fallacy"? Philosopher G.E. Moore was not the first one to raise the issue or use the term, though it is usually associated with him. There are in fact *two* crucial issues. One is whether ethical terms such as "good" or "right" can be *equated* with natural terms, such as "conducive to growth" or "admired by everyone." If we say, "that is good," is that statement equivalent to saying "that conduces to growth"? The other issue is whether normative ethical judgments about what is "good" or "right" can be *derived* from "facts"—i.e., whether they can be derived using inductive or deductive methods (Hancock, 1974, 14).[7] In Volume One of this project, the chapter on "Philosophical Approaches to Nature" described the naturalistic fallacy in detail. For our purposes here we need only to ask: Do religious traditions derive norms from facts? Does the anthropology of a religious tradition forbid alteration of what we have inherited from nature for purposes of enhancement? Does the concept of altering nature within a religious tradition imply that DNA or the human genetic code be granted special status, that it be treated as sacred and therefore unalterable by technological means? How can we forecast future public debate over the ethics of genetic enhancement, and what role might spokespersons for various religions likely play?

3.4 Altering Nature with IGF-1

3.4.1 *The Thin Line Between Therapy and Enhancement*

As we explore forecasts for the future regarding public debate over genetic therapy and enhancement, we note that genetic enhancement may in principle apply to *germline* intervention on behalf of future children or *somatic* alteration on now-living persons. Growth hormone intervention is an example of the latter. Because germline intervention is discouraged and in some cases banned, the more realistic

---

[6] "Nature needs to be resacralized," writes Muslim philosopher Seyyed Hossein Nasr (1996, 270). "The resacralization of nature stands before us as the great mission of the coming age," writes Jeremy Rifkin (1983, 252).

[7] Albert R. Jonsen (1998, 73) describes the barrier to jumping from 'is' to 'ought' as "Hume's Hurdle."

3 Religious Traditions and Genetic Enhancement

option for the near future is enhancement via somatic DNA alteration. As we try to anticipate future ethical deliberations over genetic enhancement, perhaps our present experience with growth hormone could be instructive. We will now look briefly at a genetic version of the growth hormone debate; then we will look at growth hormone case studies themselves. This will be followed by a review of the fundamental commitments of different religious traditions that might eventually be brought to bear on ethical debates in this arena.

By using growth enhancement as a template, we have the advantage of examining a medical intervention that can be either genetic or conventional, making it possible to discriminate between the *genetic* alteration factor and the *enhancement* factor. Growth hormone is available as a protein; it is being offered already. It can also be given as a genetic altering procedure. So by talking about it in two different forms, we make clear whether the fact that we use a *genetic element* makes a difference. Giving the protein is accepted widely. Is there, then, a real difference between that and giving a gene? We note that providing a protein is temporary, whereas providing a new or altered gene is permanent. Does this make a difference? This may make a difference if the persons involved cede to DNA or to our genetic code some special biological status. If DNA is thought to belong to a person's natural essence, then altering it for purposes of enhancement may seem to be a violation of the sacred. This may seem to be the case. Whether it actually is the case in light of the way various religious traditions view nature is what we will explore.

The genetic version of the growth factor debate centers on the insulin-like growth factor I (IGF-1), which plays a role in the regulation of cell growth, differentiation and apoptosis leading to multiple potential therapeutic uses. As a neurotrophic factor, IGF-1 is being evaluated for diseases such as amyotrophic lateral sclerosis (ALS), a fatal motor neuron disease in which motor neurons are progressively lost from the spinal cord and brain. IGF-1 also stimulates hypertrophy of skeletal muscle via activation of satellite cells (muscle stem cells) and increased protein synthesis, therefore it has great potential for diseases affecting muscle such as muscular dystrophies.

However, systemic delivery of IGF-1 protein has yielded disappointing results due to limited bioavailability and toxicity. Too much IGF-1 is also not desirable; it can cause detrimental cell growth contributing to cancer growth and cardiac hypertrophy. Thus, IGF-1, as a therapeutic reality, requires achieving regulated levels of the protein in the desired tissue. Recent data using tissue-specific promoters and viral vectors to deliver IGF-1 to the muscle have shown encouraging results in both ALS and muscular dystrophy models. Many hurdles still remain, such as how to effectively deliver to all the affected tissues and how to regulate expression in vivo. Toxicity has not been observed in the animal studies even after observation into old age, but mice are not men (or women). Ultimately, gene transfer may not be necessary if small molecule approaches can be used to selectively activate the IGF-1 pathway in affected tissues. Either way, any progress is welcomed by doctors, patients and families in the devastating diseases for which IGF-1 may be therapeutic and no therapy currently exists.

Rodent experimentation is showing promise. IGF-1 was shown to increase muscle mass and strength in muscles of rats injected with an AAV vector expressing IGF-1 under control of a muscle-specific promoter. The intervention enhanced the effectiveness of resistance training (ladder-climbing) and diminished the loss of muscle mass and strength after training was ceased. Many clinical applications can be envisioned for such an approach such as facilitating recovery from muscle atrophy caused by injury.

Now, with this in mind, let us try this possible scenario. Photographs of California governor Arnold Schwarzenegger, as the editor of *Muscle and Fitness*, appear in an advertisement for a new drug to enhance muscle build-up requiring less training time and only a single administration. Another ad appears in AARP magazine for a new drug for the elderly to improve strength and reduce the risk of falls, which does not add another pill to the medicine cup or another thing to remember, just a single injection. Now, how do we compare the young athlete to the grandparent?

Researchers and clinical physicians worry about the potential for abuse by athletes (Minuto et al., 2003; Kniess et al., 2003; Lissett and Shalet, 2003).[8] They are concerned about safety and about the ethical issue, namely, unjust advantages in athletic competition. Yet, they wish to improve the quality of life for aging patients. Where does one draw the line? Aging is actually accompanied by a reduction in IGF-1 and growth hormone levels. Perhaps replenishing these factors by the insertion of genetic material up to the levels of younger people could be a way to retard age-related muscle wasting. Would this be considered therapeutic or enhancing? Would it be any different if it was delivered as a protein or any other non-genetic chemical?

Some in the medical fields use the safety issue as the focus to avoid dealing with the difficult ethical dilemma. However, this may not pass the scrutiny of the expert eye. Although these genetic therapies may appear less safe because they involve gene transfer, the safety concerns are not much more risky than those that have arisen from protein therapies. For example, erythropoietin may increase the risk of cancer growth, and may also significantly improve quality of life. Hormone replacement therapy is another example where long-term use may provide improved quality of life, but it may also lead to increased risk of cancer or cardiovascular disease. Thus, the long-term use of a protein therapy is not so unlike the semipermanent introduction of a gene into a post-mitotic tissue to generate a local in vivo protein factory.

As with any new drug in development, safety must be a primary concern and needs to be evaluated carefully. "Safety" is the watchword for secular deliberations over public policy, and it covers over, and in some quarters replaces, direct

---

[8] These data illustrate that although activity of the GH/IGF-I axis declines with age, peripheral responsiveness to GH is not attenuated. This suggests that a decrease in GH responsiveness does not contribute to the age-related fall in circulating GH-dependent peptides. Thus, for those embarking on trials of GH therapy or GH secretagogues in the elderly, the capacity to generate IGF-I will not limit potential efficacy. Furthermore, the dose of GH replacement required for patients with organic GH deficiency is likely to be lower in the elderly compared with young adults.

confrontation with the ethical issues. Yet, we need to look the ethical issues straight in the eye. One way to do this is to turn from athletes and seniors to children, and look at four imaginary scenarios based upon what is now available.

3.5 Four Case Scenarios: The Ethics of Growth Hormone

As we anticipate the medium- and long-range futures regarding possible advances in gene splicing and genetic intervention, we might benefit from a template to sort out expected religious reactions to these advances in medical science. Our society has already had a taste of enhancement in its experience with growth hormone, so perhaps this partially understood phenomenon might provide insights regarding what might happen ethically when religious traditions confront the new genetics.

Our society places great emphasis on height, particularly for men. Statistically, taller men are perceived as more "successful" and earn more than shorter people. This may be a consequence of wider social values or of family upbringing. Children who are short for their age often suffer from being treated as though they are younger rather than just smaller. This often leads to decreased expectations from and by the child. These children may consequently suffer psychological harm—and, in some cases, serious physical limitations. Thus, there are physiological and psychological reasons for parents to desire normal, and in some cases supra-normal, growth for their children.

There are many conditions and diseases that can cause poor growth. Administration of growth hormone (GH), also known as somatotropin, has many effects in addition to growth stimulation during childhood. Most commonly touted in the general press recently have been its "anti-aging" and "muscle-building" effects. However, here we will only deal with its growth-increasing effects in children.

Most children with short stature grow at a normal rate and reach an adult height that is about the same as that of their parents. These children do not have a growth hormone deficiency. However, a child who grows at a slower than normal rate may have a growth hormone deficiency, regardless of his or her final height. In other cases, children do not have a deficiency of growth hormone, but may nonetheless be unusually short in stature.

Historically, treatment for GH deficiency has been limited to therapy by injection with GH protein four to seven times per week. This therapy usually lasts for years until the end of the child's growth period. GH protein was originally isolated from pituitary glands obtained at autopsy. In April 1985, pituitary-derived GH was removed from distribution in the United States and many foreign countries following the deaths of several patients from a very rare viral disease that may have been transmitted through the pituitary growth hormone they had received. In October of that same year the U.S. Food and Drug Administration approved the use of GH produced using recombinant DNA technology (rGH) for children with growth hormone deficiency. Recently, the FDA approved a supplemental application for its use to treat children who are healthy but unusually short (defined as an adult height of less than 5 feet 3 inches for men and 4 feet 11 inches for women) without a known cause. This approval makes GH available to short children who may suffer from, but cannot demonstrate, a medical condition that is responsible for short stature. The FDA based its decision on studies that found the biosynthesized hormone Humatrope added between 1 and 3 inches in height to children who took it for four to six years, and that there were no significant health risks for the children.

Human GH protein therapy is now quite safe but still requires many years of repeated injections and is very expensive with a cost of $30,000 to $40,000 a year. If a genetic transfer approach could attain similar or even more dramatic effects, without added risks, the morbidity and cost of the therapy may be significantly decreased. This could significantly increase the number of children who receive exogenous GH. The decrease in monetary and intervention costs may also make this therapy more available to those with less economic and educational resources, although critics may argue that the expanded use of GH is likely to be misused for non-health reasons.

Suppose a gene therapy approach to treat growth hormone deficiency with a viral vector that requires only a single application per year is approved by the FDA. It would be much less costly and intrusive than the daily injections of recombinant protein previously required. This would seem as a godsend to many families, especially for those without previous access. However, the increased access and the use of genetic transfer would also raise potential societal ethical concerns.
Consider the following four scenarios:

1. Johnny is a child with a severely retarded growth rate diagnosed with Somatotropin Deficiency Syndrome, low levels of human growth hormone, and an estimated final adult height of only 3 feet 6 inches. He was to receive an injection of the engineered virus to transfer a functional human growth hormone (hGH) gene into his muscle and allow him to attain a normal growth rate. He may still be short, but could gain an extra 12–24 inches in final height.
2. A second boy, Bobby, could receive rGH protein injections based on an unusually short predicted stature. However, Bobby is not diagnosed with hGH deficiency. Although he is below the 25th percentile for his age group and has a predicted adult height of only 5 feet, he seems to have a normal growth rate. Like his parents, he is of unusually short stature. Gene therapy for this boy may add 6–12 inches to his height.
3. A third boy, Tom, is displaying a slightly slower than normal growth rate, and has a possible growth hormone deficiency. However, because of his genetic composition, he is within the norm of height for his age group although predicted to be shorter than his unusually tall family members. He is predicted to achieve an adult height of 5 feet 7 inches, compared to his six-foot-six father. Gene therapy may bring Tom to a height similar to that of his family members.
4. A fourth case is that of Erwin, a boy with normal growth rate and above average height. Erwin was well above the 97th percentile in the growth chart with a predicted adult height of 6 feet 4 inches, similar to that of his father. However, his

parents feel strongly that if he were 6 inches taller he would be a happier and more successful person. Gene transfer could achieve his parents' wishes for Erwin.

These four cases present distinct therapeutic and enhancement scenarios for a genetic drug that makes the patient's cells the manufacturing site for the final effector, the GH protein. As noted above, many contemporary theorists would try to assess these cases by making the distinction between "therapy" and "enhancement." Where there is a clearly measurable medical deficiency, administration of growth hormone, or use of a genetic intervention, would be considered "therapy." Thus, for example, use of genetic intervention in Johnny's case would be considered acceptable, and use in Tom's case possibly acceptable, since there is the possibility of a physiological abnormality. But the scenarios involving Bobby—and certainly that involving Erwin—do not easily fit the model of therapy. Administration of the hormone, or genetic intervention, would be seen as "enhancement" rather than therapy.

In our view, it is difficult to justify such a distinction. Although Bobby may not have a measurable physiological abnormality, he will surely suffer psychologically as much as Jimmy will. If it is suffering that justifies intervention, it is hard to see why intervention should be acceptable in the one case but not in the other. Here is one place where understandings of "nature" might become crucial: must there be some deviation from a natural norm in order for genetic intervention to be acceptable, or is suffering alone sufficient justification? Different traditions will answer this question differently, as we shall see.

Among the complexities to be encountered here are the "genetic exemption", the potential changes of what is self; the perceived or real permanency of introduced changes; the potential for horizontal transmission of the drug (infection); the potential—however remote—of vertical transmission (to unborn offspring); the "slippery slope" of using genetic manipulations leading to psychological, moral or even species changes; the differential access to interventions that could provide unfair advantages to elite members of a society, and many others. In the following sections, we look to see whether the way that different traditions interpret nature has a direct impact on how they understand the ethical acceptability of genetic intervention, and whether such intervention is considered therapy or enhancement.

3.6 Contemporary "Naturalism" and Genetic Arguments

As we turn to alternative religious visions that serve as foundational for ethical visions, we will look first at *Naturalism*. This school of thought has developed primarily in Western Europe during the post-Enlightenment period. When fully articulated, it proffers the view that the only reality we can know is nature, and that nature is exhaustive of reality. Further, our cosmos and our bodies are organized according to natural laws, and these laws never go on a holiday. The laws are always in effect. No divine interruption (as in miracles) or even divine guidance (as in providence) is permitted. Naturalists see the world as a closed system, closed to supranaturalism, sometimes closed to all forms of transcendence. Although naturalism does not have a church or other form of institutionalization, it is a strong force in Western culture and influences public policy.

Naturalism and science are frequently associated. Science is seen by many as justifying the philosophy of naturalism. David Ray Griffin has provided insightful studies of the intellectual roots of naturalism, its marriage to modern science, and its conflict with religion. He provides a most illuminating definition for two levels of *Scientific Naturalism*.

> In the *minimal* sense, scientific naturalism is simply a rejection of the world's most fundamental pattern of causal relations. Understood *maximally*, by contrast, scientific naturalism is equated with sensationalism, atheism, materialism, determinism, and reductionism. Thus construed, scientific naturalism rules out not only supernatural interventions, as just defined, but also much more, such as human freedom, variable divine influence in the world, and any ultimate meaning to life (Griffin, 2000, 11).

What this definition in two levels tells us is that naturalism is a worldview, which sees the world as a closed causal nexus, exhaustively governed by laws discernable by the scientist.

Because naturalism does not have a church, it could take up alliances with those in churches; and in fact naturalism in the minimal sense is seen as compatible with biblical religion, especially Christianity. "Scientific naturalism and Christian faith, properly understood, are both true. Truth is one, so all truths are compatible with each other. As far as both scientific naturalism and Christian faith are true, therefore, they cannot be in conflict" (Griffin, 2004, 9). Judaism can make alliances with naturalism as well. This is important for the contemporary public controversies surrounding genetic research in general and enhancement in particular. Naturalism comes to expression not so much as a tightly argued ideology, but more as an intuition in partnership with established religious intuition. What distinguishes the naturalistic influence is the crudely articulated maxim that nature provides moral guidance.

This may not be obvious because of widespread acceptance of instrumentalism. Instrumentalism is the view that science and technology are value-neutral. If we make instrumentalist assumptions, and if we ask about the source or ground of values which we want to bring to bear on the question of altering nature, curiously, values appear to come *from humanity* or *to humanity*. If we begin with technology, it appears that values must come from human subjectivity. Technology in itself is value-neutral, so those whom we label *instrumentalists* assume; therefore, we need ethical direction to guide our use of technology. Moltmann looks at it this way. "Technological reason must be freed from ethical, practical reason, but must at the same time be integrated into it. It is only in their continual interplay that the two can grow from one another. Taken by itself, technological reason offers no criteria for the goodness of what can be done with it. Taken by itself, ethical reason has no means with which it can fulfill its purposes" (Griffin, 2004, 9). What is being said here is that the technology by which we alter nature is value-neutral, and therefore, we need to appeal to human commitment for our ethical norm.

The naturalist position, in contrast, holds that human subjectivity is prompted by an essence within nature, so that we value nature as nature. Opposition to the instrumentalist view can be found among scholars who honor the public outcry against genetic technology. They see the outcry as an intuitive insight into a natural link between fact and value. These scholars are dissatisfied with the disjunction between fact and value presupposed by analytic philosophers, and dissatisfied with the removal of teleology from nature. Celia E. Deane-Drummond, a British Roman Catholic bioethicist, and Leon Kass, former Chair of the U.S. President's Council on Bioethics, both appeal to what is natural for our essence and hence ethical norm. For these ethicists, values come to humanity from nature.

Celia E. Deane-Drummond tries to perceive purpose within nature while avoiding superimposing purpose from the outside. To accomplish this she retrieves the concept of latent dispositions toward essence.

> Aristotle and Aquinas approached moral theory very differently, using an ontology of dispositional essence. Such a scheme breaks down the fact/value distinction, so that it is the disposition of the natural process towards some goal that forms part of its nature. In other words, the good is not added to the fact about the natural world, but is inherent in it as it progresses towards its goal. The value is the end of the natural process (Deane-Drummond, 2002, 212).

She retrieves Aristotelian categories such as final cause to buttress her position. Deane-Drummond is well aware that modern science rejects final causation; science has explicitly expunged purpose and design and especially progress from biology. Yet she presses on to apply her theory of inherent value to an ethic that draws limits to genetic engineering. She applies the notion of latent essence to genes. "While teleology is officially rejected by science, I suggest that the idea of latency is not, especially in the light of our current knowledge that different genes are switched on and off at different stages of development of complex organisms. This dynamic understanding of creaturely being is far closer to Aquinas' notion of essence compared with the fixed ontology of Moore" (Deane-Drummond, 2002, 213). This leads to sympathy for the public intuition that we should refrain from xenotransplantation and cloning.

A connection is here presumed to exist between nature's essence and our human intuition. This becomes grounds for an argument raised with considerable passion and drama by Leon Kass. "There is something deeply disquieting in looking on our prospective children as artful products perfectible by genetic engineering, increasingly held to our willfully imposed designs, specifications, and margins of tolerable error....a major violation of our given nature" (Kass, 1998, 23). Nature communicates its essence to our intuition, believes Kass, and we need to listen to the wisdom of our repugnance when that essence is violated.

What causes anxiety and concern for Kass is the danger inherent in Promethean hubris, the temptation for human beings to play God, what some call "Brave New World" or the "Post-human future." This danger derives from the human pursuit of perfection. We see it in enhancement, not therapy. Medical science will be capable of providing significant improvement in human health through genetic therapy, notes Kass, and for this we should be grateful. Medicine is not, despite what some people think, a form of mastery over nature. Rather, medicine acts as a servant to aid nature's own powers of self-healing to overcome a deficiency in natural wholeness. Yet, this same medical science may tempt us to enhancement, and the problem here is that enhancement is based upon unrealizable fantasies about human perfection. The seductive lure of perfection appears in our projection of ends such as ageless bodies, happy souls, better children, a more peaceful and cooperative society, and such.

The fantasy of perfection provokes two philosophical problems, according to Kass. The first is the problem of ends and means. The present generation does not have a clear focus on the good, on the proper end or goal of pursuing perfection. We risk becoming homogenized at a level short of excellence. In a moment of eloquent flourish, Kass says,

> We are right to worry that the self-selected non-therapeutic uses of the new powers, especially where they become widespread, will be put in the service of the most common human desires, moving us toward still greater homogenization of human society—perhaps raising the floor but greatly lowering the ceiling of human possibility, and reducing the likelihood of genuine freedom, individuality, and greatness (Kass, 2003).

The second problem on Kass' list is the loss of normativity determined by our nature. The natural normativity Kass is referring to here is our finitude. As finite creatures, we can expect to age, deteriorate, and die. Recognized finitude spurs aspiration; and aspiration acted upon becomes the core of happiness. Such happiness cannot be gained through technological intervention, but only through expression of the essential soul which nature has endowed to each of us. "I have tried to make a case for finitude and even graceful decline of bodily powers. And I have tried to make a case for genuine human happiness, with satisfaction as the bloom that graces unimpeded, soul-exercising activity" (Kass, 2003).

What we have here are ears straining to listen to what nature might be saying to us, asking what if any limits we might consider when contemplating intervention into the human genome. Deane-Drummond would like to avoid committing the naturalistic fallacy while still incorporating sensitivity to nature into her theologically informed vision. Kass listens to nature alone. On the bases of what nature whispers in their ears, both Deane-Drummond and Kass end up closing the door to enhancement even while leaving it open for some therapy.

What this type of naturalism opposes is human hubris in pursuit of perfection. We observe that perfection may not apply to what is at stake in the enhancement controversy. Rather than perfection, many families who entertain the possibility of genetic enhancement are simply looking for betterment. Short of ultimate perfection, medical enhancement alters our inherited nature by offering a better life.

3.7 Roman Catholic Views on Therapy and Enhancement

We now move from *Naturalism* to *Natural Law* squarely within Christianity. Any contemporary Roman Catholic approach to ethical considerations surrounding therapy and enhancement will almost inevitably draw on a long tradition of appeals to natural law combined with commitment to human dignity in community. In this section, we will provide some background discussion of Roman Catholic theological precedents and methods, and we will attempt to forecast how Roman Catholic moral theologians are likely to assess genetic therapies and possibilities for enhancement.

In Roman Catholic tradition, "the words of scriptural revelation alone do not suffice to establish ethical norms, and therefore recourse must be had to the being or nature of man" (Grundel, 1975, 1017). According to Roman Catholic natural law theory, there exists a human essence, here called the "nature of man." This essence is discernible through human reason apart from special revelation. This essence sets limits on what people may do to themselves and to others.

The nature in natural law theory is not the laws of nature that would be described by a physicist or biologist. Rather, it is a natural *moral* law, built into the order of things as established by God the creator. Thus, nature is already graced. Nature points beyond the mere physical toward the end or *telos* of the creature. The natural law, then, incorporates an element of transcendence. It is a way of speaking about God's intentions for human beings, and about our proper ends, toward which human activity should be oriented. It is the participation of the eternal law in the rational creature that is the natural law (*Summa Theologica* II.1.Q.91, A2).

Because the natural law is discernible by reason, it is universal. It is not limited to believers or those with special revelation. In Catholic tradition, therefore, this natural law should be the basis for all positive law and for all human ethics. In this way, Christians and non-Christians have grounds for dialogue about ethics.

Contemporary Roman Catholic moral theologians affirm that nature changes, both through evolution and through human design (Grundel, 1975, 1020). The latter raises the question of deliberate change: should we alter nature? What alterations are ethical, and what alterations represent an unethical violation of the natural law? "Interventions in the course of nature are in fact to a great extent desirable for [human] existence. Hence, they are not *ipso facto* immoral. But they must not be detrimental to [human] dignity" (Grundel, 1975, 1019–1020).

Human dignity therefore emerges as a central norm that sets limits on human interventions into nature. Thus, the natural law norm does not stand alone; rather, the role played by human dignity in contemporary Roman Catholic ethical thought is decisive. The role of dignity emerges in several ways.

First, as with Enlightenment culture in general, dignity means that we treat each human individual as an end and not merely as a means to some further ends. When coupled with the understanding that life is human "from the moment of conception to death" (McKenny, 2000, 303), this norm prevents destruction of the embryo, even the very early conceptus or blastocyst. Thus, Roman Catholic moral reflection generally rejects any and all research that might involve the destruction of an embryo, such as stem cell research or attempts at cloning. Any genetic intervention that might destroy an embryo, even if done for purposes of developing genetic therapies, would be forbidden in this view.

Second, dignity involves an understanding that humans must be seen as "one in body and soul" (McKenny, 2000, 303). To respect in this way means to reject any technologies that separate the act of procreation from the biological and spiritual union of husband and wife. For this reason, the Roman Catholic Church has consistently rejected interventions such as in vitro fertilization or artificial insemination. Any new technologies that involve manipulation of the embryo, or creation of the embryo, outside the womb will be seen to violate this understanding of human dignity.

Finally, human dignity includes an affirmation that human life must never be reduced to the status of an object (McKenny, 2000, 303). Hence, for example, the life and health of some may not be pursued at the expense of reducing others to objects to be manipulated. Since personhood begins at conception, this principle extends to research using human embryos (Catholic Bishops' Joint Committee on Bioethical Issues, 1995). Enhancement would be prohibited if it objectifies the person.

These affirmations about human dignity, respect for life, and human liberty mean that therapeutic interventions can be accepted when they are designed to restore health or improve one's condition. However, most germline gene therapy would be excluded because it involves direct manipulation of the human embryo, in a situation where the embryo cannot exercise its liberty to consent. While there might be circumstances in which enhancements could be countenanced, the general tendency is to stress the correction of maladies and this casts "a general suspicion on enhancements and eugenic efforts" (McKenny, 2000, 303).

While several contemporary Roman Catholic theologians, notably Richard McCormick, Thomas Shannon, and James Keenan, extend natural law theory in different directions, none rejects the basic Roman Catholic ascription of value of human life from the moment of conception. The official Vatican teaching, however, is extreme:

The Church has always taught and continues to teach that the result of human procreation, from the first moment of its existence, must be guaranteed that unconditional respect which is morally due to the human being in his or her totality and unity in body and spirit:

> The human being is to be respected and treated as a person from the moment of conception; and therefore from that same moment his rights as a person must be recognized, among which in the first place is the inviolable right of every innocent human being to life (John Paul II, 1995; see also Congregation for the Doctrine of the Faith, 1987).

Any efforts at enhancement that would require manipulations of the embryo or possible loss of embryonic life are prohibited by this understanding of the guarantee of respect due to the human beings at all stages of life.

It is clear, then, that the natural law method, with its stress on human dignity, will set limits on what medicine and technology may do. Technologies that separate lovemaking from procreation will be rejected. Technologies that involve destruction of the early embryo will be rejected.

It is less clear, however, what the implications might be for gene therapy and enhancement. On the one hand, it seems consistent that most forms of gene therapy would be accepted. Since therapy is designed to fulfill human purposes and enable one's proper *telos*, therapeutic interventions are generally accepted in Roman Catholic tradition. Of our case scenarios, therefore, it seems very likely that

Roman Catholics would approve genetic intervention for Johnny, who is diagnosed with Somatotropin Deficiency Syndrome.

But what about the other case scenarios? What about Bobby, who will be very short but has no diagnosed deficiency? Or Tom, who may have a deficiency but again it is not diagnosed? Or Erwin, who has no diagnosed deficiency and who will already be fairly tall, though not as tall as his parents would like? There is nothing in principle in Roman Catholicism to prevent efforts at enhancement, if these efforts could also be seen as moving humans toward our proper ends. For example, if certain forms of excellence are among the ends of human beings, then it is possible that drugs that help people to focus mentally and to achieve such forms of excellence might be accepted.

At the same time, it is to be expected that most Roman Catholic theologians will not find themselves very comfortable with enhancement. The problem lies not in the technologies themselves (as it would for stem cell research or cloning, which involve the destruction of embryos), but in an understanding of what "excellence" means. The Catholic tradition has stressed the development of certain virtues, such as courage and fortitude, justice and temperance. Would temperance cease to be a virtue if it is enabled by administration of a drug, rather than by development of character over time?[9] As Donal O'Mathuna, a Protestant writing in a Roman Catholic organ, puts it, "Drugs can enhance people's sex lives, relieve baldness, increase height, improve concentration and hence take away conditions that many feel cause them to suffer needlessly. But is it appropriate to use the powerful tools of medicine to relieve these forms of suffering?" (O'Mathuna, 2002, 278).

Moreover, Catholic moral theology would always raise the question of whether true human ends, such as justice and mercy, are served by technological interventions. Mathuna notes that Catholics are leery of using medicine for enhancement because efforts at enhancement are so obviously bound up with value choices: "In making a person taller medicine promotes the belief that short people are of lesser value and that height is significant in achieving the good life" (O'Mathuna, 2002, 283). Some of these values are questionable from a Christian perspective: "Physical health is not the ultimate priority in Christian eyes" (O'Mathuna, 2002, 284). In short, if the *telos* or proper end of human life is the development of a relationship with God and the cultivation of those excellences or virtues necessary for living in community, then even some forms of suffering can be seen as serving those ends. Not all suffering should automatically be removed from human life. "Attempting to remove the trials and difficulties of life by genetic enhancement might derail the very ways in which God wants to shape our characters" (O'Mathuna, 2002, 295).

This position would not immediately posit a mandate to provide gene therapy for all forms of suffering, and it certainly would raise questions about the legitimacy of genetic enhancement. Thus, it seems very unlikely that any of these three scenarios would be considered acceptable in Roman Catholic tradition. Although increased height may be important for social standing or even for psychological health, Catholic tradition is much more likely to stress the virtues or excellences of

---

[9] The President's Commission on Bioethics suggests as much. See Kass (2003).

character. Hence, if Bobby, Tom, and Erwin suffer to some extent because of their lack of height, this may simply be an opportunity to develop the strength of character to deal with adversity.

In summary, Roman Catholic tradition is long and complex. It is probably more accurate to speak of natural law *traditions* than of a single tradition. Thus, it is risky to generalize in a few paragraphs. Nonetheless, the central emphasis on human dignity, the affirmation that human life is to be seen as inviolable from the moment of conception until death, and the stress on development of virtues or excellences in human life all tend toward a conservative stance when it comes to genetic interventions for enhancement purposes. Nonetheless, we note that there is no automatic rejection of technology. What is rejected is any action or intervention that disrespects human dignity, understood as the unity of body and soul, the value of each human life, and the importance of human freedom and character.

3.8 Protestant Views on Therapy and Enhancement

Like Roman Catholic moral theologians, Protestant ethicists are conscientiously searching their respective traditions for precedents and core convictions to guide them through the perplexing array of new challenges posed by the frontier of scientific research. Coming from the sixteenth century Reformation in northern Europe, the core convictions of these protestors within the western Latin church became salient through appellations such as *solus Christus* (Christ alone), *sola scriptura* (Scripture alone), *sola fide* (faith alone), and *sola gratia* (grace alone). The overarching ethical commitment common to Reformation theology is a commitment to neighbor love springing spontaneously out of the freed or liberated human heart. Sometimes called *agape*, this love is a free expression of one's Christian faith rather than an appeal to a commandment or law. Ethics for the Protestant is closely connected to freedom, and freedom when exercised is creative.

Turning to the specific matter at hand, this freedom is tested by the question of enhancement. To date, few Protestant thinkers have weighed in on the question; yet, we do have some testimony by thoughtful Protestant ethicists.[10] While individual differences among these ethicists are notable, the spectrum of views gives grounds for some broad generalizations. In this section, we have organized these around a number of tensions or paradoxes inherent in Christian anthropology that will influence the morality of genetic enhancement. Five such tensions can be identified, and set the stage for understanding the range of responses in the Protestant Christian community. Audrey Chapman observes that few ethicists carefully and explicitly tell

---

[10] Although Catholics and Protestants freely utilize one another's resources, we can expect lower reliance among Protestants on precedents set by natural law theory and an increased reliance on appeals to scripture (*sola scriptura*). Interestingly, however, most contemporary Protestants reflecting on genetics make scant reference to Scripture.

us how they draw on their theological or anthropological assumptions when making ethical pronouncements (Chapman, 1999, 40).[11] Thus the link between underlying anthropology and specific judgments about enhancement may be present but weak.

The first tension is the dialectic between creation and "fall," which plays a significant role in Christian anthropology generally, and becomes especially significant for Protestants as we move from anthropology to ethics. Such anthropology affirms both that humans are part of God's good creation and that humans live in a state of sin in which that goodness is distorted. In classical theology, this tension is represented by doctrines of Creation and Fall. This tension leads to alternative frameworks. Some Protestants will stress the goodness of creation and see genetic enhancement as a way to affirm and celebrate this goodness. For example, the World Council of Churches in 1989 took an overall positive view toward genetic engineering and biotechnology.[12] So did the Church of Scotland's "Religion and Technology" project (Chapman, 1999, 54). An overall positive view toward genetic interventions might accept a rather wide range of such interventions, including all four of our case scenarios.

Other Protestants will be more cautious and inclined to stress the problematic nature of human existence and the limits of human knowledge. They will tend to see enhancement as either a usurpation of God's authority (often called "playing God") or as an exercise fraught with risks because of the limits of human wisdom. In 1982, for example, a Working Committee of the World Council of Churches cautioned that genetic manipulation "amplifies and accelerates the tendency toward total reductionism" (Cole-Turner, 1993, 70). In fact, Chapman finds that the World Council of Churches generally exercises more of a "hermeneutic of suspicion" than do American denominations (Chapman, 1999, 70). As Ronald Cole-Turner notes in his summary of a number of church documents, some writers believe "that our sinful human nature will prevent us from fully seeing the misuses of our technology" (Cole-Turner, 1993, 78). The phrase "playing God," in particular, is often invoked to connote arrogant interference with nature; thus, many theologians would caution against enhancement if it is understood as contradicting the goodness of God's creation (Chapman, 1999, 53). Such views would tend to reject most genetic enhancement, as it would be seen either as a violation of God's good creation or as unnecessary in light of the fact that God's creation is good.

Cole-Turner argues that if we see creation or nature *only* as good, then we tend to overestimate our own goodness (e.g., our powers of reason and ability to know what God wants) or we find it hard to name defects in nature that should be corrected. However, if we see creation *only* as flawed or fallen, we then assume that it is open to any and all manipulations and we fail to see that it has value independent

---

[11] Chapman reviews critically extant church documents and individual authors' arguments prior to 2000; in addition she offers her own critique and theological perspective, which is largely liberal.

[12] See Cole-Turner (1993, 71). Teaching in a Christian seminary, Cole-Turner writes with explicit attention to a theological framework in which Jesus as redeemer figures prominently. Cole-Turner represents a moderate liberal approach.

of us. He urges a stance that holds in tension both the goodness of creation and the fallenness of our situation in the world; this stance reflects the classic tension of Christian traditions. Perhaps the general tendency is summed up by a Lutheran pronouncement that urges critical engagement, affirming in general the goodness of knowledge and new technologies, but being poised to offer criticism of particular applications (Chapman, 1999, 38).

A second tension is between the two concepts of stewardship and co-creation. Protestants, like Roman Catholics and Orthodox Christians, affirm that humans are not the world's Creator (God is creator), but we are charged with responsibility to carry on God's work in the world. The traditional language for this responsibility is stewardship (Chapman, 1999, 42). Stewardship generally implies that God sets limits on human activities. As stewards of God's creation, all decisions must be made with a view to fulfilling God's purposes rather than human desires. Christians have tended to see in nature some *logos* or divine purpose that stands as normative and sets limits on human intervention (National Council of Churches in the USA, 1980). For example, Princeton ethicist Paul Ramsey argued that we "ought not to play God" before we learn to be human, and that once we learn to be human we will not play God. Ramsey opposed genetic manipulations. Thus, Chapman finds that stewardship often evokes a static view: things were created a certain way and it is our responsibility as humans to ensure that they remain that way (Chapman, 1999, 43). In such a view, genetic intervention would be likely to be approved only in cases where there is a diagnosed deficiency, as in Johnny's case.

Some theologians have abandoned the language of stewardship in favor of co-creation. Renowned Catholic theologian Karl Rahner argued that humans are the creatures who "freely create" ourselves. A Lutheran leader in the dialogue between faith and science, Philip Hefner, coined the phrase "created co-creator" to emphasize that we human beings, created by our creator God in the divine image, are responsible for the ongoing creativity of the natural world. The concept of the created co-creator is both descriptive and prescriptive—that is, it provides us with a theological anthropology and with an ethical ideal (Hefner, 1993, 1998).

Those who see humans as co-creators generally see creation as continuously evolving and therefore do not see genetic intervention as an inappropriate use of human power. As early as 1980, a task force of the National Council of Churches argued that life is a gift from God but that humans are to show creativity in exploring life's possibilities (Cole-Turner, 1993, 71). More recently, Ted Peters has argued that the term "Playing God" can be understood in two ways: as a Promethean attempt to design an ideal or perfect future, or as an appropriate assumption of human responsibility. Playing God in the latter sense is appropriate because our ethical responsibility "*includes* building a better future through genetic science" and this exercise of responsibility is "a form of human creativity expressive of the image of God...." (Peters, 1997, xvii).[13] The United Church of Canada has been

---

[13] Lutheran theologian Ted Peters is largely favorable toward genetic science and suggests that there are interpretations of playing God that make genetic interventions acceptable.

most emphatic and clear on this point: "We are called to be co-creators with God..." (Chapman, 1999, 43).

Some who use the language of stewardship attempt to include within it notions that resonate with co-creation. Evangelical Protestants Bruce Reichenbach and Elving Anderson suggest that stewards are charged with filling, ruling, and tending the land. Hence, they believe that humans *must* "play God," and that genetic intervention can be a part of responsible stewardship (Reichenbach and Anderson, 1995).[14] From this perspective, "the effort to improve on nature is not inherently wrong," as the National Council of Churches notes (Cole-Turner, 1993, 72). Several individual denominations have taken a similar stance. For example, the United Methodist Church in 1991 adopted a report that declared: "Humans are to ... employ, *develop and enhance* creation's resources in accordance with God's revealed purposes" (Cole-Turner, 1993, 76, emphasis added). Under such a view, improving on what nature gives us is not necessarily wrong. Hence, there would not necessarily have to be a specific diagnosed deficiency in order to approve genetic intervention to increase height.

Thus, the tension for Protestant churches is a tension between the sense that human beings may—indeed, sometimes must—manipulate and change nature but that in so doing we are held to standards established by God's purposes in creation. Depending on whether these purposes are seen to reside in physical creation or in some notion or vision of a future perfect or redeemed world, there may be more or less latitude for interventions that enhance the givenness of human genes. For most Christians, interventions are permissible so long as "the researcher follows God's design" (Pope John Paul II cited by Cole-Turner, 1993, 77). However, Chapman is correct to charge that neither the stewardship model nor the co-creation model by itself gives clear guidelines as to what interventions would be permitted and what would be prohibited (Chapman, 1999, 44).

A third tension occupies us in this essay, namely, the difference between therapy and enhancement. How does what we have said about theological anthropology apply to questions of genetic alteration for purposes of therapy and enhancement? Hidden within the tension between stewardship and co-creation are assumptions as to whether God's creation, including human nature, is continuously evolving and open-ended or whether creation has a fixed character in the sense of limits that should not be transgressed. The question of whether human nature, and creation in general, is fixed or infinitely variable becomes important in discussing genetic enhancement. Several ethicists note that it is difficult to draw the line between therapy and enhancement. But most ethicists do believe that it is possible to specify a base-line of human functioning that is part of the intended order of creation. When someone falls below this base-line, genetic interventions are called therapy or correction. When an intervention would move someone above the base-line, it is called enhancement. For example, Paul Ramsey argued strenuously for an "exact"

---

[14] Writing from an explicitly evangelical Christian perspective and drawing carefully on biblical warrants, Reichenbach and Anderson argue for a stewardship framework.

and limited meaning of the term "genetic therapy," precisely so that it could not be used for enhancements that really treated parental desires (Ramsey, 1972).[15]

Similarly, LeRoy Walters and Julie Gage Palmer assert that "disease and disability" are "evils that should be overcome as quickly and efficiently as possible" (Walters and Palmer, 1997).[16] While all short children may experience pain or discrimination, the question of whether there is a physiological basis for the short stature is taken to be morally relevant for policy purposes. "We are attempting to draw a sharp line between *bona fide* illness... and physical traits that can lead to discouragement or discrimination or both..." (Walters and Palmer, 1997, 113). They approve genetic enhancement for children of short stature who have hormone deficiencies, but not for children of short stature who do not have hormone deficiencies.

Both Ramsey and the team of Walters and Palmer would therefore appear to join the Roman Catholics in approving genetic intervention for Johnny, since Johnny has a diagnosed deficiency in human growth hormone. However, it does not appear likely that they would approve genetic intervention or 'enhancement' for Bobby, Tom, or Erwin, where there is no clearly diagnosed disorder.

However, the picture may be more complicated than that. In general, Walters and Palmer adopt Norman Daniels' concept of "species-typical functioning" as their base-line. Nonetheless, they approve some enhancements, such as improvements to the immune system, a reduction in the number of hours that people need to sleep, and memory enhancement for people with senile dementia, even while acknowledging that such enhancements go beyond species-typical functioning. It is possible, therefore, that they would approve 'enhancing' height for Bobby or Tom, if it is predicted that either boy will grow up to be of very short stature. Even though neither has a clearly diagnosed disorder, such enhancement might fit Walters and Palmer's understanding of improving general functioning (Walters and Palmer, 1997, 121).[17]

Retired Union Seminary ethicist Roger Shinn notes the problem with standards such as "species-typical functioning." "Terms like disease, ailment, defect, liability, and anomaly all imply some departure from a state of health regarded as normal," he asserts, but he cautions that the very idea of normality "needs critical investigation" (Shinn, 1996, 96, 101).[18] Shinn further notes that culture has a great impact on what is considered biologically normal, for example, the availability of eyeglasses makes myopia trivial and it is culture that makes dark skin a liability.

---

[15] Ramsey was an early critic of genetic interventions and argued strongly for limits that should not be transgressed. A Methodist by affiliation, he is generally understood to fall on the conservative end of the theological spectrum.

[16] In this work, Walters and Palmer do not argue out of an explicitly Christian or theological perspective, but we include them because their book-length treatise has become a standard for assessing ethical issues in genetic interventions and Walters has a background in Christian ethics.

[17] The discussion by Walters and Palmer is full of contradictions. They appear to establish physical norms, such as species-typical functioning, and then immediately violate those norms in what they propose. This suggests to me that their anthropology is more fluid than they acknowledge.

[18] Shinn's career as a Christian ethicist spans several generations. His approach, though not heavily theological, is informed by liberal theology.

Nonetheless, when Shinn lists six insights that help us formulate norms, he begins with the understanding that health is a good and so is healing. Such statements presume a base-line for making judgments regarding wholeness or health. Chapman notes, however, that the base-line may be shifting already and will certainly do so in the future, with the result that interventions that are considered radical today will become acceptable in the future. It will in fact, she argues, become increasingly difficult to distinguish between prevention and enhancement. Interventions that today would be considered enhancement may some day be considered simple therapy, and there is likely to be "creeping enhancement" (Chapman, 1999, 75). If Chapman is correct, then even the scenario involving Erwin, who is not predicted to be short but whose parents would like him to be taller than predicted, would some day be considered therapy rather than enhancement or at least would fall within the acceptable range of genetic interventions.

At stake for many Protestants, as for Catholics, is the fundamental value of human life in all its conditions. To love our neighbor with *agape*—that is, to love a person as an end and not merely as a means—is to confer dignity on that person, to treat each person as having dignity. Children and adults who, due to genetic reasons, deviate from the norm or suffer from disease, have dignity. All Protestants, like their Catholic colleagues, will agree on this fundamental point. The United Methodist Church declared boldly: "we understand that our worth as children of God is irrespective of genetic qualities, personal attributes or achievements" (Cole-Turner, 1993, 76). The National Council of Churches had earlier also declared that human life is to be valued in relation to God, not in relation to human standards of genetic health (Cole-Turner, 1993, 73). Human dignity is God-given and not defined by human norms or standards. Several denominations have raised cautions about possible eugenic efforts associated with new genetic technologies. For example, the Christian Life Commission warns against eugenic applications or efforts to remove "undesirables" (Chapman, 1999, 47). The United Methodists in 1992 adopted a resolution that also raised concerns that efforts at genetic screening may have eugenic consequences. Ted Peters expresses very plainly a principle that many Christians would affirm: "God loves us regardless of our genetic make-up and we should do likewise" (Peters, 2003).

What Protestant spokespersons and theological ethicists try to hold together here is the paradox of saying that every human life, no matter what its genetic condition, is equally and ultimately valuable and at the same time affirming that it is acceptable to intervene in that life to correct defects. This leads Cole-Turner to ask pointedly whether illness is seen as natural or as a defect of nature (Cole-Turner, 1993, 78). If illness is natural, then what justifies intervention at all? If it is a defect, then intervention may be justified, but how does one then applaud the fundamental value of that defective life? This is the paradox for Christian theology, which seeks both to love and affirm all of God's creation and yet to accept a mandate to reduce suffering and improve the human condition.

A fourth tension arises when we try to put together freedom and justice. Protestants, especially liberal Protestants, are acutely concerned about the relationship of freedom to justice. Responsibility, whether as stewards or co-creators, implies freedom.

Protestant Christians usually believe that freedom is one of the greatest gifts given to humans by God. A major part of being created in the image of God is being free to choose and act. It is therefore a violation of the dignity of human beings to take away their freedom or to interfere in their right to make decisions about things that affect them. Most Protestant ethicists hold strongly that who we are is not determined by our genes and that genetics may shape our predispositions but does not remove arenas of freedom of choice. Peters, for example, holds that determinism at the level of *genes* would not remove freedom at the level of *persons*. Chapman argues that a genetic predisposition may place a greater rather than lesser burden of moral responsibility on a person. Hence, most authors do not see genes as reducing human freedom and responsibility. Freedom remains a core value.

At the same time, it is possible to use human freedom in such a way that it negates human dignity. Most Christians believe that the God who created humans is a God of justice and love. Humans are to use our freedom in loving service of others and in an effort to bring about right relationship or justice. In other renderings, such as the United Church of Christ, compassion is the overriding theme and freedom is oriented toward compassionate healing. As the working group of the Church of Scotland emphasized, the biblical mandate is to serve human welfare. The measure of justice or service is often understood to be the plight of the poor or dispossessed (the widow, the orphan, and the stranger in the land are the paradigmatic cases).[19] Any use of freedom that undermines the position of the poor is therefore understood as wrong. For example, the National Council of Churches raised cautions about eugenic programs (Cole-Turner, 1993, 73). It also queried whether the poor would have equal access to any genetic technologies that are developed. Similarly, the United Church of Christ in 1989 welcomed the development of genetic engineering *provided* there was appropriate regulation and "justice in distribution" (Cole-Turner, 1993, 76). Philip Hefner cautions that use of our freedom to shape our own future may result in interventions being manipulated in accord with the interests of dominant classes. Karen Lebacqz has also cautioned that the genome project may be "no deal for the poor" (Lebacqz, 1997b).[20] The United Church of Canada expressed as a general principle that the "rights of the weaker and the needy" must be protected in any genetic interventions (Chapman, 1999, 60).

The combined stress on freedom and on justice sets a possible tension: to take seriously the plight of the poor could set limits on human freedom. We may be obligated to refrain from genetic interventions that might damage the situation of the poor. This tension is well illustrated by the one place where Walters and Palmer disagree with each other: in discussing the distribution of genetic technologies, Palmer accepts few limits on freedom, whereas Walters believes that the long-term goal of

---

[19] None of the ethicists considered here would qualify as a liberation theologian, but liberation theology out of Latin America has been particularly strong in arguing for a preferential option for the poor.

[20] A liberation Protestant theologian, Lebacqz frames her discussion of genetics within the language of justice concerns.

reducing the gap between those who are best-off and those who are worst-off is preeminently important.[21] Roger Shinn suggests that freedom must always be understood as "freedom in community," not simply freedom to do what one will. All of these concerns point to the possibility that human freedom would be limited in the interests of welfare, compassion, and justice. With regard to our case scenarios, for example, the acceptability of genetic intervention would not rest simply on what parents want (for Erwin, e.g.). Interventions would have to be measured by their overall social impact. If genetic interventions to make people taller in the West could be shown to damage the opportunities available for others, those interventions might fail the test of justice. However, Chapman notes that there is little development of an explicit "justice trajectory" among Protestants examining genetic interventions even though justice may be given lip service (Chapman, 1999, 48).

Finally, the fifth tension through which we discern the profile of Protestant ethics is the tension between reason and revelation. A perennial problem in Christian theology is the relative weight to be given to reason or to revelation. Do norms come from nature or rational discourse reflecting upon nature? Or do they come from Scripture or extra-rational authoritative sources? In the discussions of enhancement, this tension is reflected in differing stances on the precedence given to science or theology. Does science provide facts to which theological anthropology must respond? Or does theology provide a framework within which any scientific discoveries must be understood?

While many of the Protestants considered here agree broadly on policies regarding genetic enhancement, they nonetheless may approach this question from noticeably different perspectives. The most common approach in Protestant theology is to find some general theological affirmation (e.g., that humans are created in the image of God) and move to assess science and technology from that affirmation. Most begin with the language of *creation* and ask whether God's activities as Creator set any limits on human genetic intervention. However, Cole-Turner frames his response to technology in the language of *redemption*. For Cole-Turner, since Jesus came to redeem all of creation, Jesus' acts of healing are the norm by which judgments can be made regarding what needs to be corrected and what does not. A collection of Presbyterian essays entitled *In Whose Image?* (Burgess, 1998) draws explicitly on the notion of *imago dei* for its theological grounding. Thus, ethicists may depend on different theological affirmations—on God as creator, God as redeemer, or humans as made in the image of God—but they all share the strategy of drawing explicitly on theological affirmations as grounding for ethical evaluation.

Other theologians, however, do not use the story of creation, the life of Jesus, or notions of *imago dei* as guides for understanding what needs healing. Shinn, for example, simply turns to "widely accepted" views in order to argue that some interventions are acceptable and some are not. This seems a more anthropological appeal than a

---

[21] It appears to us that this disagreement reflects Walters' training in Christian ethics, with its stress on a preferential option for the poor, whereas Palmer adopts the liberal values of American law and philosophy.

theological appeal. Similarly, Chapman charges that most church documents review theology only in general and rather traditional terms, and fail to engage theological resources in a deep manner when trying to discern ethical issues or argue for ethical positions: "theological affirmations once made do not become the grounding for subsequent ethical and policy discourse" (Chapman, 1999, 40). Serious theological work and its connection to the ethics of genetic interventions has yet to be undertaken.

Chapman's own constructive work on human nature takes yet a different tack. She permits science to trump theology: the discussion turns on what genetic science would require by way of *revisions* of Christian anthropology. In other words, science sets the stage and theology must adapt. Walters and Palmer also appear to do this, albeit somewhat indirectly, insofar as they begin every chapter with a careful review of the science and tend to accept scientific ways of dividing the world into categories. For these theologians, then, it is not revelation but reason that sets the stage, and revelation must follow and adapt appropriately to what science or reason tells us. Even though they might not differ much when it comes to concrete policy proposals, therefore, Cole-Turner, Shinn, and Chapman reflect differing views on the precedence of reason or revelation, science or theology.

At issue here is whether Christian theology can enter the public arena, and if so, how. Any strict dependence on theological norms such as views of creation or redemption limits the audience for the Christian theologian: only those who hold similar views will be persuaded. To enter into public dialogue, therefore, theologians often turn to generally accepted views or give priority to the scientific views that dominate in the contemporary dialogue.

In summary, it is impossible to capture the richness of Protestantism within the wider Christian tradition and reflection in a short section. However, the five tensions or paradoxes outlined here give at least some indication of why Protestants will disagree with each other on questions of genetic enhancement and also of the basic values that must be held together in any Christian view. Christians attempt to hold together both the goodness of creation and the distortions of human history, both the excitement of human creativity and the sense that this creativity must be in God's service. Protestant Christians will therefore struggle to affirm the goodness of every human life, no matter its genetic constitution, and at the same time to find a permissible range for interventions that "correct" or "improve" that life. Every correction will be seen as potentially dangerous, however, because of the sinfulness of human life. In short, there is a circle here: from affirmation of the goodness of life, to the call to intervene and make it better, to the recognition that such intervention is fraught with the dangers of self-deception and injustice.

3.9 Jewish Medical Ethics and Genetic Enhancement

"For Judaism, God owns everything, including our bodies," writes Elliot N. Dorff. "God lends our bodies to us for this duration of our lives, and we return them to God when we die." What this implies is that "God can and does assert the right to restrict how we use our bodies" (Dorff, 1998, 15). This leads directly to the mandate to heal, to moral support for the practice of medicine. "Because God owns our bodies, we are required to help other people escape sickness, injury, and death" (Dorff, 1998, 26). Support for clinical medical practice implies, in addition, support for scientific research on behalf of human health and well being.

The most frequently appealed-to method for Jewish theology and ethics is one of interpretation, one of interpreting the Torah through the history of texts that make up the Hebrew and Jewish traditions. Moral laws are derived through application of interpretations, through *halakhah*. Are such moral laws strictly positive—that is, are they strictly grounded in the religious legal system?—or are they rooted in the natural condition of which the legal system is a cultural expression? Even if moral laws derive ultimately from God, do they also adhere to what can be discerned as natural law? Some Jewish thinkers affirm the latter. "Natural law theory is necessary for an adequate essential characterization of *halakhah*," writes David Novak (Novak, 1995, 40).

The appeal to tradition combined with appeal to natural law shows a kinship between Jewish ethics and Roman Catholic ethics. Aaron L. Mackler sees more convergence than contrast between the two approaches. On method, he says, "Jewish approaches generally are based on tradition, especially halakhah—a term meaning 'path' or 'way' and denoting Jewish law. Although Catholic moral approaches accord significant weight to tradition, more commonly they are centered on natural law, together with magisterial teaching" (Mackler, 2003, 1). On shared values, he says, Jews and Catholics share a commitment to "the intrinsic dignity of human persons, created in the image of God; the responsibility of a just society to offer needed support to its members; and a divine mandate to provide healing to people in need" (Mackler, 2003, 190). The second in this list of three, the commitment to justice, would likely place some Jewish ethicists along with some Roman Catholic colleagues in league with Protestants regarding problems of equal access associated with enhancement.

Let us turn here to the relationship of tradition to natural law. Within the context of *halakhah*, we can find some movement from *is* to *ought*. Dorff makes this move, but only when the concept of nature is theologically understood. Spcficially, human nature includes our creation in the image of God. Our nature includes an essence and a *telos*, namely, to fulfill the divine image within us. "Locating the divine image within us may also be the Torah's way of acknowledging that we can love, just as God does, or that we are at least partially spiritual and thus share God's spiritual nature. Not only does this doctrine *describe* aspects of our nature; it also *prescribes* behavior founded on moral imperatives" (Dorff, 1998, 19). This is a theological, not a naturalistic, understanding of human nature. Because it includes a *telos* or purpose, such a view of human nature created in the divine image is compatible with a transformatory ethic.

Both Jews and Christians inherit from the Bible a future orientation, a divine promise of transformation. This seems incompatible with the naturalistic fallacy—that is, incompatible with an ethic projecting a vision of *what ought to be* in the future on *what is* the state of the human make-up in the present. Rather, biblical theologians look forward to a future that differs from the past, that cures the ills of the past. Christians might emphasize an eschatological vision of a future new creation, a vision in which there will be no more suffering or pain and God will wipe away our tears (Rev. 21:1–5). This tends to support strongly efforts toward healing as anticipatory of God's redemption. What ought to be done is determined by a vision of what lies beyond the present state of nature. Jewish medical ethicists might emphasize rather the mission to heal, the God-appointed task for the human race to take care of our bodies. The Bible which Christians and Jews share is filled with God's promises that all things will be new (Isaiah 65:17), so human endeavors to transform the ills of present existence in light of a future healthier life is based on a ethic that relies upon divine promise.

Despite good theological reasons for supporting an ethic of transformation, listening to nature still provides a level of caution for some Jewish bioethicists. Elliot Dorff illustrates both the vision and the caution on the topic of genetic therapy and enhancement:

> The potential of stem cell research for creating organs for transplantation and cures for diseases is, at least in theory, both awesome and hopeful. Indeed, in light of our divine mandate to seek to maintain life and health, one might even contend that from a Jewish perspective we have a *duty* to proceed with that research. As difficult as it may be, we must draw a clear line between uses of this or any other technology for cure, which are to be applauded, as against uses for enhancement, which must be approached with extreme caution. Jews have been the brunt of campaigns of positive eugenics....so we are especially sensitive to the dangers in creating a model human being that is to be replicated through the genetic engineering that stem cell applications will involve (Dorff, 2001, 92).

Within the divine mandate to heal and even transform, we find a caution against enhancement.

This caution goes so far as to appeal to the problem of playing God in medicine. Ordinarily, the risk of playing God is not a large factor in Jewish ethical deliberation. Laurie Zoloth writes, "whereas moderns are worried lest we 'play God', the rabbis were concerned that we act *more* like God might in many ethical and social-political arenas, as in helping the poor, creating justice, and healing the sick" (Zoloth, 2001, 96). Yet, appeal to the commandment to avoid playing God arises when the question of enhancement arises. Elliot Dorff asks, "How do we determine when we are using genetic engineering appropriately to aid God in ongoing, divine acts of cure and creation and when, on the other hand, we are usurping the proper prerogatives of God to determine the nature of creation? More bluntly, when do we cease to act as the servants of God and pretend instead to be God?" (Dorff, 1998, 162).

The risk of playing God arises when considering genetic enhancement. Elliot Dorff goes so far as to endorse germ line intervention on the grounds that it serves the divine mandate to heal and he applies it to future generations, but he shies away from enhancement because it goes beyond healing.

> Since sickness is degrading, it would be our *duty* to cure the disease at its root if we could, so that future generations will not be affected. But the more powerful our abilities to intervene in preventing genetic diseases, the more urgent it becomes to accomplish the philosophical and moral tasks of defining the line between therapeutic and non-therapeutic uses of this technology and, in so doing, the boundary between us and God (Dorff, 1998, 164).

What we see here is reaffirmation of the Jewish commitment to heal combined with a cautious reluctance to go beyond healing to enhancement. The difficulty is in finding the right place to draw the line.

With this discussion in mind, we could easily imagine Jewish bioethicits strongly prescribing growth hormone for Johnny on the grounds of the divine mandate to heal. This would likely apply to Bobby and Tom as well, seeing promotion of social fitness as itself a form of therapy. When it comes to Erwin, however, the question of justice would be raised. If growth hormone would make Erwin superior to his peers at school or in the neighborhood or give him an edge in his profession, then we would want to ask: does it matter that Erwin has access to this enhancement while others do not?

In summary, Jewish ethicists appeal first and foremost to *halakhah*, to making application to present circumstances based upon interpretation of the Torah through historical commentary. The result is a system of moral laws. In addition, some Jewish ethicists are willing to incorporate a hint of natural law, especially if appeal to what nature teaches us aids in our *halakhic* deliberations. What all this yields is a solid reaffirmation of our moral obligation to heal—in this case supporting genetic intervention for the purposes of therapy—but a reluctance to engage in enhancement. Enhancement risks playing God.

3.10 Islam: What Might We Expect?

When we turn to Islam, we turn away from the shared commitment to natural law that we found in Roman Catholicism and the Enlightenment of western Europe. This is illustrated in the controversy over reproductive cloning which broke out in 1997. At that time, the U.S. National Bioethics Advisory Commission invited religious leaders to provide guidance for what might become government policy. Aziz Sachedina, a Muslim bioethicist at the University of Virginia, testified that the Qur'an and subsequent Islamic tradition do not provide background or principles that anticipate modern biological knowledge about the embryo or genetic inheritance (Sachedina, 1997). Decisive ethical guidance must await a process of interpretation which will involve application of past tradition to present circumstances. In Islam we appeal not to nature plus scripture, but to scripture alone, to the Qur'an.

In actual practicē, the Qur'an does not stand alone. It is accompanied by its tradition of elaboration in the Sunna (meaning "trodden path"). To these two, the Qur'an and the Sunna, are added two other sources, consensus (*ijma*) of the early Muslim community plus the principle of analogy (*qiyas*) (Sachedina, 2003, 14). The latter, analogy, is a method of reasoning from data furnished by the Qur'an and the Sunna in which the unknown is approached via analogy of what is known. Relying upon ancient sources for moral reasoning in an era of fast-moving medical science is now giving birth to a nascent and as yet undeveloped field, Islamic Bioethics.

As Islamic ethicists and jurists confront new and unprecedented scientific challenges to human self-understanding, appeals to the Qur'an and ancient tradition pervade their analysis. The conceptual apparatus inherited from the tradition provide the matrix for deliberation. One can expect that much of the vocabulary employed to sort through biomedical conundrums will be classical theological vocabulary, with only carefully filtered additions of contemporary scientific terminology. Because issues formulated by the contemporary scientific situation can be addressed only indirectly rather than directly, the method will necessarily be one of analogy (*qiyas*). Analogs to past juridical deliberations will be retrieved and contemporary applications sought.

Part of the tradition that makes up Islamic thought are philosophical debates regarding such subjects as the human soul, especially the relationship of the soul to the body. We can safely forecast that ancient debates on this topic will resurface as influences on contemporary jurists (*fuquaha*) as they ply their craft, the science of jurisprudence (*usul al-fiqh*). We can almost expect a twenty-first century extension of a fissure that opened up in the eleventh century, namely, the split between Ibn Sina (Avicenna, 980–1037) and Abu Hamid al-Ghazali (1058–1111).

Ibn Sina's anthropology comes close to that of Plato's. The soul is incorruptible and does not die with the body. Consciousness can perdure in a disembodied state. Attachments to the body which involve temperaments are accidental, not essential, to the soul. The body is not the form of the soul, nor does the soul imprint itself onto the composite parts of the body. The soul is not intrinsically dependent on the body; rather, the soul's fundamental relationships are with eternal principles that escape change or corruption. The result of this dualism is a form of everyday naturalism, according to which what happens to the body is exhaustively explained by its place in the physical causal nexus. The body's natural nexus of activity drops into near insignificance compared to the soul's destiny (Sina, 1952).

Ghazali, in contrast, is more Aristotelian. Because all humanly initiated action requires an act of the will in the soul, Ghazali argued against Ibn Sina, the soul cannot avoid an inextricable attachment to bodily movement. The natural causal nexus and the social nexus include the soul's activity. A continuity exists between the spiritual substance of the soul and the physical substance of the material world, though they are not identical. Ghazali describes the body as the camel the soul mounts to ride toward God. Without the body, the soul cannot reach its destination. Whereas dualists such as Ibn Sina could think of eschatology in terms of a disembodied soul, Ghazali holds out for a bodily or corporeal resurrection (Ghazali, 1988).

In an essay describing alternative views regarding the place of brain death and organ transplantation in Muslim bioethics, Ebrahim Moosa demonstrates that these two anthropologies influence two contrasting positions. He compares two *fatwas* or non-binding juridical opinions, one from Pakistan and one from Egypt. In the case of the Pakistani *fatwa*, organ transplantation violates human dignity (*karam wa hurma*). The person declared brain dead is the one whose dignity is being protected here; dignity is preserved by not dismembering the corpse. Repelling harm to the body—even the dead body—takes precedence over potential medical benefit to someone else. Moose concludes that the anthropology of Ghazali is at work here.

The Egyptian *fatwa*, in contrast, permits organ transplantation. Here, it is assumed that the soul's presence is the source of animation, and hence it is tied to

3 Religious Traditions and Genetic Enhancement

brain function. Because a brain-dead person is no longer animated by the soul, and because the soul has been released to eternity, what remains is a body subject solely to the physical causal nexus. It is solely natural, with no supernatural component remaining. Organ transplantation is permissible. No precedent in Islamic law on this matter makes the question open to juristic discretion (*itjihad*), and the discretion employed here relies on the dualist assumption.

> Once it could be argued that the locus of the soul is the brain and that consciousness is an indicator of brain function, brain death can easily be justified. Those jurists who opposed brain death and organ transplantation used the same texts and sources as their fellow jurists but arrived at an opposing and differing position. Their emphasis was on the social imagery of the body as inviolable in its dignity (Moosa, 2002, 344).

With these precedents in mind, we can offer forecasts regarding likely ethical trajectories in Islamic thought regarding issues such as genetic therapy and genetic enhancement. Ethical postures deriving from the dualistic tradition may find it easier to engage in genetic engineering, both therapy and enhancement, on the grounds that the body belongs solely to the physical nexus. If the soul can be distinguished sharply from the DNA, then engineering changes in the DNA should leave the soul uncompromised. Ethical postures deriving from the Ghazali tradition, however, may be looking for physical elements to identify with a person's dignity. The genome would become a likely candidate, especially if it is imagined that the soul would have an essential connection to the person's genome. Then engineering of the genome could be construed as a violation of the soul, not just a healing for the body. The commandment, "thou shalt not play God," might be heard more frequently in Muslim circles.

No doubt therapy will count for more ethically than enhancement will. Preserving human life from suffering will play a role in Muslim deliberation, to be sure, trumping enhancement. Yet, even therapy may not provide sufficient ethical warrant to approve genetic engineering if by changing our physical nature it is believed we are violating the dignity of the soul.

Let us return for a moment to our four cases, which raise the difficult question as to whether there is justification for rejecting protocols solely on the basis that they extend an enhancement technique that is proven safe and effective, to an area other than the treatment of the disease for which it was originally devised. The cases of the boys offer some pointers. The use of gene therapy for Johnny, with hGH deficiency, appears straightforward in principle, because it can be viewed as treatment for a disease. Bobby, on the other hand, is short, not diseased; therefore his case is more controversial. However, differences in the cases of Johnny and Bobby are not so clear-cut if disease is defined as a condition that impairs human functionality below species-typical levels. Even if Bobby's short condition failed the classification of a disease, the gene therapy could be justified as an upgraded form of the hGH protein treatment he is already receiving, which is non-controversial, and which is cheaper and less cumbersome.

Tommy is normal, though slow in growth rate. This is not yet remarkable. Erwin is remarkable, because, although he is normal, his parents are considering enhancement. Are we crossing a line? On the right side of the line, it is argued that gene therapy is developed for those who fall below the normal range of some characteristic or function. That line seems to be crossed when the treatment takes someone beyond the normal range to a higher level. There are two problems with this analysis. One: notions of normal and abnormal are technologically and culturally conditioned, and vary in time and place. Two: even if there were some universal standards, some functions can be greatly enhanced over the normal range by treatment with interventions designed for treating or preventing diseases.

Thus much of the debate boils down to what we mean by enhancement. In medico-ethical circles the major trend is to contrast it with the curing of a disease, and hence it is considered a procedure or intervention that is "nonessential at best and suspect at worst." Again, because the line between therapy and enhancement cannot be precisely drawn, we simply note that it has ethical significance even when it remains medically ambiguous.

3.11 Hindu Perspectives on Genetic Enhancement

"Hindu bioethics believes there is a medical and moral divide between *somatic cell gene therapy* and *enhancement genetic engineering*, which must not be crossed, and which serves as a marker for how far genetic engineering should go at this stage of development," writes Cromwell Crawford (Crawford, 2003, 153). What is the religious background to this conclusion?

The Hindu concept of nature and the place of humans in it is vastly different to those of the Jewish, Christian, and Muslim traditions, and therefore we can expect some theoretical and practical differences, yet important commonalities also exist. Since Hinduism is a pluralistic tradition that allows for many expressions, we choose the view found in the classical Vedanta philosophy of Sankara (8–9 c.). Our purpose is to show that certain cognitive and moral insights that are necessary for a productive relationship between humans and physical nature, with implications for therapy and enhancement, find their locus in Vedanta philosophy.

The first insight is interconnectedness. You do not have to be a religious person to affirm this. Environmentalists have been telling us for decades that everything in nature is intrinsically connected with everything else—humans, plants, animals, and all sentient beings are part of the web of life. Nature is so structured that for every act there is a corresponding reaction, the full effects of which are not immediately visible.

The idea of nature as a connected whole has been cultivated by Hindus for millennia, and is capsulized in the doctrine of karma. The word is derived from the root *kr-*, "to act." The Brhadaranyaka Upanishad (c. 1000 BCE) declares: "According as one acts, according as one conducts himself, so does he become" (1V.iv.5, in Hume, 1931, 1031). Actions that are born of ignorance and selfishness entangle the doer within the cycles of birth and rebirth. Thus the wheel of life is kept in motion, because every deed both impacts the world of nature, and also produces "tendencies" (*samskara* or *vasana*) in the doer in the form of habitual patterns of behavior. Deutsch states, "whatever one does will have effects not only in the immediate

3 Religious Traditions and Genetic Enhancement

present but in the long future as well: any act, in short, will have consequences that reach far beyond the act itself. And everything in nature is so interconnected through causal chains and relationships that we ourselves become part of the natural process and are conditioned by it" (Deutsch).

The second Hindu insight is linked to the first and states there is a natural kinship between humans and all of physical nature. By contrast, the West has held to a religious belief in an original creation of fixed species until recent times (and many still do). Darwin demolished this religious worldview with his theory of biological evolution. Details of Darwin's theory have since been revised, but his principle of evolution is universally accepted by the scientific community.

For Vedanta, the evolutionary orientation of science affirms its position that fundamentally all life is one; that in essence everything is reality; and that this oneness finds its spiritual expression in a reverence for all living things. These sentiments are embodied in the well-known concept of *ahimsa* (non-violence). *Ahimsa* is the primary virtue Hindus must observe in their relationships with all facets of nature of which they are a part. Gandhi interpreted *ahimsa* as 'non-violence' in a universal sense and ranked it as the foremost human quality.

A third Hindu insight is related to the doctrine of creation. Western attitudes toward nature have been shaped by the church's historic interpretation of the Genesis narratives of creation. Correctly or incorrectly, the church has historically believed the world was created by God for human ends, and that he has given them absolute dominion over nature. Only humans are made in God's "image." One derivative of this interpretation of the biblical doctrine of creation is that it introduces radical splits in life – splits between humans and the Creator, the Creator and physical nature, and between physical nature and humans.

Hindus understand "creation" through an emanationist theory. They see a natural unfolding of spirit in the world, and believe that the presence of spirit in matter invests the whole created order with spiritual worth. Nature is thus connected both horizontally and vertically, disallowing any sharp division between body and spirit in humans. Both body and spirit contribute to the whole, and express the full integrity of the whole. Thus the dualism that has dominated Western thinking finds no room in the philosophy of Vedanta.

In the next part of this paper we shall suggest the implications of these three core ideas—creation, continuity, and interconnectedness—for a proper evaluation of gene transfer for therapy and enhancement.

Hindu bioethics distinguishes between (1) somatic cell gene therapy and (2) enhancement genetic engineering. In terms of somatic cell gene therapy, many diseases such as ADA deficiency (an immune deficiency disease of children), sickle cell anemia, hemophilia, and Gaucher disease, are caused by a defect in a single gene and are treatable. Hinduism responds positively. It believes that all living beings have been created for health; that all living organisms are regulated by the principles of pleasure and pain; and that disease is an impediment to the fulfilment of all human goals, including spiritual fulfilment. Hence, the response to disease must be with daya (compassion). *Daya* is not pity but empathy—empathy that is based on the realization of our interdependence and interconnectedness.

Gene therapy must be given to persons with these sicknesses, for it is their only hope. The ethical principle this context is ahimsa—do no harm (*himsam ma kuru*). It aims to balance risks and benefits in specific interventions. We know that gene therapy is a risky business in this early stage of development. Cutting-edge medical research is always risky, but relative to the severe privations and threat of death, the risks and uncertainties of gene therapy are at acceptable levels for many of these patients.

However, somatic cell gene therapy also has the potential for enhancement genetic engineering—for supplying a specific characteristic that individuals might want for themselves (somatic cell engineering) or for their children (germline engineering) which would not involve the treatment of a disease. The slide from correction to perfection is already underway. The human-growth hormone was devised for children with prospects of growing up the size of dwarfs, but it was soon used by children who only thought they were "dwarfs," and who were blessed with wealthy parents who could pay $30,000 for a year's treatment of growth hormones.

What is the Hindu position on enhancement? Much depends on particular cases. First, in some cases enhancement could mean the use of biotechnology for the purpose of "self-improvement." Self-improvement is as much of an American religion as being Baptist. Hindu bioethics has no problem with that, as long as one has a clear idea of the nature of the "self" that is to be improved. Arguing from one view of the self, a person can say, there is absolutely no difference between getting one's child the best school and getting one's child a perfect gene. What is the big fuss? Hinduism would answer that there is a difference and it has to do with *buying* "personal" traits versus *cultivating* those traits. The two approaches differ radically in their means, which then transform the end. For example, a parent could buy a Harvard education for a child, and the quality of the education could very well enhance the child's natural gifts, but this type of enhancement is quite different to the purchase of those capacities. Ethicists like Erik Parens make the same point today that the Gita made a long time ago: "let a man lift himself; let him not degrade himself; for the Self alone is the friend of the self [person] and the Self is the enemy of the self" (V1.5). Both voices unite in a single message: personal transformation is a function of the inner life. Seeing with the "third eye" is not an acquisition of reconstructive laser surgery. To uplift himself an individual must engage creative forces that are within, and not simply rely on appendages that can be purchased at a price. In fine: *self-improvement is improvement of the self*.

Second, the use of biotechnology for enhancement raises questions of social fairness. Do we wish to usher in a society where only the rich become smarter? Who will have the right to access the technology once it becomes financially out of the reach of the common person? Every parent would want his or her child to be intellectually enhanced, but only a minority would be able to afford it. Would this not create a new 'caste system' in which the wealthy Brahmins of society constitute a new intellectual aristocracy that looks down upon children who are not enhanced, because they have lower IQs? Thus, Hindu bioethics appeals to its principle of justice, based on our common spiritual heritage and the connectivity of human life. The Gita says: "When one sees Me everywhere and everything in Me, I am never lost to him and he is never lost to Me" (V1.30). Such thoughts invest each individual with equality, and evaluate all attempts at enhancement with the demands of social fairness.

Third, even if the questions of social fairness were resolved, is the enhancement of human capacities medically sound? The position of Hindu bioethics is that in situations of life-threatening disease, taking risks may be justified by the potential benefits of gene therapy; but in the absence of life-threatening disease, the risks may outweigh benefits, and enhancement is no longer an option. Five areas of medical concern stand out, all of which are addressed by Hindu principles derived from the core ideas presented above.

The primary medical principle of Hindu bioethics is *ahimsa*—do no harm. This is the first flag that goes up. Medicine is not an exact science, and when we stand on the medical frontiers of enhancement engineering, knowledge recedes while risks increase. For instance, we have preliminary ideas of how genes run a cell, yet what do we know about how the configuration of an organ takes shape? We know how the central nervous system works through electric circuits, memory storage, etc., but what do we know about "thought," about "consciousness," about "spirituality"?

The following are some of the areas of concern raised by pioneer geneticist French Anderson. He gives us a glimpse of the genie in the bottle. He says:

> Even though we do not understand how a thinking, loving, interacting organism can be derived from its molecules, we are approaching the time when we can change some of those molecules. Might there be genes that influence the brain's organization or structure or metabolism or circuitry in some way so as to allow abstract thinking, contemplation of good and evil, fear of death, awe of a 'God'? What if in our innocent attempts to improve our genetic make-up we alter one or more of those genes? Could we test for the alteration? Certainly not at present. If we caused a problem that would affect the individual or his or her offspring, could we repair the damage? Certainly not at present. Every parent who has several children knows that some babies accept and give more affection than others, in the same environment. Do genes control this? What if these genes were accidentally altered? How would we even know if such a gene were altered? (Anderson, cited by Gibbs, 1999).

A second area of medical concern has to do with side-effects, which Hindu bioethics addresses through its principle of consequentialism. It is axiomatic to the Indian mind that everything has its own store of karma which eventually plays itself out. Anderson has made it perfectly clear that enhancement research is not at the point that all outcomes are known. It would not be a scare tactic to say that parents who are eager to give their children gene enhancement would be making decisions on behalf of their children over which they had no control and whose long-term effects would be uncertain or even dangerous. Who can predict all side effects? Can we be certain that a child engineered to become intellectually sharp could not actually turn out morally mean? What happens when the "supermice" get old? Scientists already fear that altered mice might be more prone to strokes, chronic pain, and premature death. There are many such possibilities of complications for which a karmic view of nature signals caution.

A third area of medical concern is expressed in Ayurvedas's principle of health as balance. It alerts us to the fact that changes brought about by genetic engineering in one area could adversely affect balance in other areas.

A fourth medical concern has to do with homogenization. There is a troubling prospect that we could be heading toward a homogenized society that is shaped by certain dominant traits and values representing the fashion of the day. To the contrary, the premise of Hindu medicine is biological diversity, and psychically each individual is conceived as unique by virtue of his or her karmic constitution. Hindu medicine affirms individuality in nature, which not only makes for survival but for the richness of heterogeneity.

A fifth concern is expressed in Hinduism's adoption of an inclusive approach toward humans and other forms of beings. This is completely missing in the Western approach, which limits the medical concerns of genetic engineering simply to human considerations and human wellbeing. Harold Coward points out: "Animals are genetically engineered to model some of the most devastating diseases that afflict humans. To accomplish this goal, however, requires that large numbers of animals live lives of intense pain and suffering" (Coward, *2000*). Hindu inclusivism would mandate care for animals and avoid a cruel or callous use of animals for the sake of human enhancement.

In summary, the Hindu view of nature and of the place of humans in it counsels extreme caution when questions of altering nature are considered. Because of the line drawn between therapy and enhancement, we could expect a Hindu bioethicist to approve of the use of growth hormone for Johnny but withhold it from Bobby, Tommy, and Erwin.

Yet, at the same time, since Hinduism employs reason and is scientifically open, it may eventually withdraw from making a blanket condemnation of genetic enhancement as *intrinsically evil*. Instead, it could start with the person, holistically understood, which then calls for an evaluation of all means of genetic enhancement by the moral yardstick of whether they have karmic risks which are not worth taking at this time, or whether they do indeed enhance the person for good. "Enhancement" is here defined as the natural extension of health organically conceived as physical, mental, social, and spiritual wellbeing. Its emanationist view of creation, and its philosophic recognition of change as a fundamental feature of nature, both physical and human, make Hinduism cautiously optimistic about the future.

Though this technology is in its infancy, its potential is enormous to change not just how we play, but how we pray, and pay and do everything else. Genetic enhancement could be engineering a new creation. Such prospects behoove religious leaders to dialogue with members of the scientific community to ensure that our future is not just better but good; for "better" is not always good.

3.12 Buddhist Perspectives on Enhancement

Gautama the Buddha integrated religion and medicine in his understanding of nature, both physical and human. His basic outlook was naturalistic. He left out of his teachings concepts we normally identify with religion—ideas such as God, soul, creation, judgment. He saw himself, not as a divine Saviour, but as a teacher and exemplar. He claimed that all of the salvific knowledge we need is found, not within some transcendent realm but "within this very body," and that each person has the capacity to become a Buddha.

His Asian naturalism made him a pragmatist. He said, when a building is burning, it is not the time to discuss the nature of fire, but to flee the flames. Likewise when you are shot by an arrow, you do not ask who shot the arrow, but extract the arrow and heal the wound. The point is: when people suffer, metaphysics must yield to practicality and self-reliance.

The Buddha was known as "the Great Physician," and made healing his mission. The ultimate healing is enlightenment or Nirvana, an internal realization of self-transcending achievement. His central religious teachings, embodied in "The Four Noble Truths," guide the Buddhist, especially the monk in the Samgha [the Buddhist monastic community], directly toward enlightenment. Because the ascetic monk avoids attachment to relationships such as marriage, family and children, ethical issues surrounding therapy vs. enhancement tend to involve the laity and are more subject to common secular values. Buddhist ethics thinking follows two tracks, one leading directly to Nirvana and the other indirectly via guidance for daily family and professional life. The spiritual path leading to the ultimate healing is guided by the Four Noble Truths.

The first of the four is that human life is replete with suffering. The Indian doctor first aims at an accurate diagnosis: are the complaints of the patient genuine, or is he or she only apparently ill? The Buddha's diagnosis of the human condition: suffering is universal and real. What modern philosophers have described as "anguish," "alienation," and "quiet desperation," the Buddha called "*dukkha*."

This view of life is neither pessimistic nor optimistic, but realistic. There is universal acceptance of the reality that life's passages of birth, old age, sickness, and death, are fraught with suffering—much due to our own ignorance and selfishness. Certainly, there is happiness, but happiness is subject to the laws of finite nature in perpetual change (*anitya*). Everything in nature is in a state of flux. The Greek philosopher Heraclitus, a contemporary of the Buddha, stated: "no man steps into the same river twice," because he and the river are ever changing. People are ignorant of this law of nature, because they superimpose on this ephemeral world notions of things which endure; which must be possessed in perpetuity; and which must be perfected. But the Noble Truth is: Humans belong to a single order of nature, and are creatures of change like all other sentient beings. Unlike the Hindus who taught the existence of the soul or *Atman*, the Buddha advanced the doctrine of *anatma* or non-substantiality. Just as the word 'chariot' is only a linguistic convenience to signify the combination of disparate parts, and not to signify something substantial, so too, the "I" is only a particular combination of physical and mental energies that are separable into five groups. These five aggregates (matter, sensations, perceptions, mental formations, and consciousness) are subject to the same universal law of change, and therefore they themselves are *dukkha*. Therapy and enhancement both belong to the realm of finite nature replete with inescapable suffering, the human body in a world of *dukkha*.

Turning to the second Noble Truth that suffering is due to ignorance, especially due to our ignorance about the way unsatisfied craving affects us, we will follow the model of the doctor and the patient. Having ascertained that the patient has a real illness, the Indian physician gets to the nature of the illness, and probes into its origin. The Buddha links suffering with ingrained ignorance, which is a product of the universal desire to be, to grow, to enjoy, while naively clinging to the ego. Thirst (*tanha*) is the primary cause of *dukkha*, and is all-pervasive, including craving for pleasure, power, and also attachment to ideas, ideals, theories and beliefs. The power of thirst is also described in terms of mental volition and karma.

The Buddhist understanding of karma refers to volitional action, and should not be confused with its effect, which is known as the fruit of karma. Good karma issues in good effects, and bad karma issues in bad effects. Whether good or bad, karma keeps one bound to the cycle of rebirth, because it is driven by the will to exist, to re-exist, to continue, to become more and more. Trapped in *samsara*, one endlessly suffers. But one who is freed from the false notion of a permanent self, free from thirst, and free from defilements, does not accumulate karma even though he acts, and thus frees himself from the cycle of rebirths. When death comes, the physical body ceases, but the volitional energy persists, manifesting itself in another form through the process of rebirth. There is no soul that transmigrates, only a series which continues, like the flame of a candle that burns through the night—a flame which is the same, yet is not another.

Turning now to the third Noble Truth that knowledge leads to liberation, the Indian physician is in a position to make a prognosis: *dukkha* is eradicable. The acceptance of causality in nature does not preclude freedom; rather, it gives freedom a purchase on reality. Free of any doctrines of original sin, total depravity, or predestination, the third Noble Truth states: what has been done, by that same token, can be undone. Saviours are not needed, because bane and blessing belong to the order of cause and effect in the natural world, and are self-explanatory. When the driving force for permanency and possession are recognized as delusions, the suffering consequent upon such ignorance ceases.

Nirvana is not the annihilation of the self, because there is no self in the first place; instead, it is the annihilation of the illusion of the self. It is the Absolute Truth that nothing is absolute in the world—and that all things are relative, conditioned, and impermanent.

Finally, the fourth Noble Truth, which announces the Eightfold Path or the life of Buddhist discipline, has three components: Ethical Conduct; Mental Discipline; and Wisdom. The discipline aims at uniting head and heart, intellect and emotions. Buddhism does not try to perfect persons by making them follow rules, codes, or rituals, but tries to sensitize qualities of mind and heart to produce acts of compassion. Buddhism is par excellence a religion of compassion. The function of ethical conduct in the eightfold path is to help the devotee gain control over his senses, and thereby facilitate the uninterrupted arising of enlightenment.

What might be the implications of the Four Noble Truths for our issue here, namely, the distinction between genetic therapy and enhancement? A convenient starting point for a Buddhist evaluation of gene transfer for therapy or enhancement

3 Religious Traditions and Genetic Enhancement

is the scenario described in the Hindu section which supposes the availability of gene therapy for the treatment of patients with growth hormone deficiency. Since the scenario has already been described, we will only give the Buddhist responses to each of the four cases.

In the case of the boy who suffers from hGH deficiency, Buddhism would certainly approve the treatment to cure the disease. Pain and meditation do not go together, so one must not make a virtue of pain but rather work for its elimination. A fundamental value in Buddhism is compassion. The *Dhammapada* (X.1–2) asks us to put ourselves in the place of those who suffer, and to act for the purpose of relieving suffering. Compassion is the fruit of internalizing a sense of our interconnection with all beings and our interdependence on all fellow creatures. The image of the leaning Buddha one sees in temples, bending toward all who suffer, and the countless examples of *bodhisattvas* who have dedicated their lives for the relief of others' suffering, are examples for emulation, to free the person from the world of suffering.

In the case of the boy who does not suffer from hGH deficiency but is short, there is some question whether he is eligible for treatment that is specifically intended for disease. Though his problem is not physical but mental, Buddhism would support his case because its notion of health is not limited to the body. The Buddha's view of the person allows for no dichotomy between the mental and the physical, between mind and matter. Yoga combines the two with consequences for both physical and mental health. The Buddha spoke of four kinds of food or nutriments (*ahara*). He gave full recognition to the nourishment of the physical body, and rejected all forms of ascetic lifestyles because they were painful and unproductive. At the same time he was aware of the need for psychic and social forms of nutrition, their being essential for the psychophysical personality.

In the case of the tall boy, Erwin, many reject him as a candidate for the use of gene therapy on the grounds that the therapy is intended only for someone who falls below the normal range is some function or characteristic. Buddhism is not impressed by arguments based on definitions of what is deemed "normal." We live in a world where change is order of the day, therefore it is hardly credible to make some state or condition permanent when it is intrinsically impermanent. Besides, Buddhist medical practice is not guided by what is thought to be normal, but by individual differences, learned through diagnosis.

Positively stated, Buddhism is on the side of enhancement, and is categorically supportive of biomedical progress. It is a fundamental thesis of Buddhism that all forms of ignorance which produce suffering must be overcome, including genetic ignorance. The quest for enhancement is a recognition of profound human capacities of our physical and mental powers, which the Buddha fully understood and tapped into. Once he discovered that passion and hatred are the causes of dukkha, he eradicated these causes for himself and achieved new health and happiness under the bodhi tree. For the next forty-five years he perfected his health and happiness, which served as an antidote to the severe physical pain that dogged him down the years, due to an early regimen of extreme starvation. The Buddha excelled his contemporaries, and notwithstanding his tireless labours for the welfare and happiness of all beings, and getting only three hours of sleep a day, he managed to live to the grand age of eighty.

All this is to say that from a Buddhist perspective nothing should stand in the way of a boy, even a tall boy who has big dreams of achieving the height of a basketball superstar, with the best of modern science. The problems some Jews, Christians, and Muslims face in connection with altering nature, expressed through fears of "Playing God," do not arise in the same way for Buddhists, because for them notions of sanctity belong to the natural realm and arise from our common connectivity and mutual interdependence. Nature is sacred because all life is part of a single web. The ethics of "do no harm" follows from a unitive vision of nature.

However, two strictures apply. On the side of science, the pursuit of enhancement through genetic intervention should not entail risks that would cause harm to the individual—physically, mentally, socially, or spiritually. The watch-dog principle at work is ahimsa. Secondly, existential harm follows when a person becomes so emotionally and passionately attached to the allure of an enhanced state, that he becomes transformed into its very image. This is his new identity: *I am* my height; *I am* my intelligence; *I am* my beauty; *I am* my strength. Such aspirations arise from the illusion that it is what is engineered outside us that brings us happiness. The truth is that it is this very attachment which is the root of unhappiness.

3.13 Feminist Perspectives on Enhancement

As we saw in our earlier discussions of the naturalistic fallacy and naturalism, we cannot rely on easy assumptions regarding just what we are talking about when we use the word 'nature'. Feminist interpreters of religion in the last third of the twentieth century and now in the twenty-first century have found the nature of nature to be a problematic concept. Even though feminism, like naturalism, is not itself an institutional religion, we can benefit by reporting on feminist concerns and insights into the move from nature to ethics and apply this to questions surrounding genetic enhancement.

As with Roman Catholic and Protestant theologians, few feminist texts have addressed either gene therapy or gene enhancement specifically.[22] By contrast,

---

[22] For example, neither gene therapy nor enhancement appear as categories in the indexes of several leading texts on feminist bioethics: Susan Sherwin, *No Longer Patient: Feminist Ethics and Health Care*; Mary Briody Mahowald, *Women and Children in Health Care: An Unequal Majority*; Emilie M. Townes, *Breaking the Fine Rain of Death: African American Health Issues and a Womanist Ethic of Care*; Helen Bequaert Holmes and Laura M. Purdy, eds., *Feminist Perspectives in Medical Ethics*; Susan M. Wolf, ed., *Feminism and Bioethics: Beyond Reproduction*; Rosemarie Tong, ed., *Globalizing Feminist Bioethics: Crosscultural Perspectives*. One striking exception is Rosemarie Tong, *Feminist Approaches to Bioethics: Theoretical Reflections and Practical Applications*, which has an explicit discussion of gene therapy though no explicit discussion of genetic enhancement.

3 Religious Traditions and Genetic Enhancement

"eugenics" as a category for analysis and discussion shows up with some frequency in feminist works on bioethics.[23] A cursory background on feminist bioethics will help to explain why this is so and what its importance is when we turn to genetic enhancement and to our case studies.

As feminists can operate out of a number of fundamental approaches, there is no single feminist stance on most issues in bioethics. Just as Protestants may differ considerably from each other, so may feminists. Nonetheless, there is significant agreement on some basic affirmations and concerns. These undergird feminist thought in both philosophical and theological circles.

First, feminists generally begin by asking how new technologies or social arrangements will affect *women*. (Sometimes this is extended to include women and children.) While almost all feminists would argue that attention to women's well-being ultimately is also good for men, it is important that women's concerns and interests get priority.

Second, feminists begin with the conviction that "women have historically been oppressed and that such oppression is morally wrong" (Mahowald, 1993, 85). While feminists will disagree on the remedies for oppression—e.g., whether "equal treatment" under the law is sufficient or whether laws themselves need to be changed—they all agree that women around the world have been and continue to be oppressed.

Feminists therefore bring a 'hermeneutic of suspicion' to typical arguments in bioethics. For example, in assessing new reproductive technologies, Susan Sherwin notes that most arguments *for* technologies such as IVF are based on assumptions about the importance of autonomy. From a feminist perspective, however, such assumptions are questionable: the technologies are likely to be controlled by specialists and not by the women who are the 'patients' (Sherwin, 1992a, 126). Suspicion toward structures of health care and claims of what is 'good' for women is thus the third characteristic of contemporary feminist bioethics.

Fourth, feminist ethics looks beyond the desires or situation of the individual patient to ask about how *practices* in health care will affect women (Lebacqz, 1997a). Hence, feminist bioethics includes a *political* dimension that has been largely lacking in mainstream bioethics (Sherwin, 1992b, 22; see also Roberts, 1996, 116). Feminists ask how practices 'medicalize' normal experiences such as menopause or menstruation (Sherwin, 1992a). Feminists also ask how assumptions about the 'normal' have worked to constrain and disadvantage women.

Of particular relevance here is the fact that the category of 'nature' is itself suspect in feminist thought. 'Nature' has often been used as an argument to deprive women of rights and power, and thus claims about women's "nature" are viewed with particular suspicion by most feminists. As Wendell puts it, "We are used to countering claims that insofar as women are oppressed they are oppressed by nature....We know that if being biologically female is a disadvantage, it is because a social context makes

---

[23] Eugenics is mentioned specifically in Tong (1997), Wolf (1996), Holmes and Purdy (1992), Mahowald (1993), and Tong (2000).

it a disadvantage" (Wendell, 1992, 67). In a society in which women have historically been considered by nature to be 'weaker' than men and therefore unable to participate in the 'strong' world of the public arena, any discussion of women's 'nature' is automatically suspect (Wendell, 1992, 69). Asch and Geller caution specifically that the Human Genome Project and the attention to genetics raise again the danger that anatomy will be taken as destiny, especially for women (Asch and Geller, 1992, 323). The very framing of this project is therefore problematic from a feminist point of view. Minimally, feminists would argue strongly that attention to views of 'nature' and their impact on acceptance of technology must include significant critique of how 'nature' has traditionally been understood and utilized.

While most feminists eschew the category of 'nature,' one feminist does begin her work in bioethics with an account of 'human nature.' Mary Mahowald defines women and men in terms of their typical biological maturation into human beings capable of either conceiving, bearing, and nursing children or capable of fertilizing human ova (Mahowald, 1993, 5). Further, they are typically capable of forms of rationality and autonomy. Biological maturity and rational maturity may exist independently of each other, and life experiences will influence the specific expression of either. These affirmations about human nature give some guidance for bioethics. However, Mahowald sees equality as a social goal to be actively pursued and therefore rejects the historical 'natural law' view that men and women have different 'natures' and therefore social inequalities are acceptable (Mahowald, 1993, 79).[24] Seeing human life as a developmental continuum, Mahowald argues that ethical decisions must be appropriate to the specifics of the circumstances (Mahowald, 1993, 7). True egalitarianism must attend to individual differences. This attention to the concrete other is also typical of feminist thought.

In spite of her apparent acceptance of 'nature' as a framing category, Mahowald reflects what Sherwin calls a typical feminist move: an acknowledgment of the social roots of individuals (Sherwin, 1992b, 23). Most feminists are skeptical of the extreme individualism that permeates liberal political philosophy (Asch and Geller, 1996, 327). Feminists look to the ways in which social contexts define people and structure interactions. Hence, feminists argue for an 'embedded' approach to persons, rather than for abstractions about human nature. They specifically and deliberately eschew the attention to 'private' encounters between doctor and patient, in favor of an approach that focuses on unjust social arrangements (see Roberts, 1996, 121).

Perhaps the most significant text in this regard is Emilie M. Townes, *Breaking the Fine Rain of Death*. In this "womanist" ethic of care,[25] Townes sets out precisely to analyze social location and cultural context and to provide a bioethics that takes

---

[24] She rejects natural law reasoning in general. Primarily, she objects to its presumed opposition to 'artificial' interventions into conception. Arguing that the ability of human beings to transform the world, including themselves and their capacity for reproduction, is surely part—and perhaps the most important part—of any divine design, Mahowald rejects traditional natural law reasoning (77).

[25] Many African-American women eschew the term 'feminist' as reflecting primarily the concerns of white women, and use the term 'womanist' to denote concerns arising from the Black woman's context.

history and social setting seriously. Hence, she evaluates the legacy of the Tuskegee syphilis study on Black attitudes toward health care, she reviews in painstaking panorama the barriers to health care access for African Americans, and she draws deliberately on scriptural sources and images for a framework for her womanist ethic.

Given the feminist commitment to assessing the political context and meaning of health care and reviewing the history of women's oppression, it is no surprise that feminists have not addressed some of the typical categories of bioethics. Discussions of gene therapy are missing, but references to eugenics are rather common, as feminists search for an analysis of the wider social context within which all medical practices must be evaluated.

One of the few feminist texts that does discuss gene therapy and enhancement is Rosemarie Tong's *Feminist Approaches to Bioethics*. Tong reviews non-feminist approaches to a range of bioethical issues, and then shows what a feminist view looks like by contrast. The non-feminist perspective on gene therapy, for example, assumes that so long as the purposes of somatic cell gene therapy are 'negative' and 'medical'—that is, aimed at treating genetic conditions that society widely regards as diseases—there are no moral qualms about its use. Gene therapy can simply be likened to other somatic therapies (Tong, 1997, 227ff.). "Enhancement," however, raises problems for non-feminists, because its purposes are 'positive' and 'non-medical'—that is, raising people above 'normal' functioning. Germline gene therapy raises problems because of the unpredictable risks to subjects and their offspring, because it places numerous generations in the position of being unconsenting subjects, and because it might be used to create permanent alterations to suit the powerful rather than to suit those who are altered. Advocates of germline intervention counter all three of these arguments. A common concern, however, is whether the Human Genome Project has challenged our understanding of our 'nature' and significance. For example, we may increasingly think of ourselves as the product of our genes—or, conversely, attempt to become the active manipulators of everyone's genetic destiny.

Feminists make many of these same arguments, suggests Tong (1997, 238ff.). However, their focus would be on the impact *on women* of decisions made about gene therapy for children. Are the women truly ready to accept the possible risks and burdens? Are they already being blamed for the genetic problems of their children, and would they be blamed for any risks associated with gene therapy? Will they be pushed into accepting treatments that are still experimental and for which risks are not yet well established?

Feminists also raise serious questions of justice. Will the costs be so high that only the privileged can afford such treatments? Will women who cannot afford gene therapy be pressured to abort instead? And if we extend the arguments of Roberts and Townes, we might ask whether the intent of gene intervention is truly beneficent, as its proponents might claim, or whether it will be, wittingly or unwittingly, a tool for racist and sexist oppression.

This last question is particularly germane, as feminists are quick to point out that 'disease' is already a social category. Society can manipulate the criteria for disease. What genetic conditions should count as 'diseases' for which intervention is sought? As Wendell puts it, "careful study of the lives of disabled people will reveal how artificial the line is that we draw between the biological and the social" (Wendell, 1992, 69). Sherwin, too, argues that feminists are and should be skeptical about any extensions of medical authority into "mental and social spheres" (Sherwin, 1992a, 193). This suggests that feminists might be particularly cautious about using genetic technologies for 'enhancement' purposes, since 'enhancement' is always a value judgment (Tong, 1997, 240). As for arguments that we already use techniques such as plastic surgery to 'enhance' ourselves, feminists have mounted compelling critiques of the practice of plastic surgery, and therefore would be likely to argue against gene intervention to secure desired characteristics for women.

What is crucial here is that 'enhancement' is so obviously a socially constructed category: what we find desirable today is not what would have been desirable a generation ago, nor necessarily what would desirable in future generations. Further, such standards are often racist and sexist. For example, Mahowald notes that sex selection is extremely sex biased in favor of males (Mahowald, 1993, 83). Wertz and Fletcher argue against sex selection precisely because it might be the edge of a 'slippery slope' of 'positive eugenics' that would treat children as commodities to be manipulated and controlled (Wertz and Fletcher, 1992, 245). In sum, Tong argues that "feminists question whether science and medicine should devote energy and expertise to serving ephemeral social preferences, especially those that tend to reinforce an iniquitous status quo" (Tong, 1997, 241).

Given that feminists are also skeptical about the 'medicalization' of many aspects of life, it is possible that feminists would be hesitant about gene 'therapy,' and would join Roger Shinn and others in pointing out that 'therapy' is also a value judgment. As Mahowald notes, termination of fetuses because of their anomalies may be a kind of chauvinism or social prejudice (Mahowald, 1993, 87). Most abortions for genetic conditions are not really to avoid suffering to the affected individual, but to avoid suffering and costs to others. Perhaps the strongest argument here is from Susan Wendell, who states boldly, "the idea that there is some universal, perhaps biologically or medically describable paradigm of human physical ability is an illusion" (Wendell, 1992, 66).

In sum, it is not possible to construct or assume a single 'feminist' position on gene therapy and enhancement. No doubt some Western feminists, particularly those who adhere to liberal tradition and emphasize 'rights' and autonomy, will support gene intervention for therapy and even for enhancement. But we would expect that feminists in general will be skeptical about claims for 'enhancement' and will challenge us to look at the social and political context in which these technologies are developed. Feminists will examine the *practice* of gene intervention, and will not necessarily assume that it raises no new issues or can be dealt with in the same manner as other medical practices or interventions. Even if it can be, their general critiques of medical practice will apply to this new arena as well.

When it comes to our case scenarios, therefore, we might expect the following. First, although some feminists might hold, with Mahowald, to a standard of normal or natural development as a measurement for when genetic intervention is acceptable, most feminists would reject such a standard. Thus, the finding that there is a specific, medically diagnosable disorder in Johnny's case, or in the case of Tom, is not relevant. More important is the issue of what each child may suffer from his projected height. More important than that is the question of whether the practice of emphasizing height tends to disadvantage some groups of people, especially women. Hence, we expect that feminists might agree with Wendell on the need for caution about setting any standards of "normalcy" that would alone determine the acceptability of intervention.

Given the long history of concern about eugenics and about the impact of practices, it is very likely that feminists would reject genetic intervention in the case of Erwin, where the parents simply want to make Erwin taller in order to ensure that he has an advantage in life. However, much depends on context. All of our case studies are framed around male children. Feminists would be most likely to point out the advantages that tall men already have in society, and to raise questions about practices that increase male advantages over females. If female children were involved, the judgments might be different.

Indeed, looking at context carefully suggests that we must make room for a variety of feminist responses, including possible acceptance of genetic enhancement. The "one child" policy in China has led some Chinese women to abort a fetus that they feared might be born damaged (Nie, 2000). The women generally expressed some sense of loss, grief, and regret over the aborted child. Therefore, it is certainly possible that they might have welcomed gene intervention in order to ensure *yousheng* or a "good birth." What is usually called gene therapy would be welcomed by many women in order to lighten the burden on mothers. Some of the Chinese women also stressed their support for the government one-child policy and spoke of their efforts to provide their one child with every possible advantage. Thus, it is also possible that they would desire genetic 'enhancement' were it available. This scenario makes clear how crucial is the central feminist claim that practices cannot be evaluated in isolation from social and political context and meaning.

3.14 Conclusion and Recommendations

What is the central question? Is it this: what is natural? Or, is it: what is right? Does the latter depend on the former?

We could be asking: is therapy more natural than enhancement? Both require technological intervention. The first is generally accepted regardless of religious background. The second is viewed as suspect. Yet, the grounds for suspicion are as yet inchoate and difficult to articulate. Even so, based upon the theological frameworks of living religious traditions, we can speculate. And we have.

We have noted a couple of distinctions. First, we assume that therapy and enhancement are two different things, but we hesitate to draw a sharp line between them. We work with the notion that therapy would move a person in ill health toward the average state of biological health, whereas enhancement would move a person from an existing healthy state toward a level of biological excellence above the average. Second, we distinguish between germline genetic alteration and somatic genetic alteration. The first would influence heritable traits for generations to come, whereas the second would influence only the person whose cells receive inserted genetic material. For the foreseeable future, only somatic DNA alteration has been proposed for use in enhancement procedures. This draws the focus of ethical deliberation away from influencing inheritance and places it on living individuals whose genomes may be altered by insertion of a enhancing DNA material.

With these stipulations in mind, we have considered four scenarios. In the case of Johnny with Somatotropin Deficiency Syndrome, on the basis of our review of religious traditions it appears that none would have theological reasons for denying therapeutic genetic treatment. In the case of Bobby who is short but not suffering from a deficiency, genetic alteration could bring him up to normal height. No religious precedents would prevent such medical intervention. In the case of Tom who wants only to attain the height of others in his immediate family, again nothing we can find in our review of religious commitments would render genetic intervention morally illicit. In the case of Erwin, however, who is already of normal height and whose parents want him to exceed the normal height, such enhancement is generally thought to be morally illicit.

What we see here is that it is the distinction between therapy and enhancement that is decisive. Therapy is morally licit, even if it involves alteration of the human genome. Enhancement is a moral problem for some if it provides an unfair advantage of one person over another. Enhancement is a moral problem because it is a justice problem.

As we have interrogated the fundamental anthropologies and understandings of nature in the world's religious traditions, we find no prior commitments that would lead contemporary religious adherents to think of DNA as sacred. No existing religious precedents would require that a gene or a genetic code have a status higher than any other aspect of human biology. Even though recent cultural conversation includes voices of "genetic essentialism" or implicit beliefs in the sacredness of DNA, science does not confirm such a belief nor do religious traditions necessarily poise themselves to demand special status for the genome. This leads to a principle: *DNA is not sacred.*[26]

---

[26] QUESTIONS FOR FURTHER DISCUSSION AND RESEARCH:
1. Does the plurality of religious traditions necessarily indicate an irreconcilable diversity of moral commitments to genetic therapy and genetic enhancement?
2. Why has DNA become so central in people's thinking about human nature and identity?
3. What are the theological implications of genetics? Can ancient religious resources provide what is needed for contemporary theological and ethical judgments? Must theology be responsive to new genetic advances, or must genetic advances be understood in the context of theological affirmations?
4. Is genetic intervention different from or similar to genetic selection (by prenatal diagnosis or preimplantation genetic diagnosis)?
5. Is genetic selection closer to eugenics than somatic gene transfer? Is germline gene transfer more similar to genetic selection or eugenics? What are the morally relevant differences between eugenics, germline gene intervention and somatic gene transfer?
6. Should bioethicists consider it their job to draw lines, such as the line between therapy and enhancement?

References

Asch, Adrienne, and Gail Geller (1996). "Feminism, Bioethics, and Genetics," in Susan Wolf (ed.), *Feminism and Bioethics: Beyond Reproduction.* New York: Oxford University Press, 318–350.

Burgess, John P. (1998). *In Whose Image? Faith, Science, and the New Genetics.* Westminster: John Knox.

Catholic Bishops' Joint Committee on Bioethical Issues (1995). *Genetic Intervention on Human Subjects: The Report of a Working Party.* London: Linacre Centre.

Chapman, Audrey R. (1999). Unprecedented Choices: Religious Ethics at the Frontiers of Genetic Science. Minneapolis, MN: Fortress.

Coates, Peter (1998). *Nature: Western Attitudes Since Ancient Times.* London: Polity Press.

Cole-Turner, Ronald (1993). *The New Genesis: Theology and the Genetic Revolution.* Louisville, KY: Westminster/John Knox.

Congregation for the Doctrine of the Faith (1987). *Instruction on Respect for Human Life in Its Origins and on the Dignity of Procreation (Donum Vitae).*

Coward, Harold (2000). "Ethics and Genetic Engineering in Indian Philosophy." Unpublished paper presented at the Eighth East-West Philosophers' Conference, University of Hawaii, January 19.

Crawford, S. Cromwell (2003). *Hindu Bioethics for the Twenty-First Century.* Albany, NY: SUNY.

Deane-Drummond, Celia E. (2002). *Creation Through Wisdom.* Edinburgh: T. & T. Clark.

Deutsch, Eliot (n.d.). "Religion and Ecology, Ancient India and Modern America." Unpublished paper.

Dorff, Elliot N. (1998). *Matters of Life and Death: A Jewish Approach to Modern Medical Ethics.* Philadelphia, PA/Jerusalem: The Jewish Publication Society.

Dorff, Elliott N. (2001). "Stem Cell Research—A Jewish Perspective," in Suzanne Holland, Karen Lebacqz, and Laurie Zoloth (eds.), *The Human Embryonic Stem Cell Debate.* Cambridge, MA: MIT.

Duster, Troy (1990). *Backdoor to Eugenics.* New York: Routledge.

Fukuyama, Francis (2002). *Our Posthuman Future.* New York: Farrar, Straus & Giroux.

Ghazali, Abu Hamid al- (1988[1409]). "al-Munquidh min al-Dalal," in Ahmad Shams al-Din (ed.), *Majmu'a Rasa'il al-Imam al-Ghazali.* Beirut: Dar al-Ktuub al-'Ilmiyyah.

Gibbs, Nancy (1999). "If We Have It, Do We Use It?," *Time* (September 13).

Griffin, David Ray (2000). *Religion and Scientific Naturalism: Overcoming the Conflicts.* Albany, NY: SUNY.

Griffin, David Ray (2004). "Scientific Naturalism: A Great Truth That Got Distorted," *Theology and Science* 2(1), 9–30.

Grundel, Johannes (1975). "Natural Law," in Karl Rahner (ed.), *Encyclopedia of Theological: The Concise Sacramentum Mundi.* New York: Seabury.

Hancock, Roger N. (1974). *Twentieth Century Ethics.* New York: Columbia University Press.

Hefner, Philip (1993). *The Human Factor.* Minneapolis, MN: Fortress.

Hefner, Philip (1998). "Determinism, Freedom, and Moral Failure," in Ted Peters (ed.), *Genetics: Issues of Social Justice.* Cleveland, OH: Pilgrim Press.

Holmes, Helen Bequaert, and Laura M. Purdy (eds.) (1992). *Feminist Perspectives in Medical Ethics.* Bloomington, IN: Indiana University Press.

Hume, Robert E. (trans.) (1931). *Thirteen Principal Upanishads.* London: Oxford University Press.

John Paul II (1995). *Evangelium Vitae.*

Jonsen, Albert R. (1998). *The Birth of Bioethics.* Oxford: Oxford University Press.

Juengst, Eric T. (1998). "What Does *Enhancement* Mean?," in Erik Parens (ed.), *Enhancing Human Traits: Ethical and Social Implications.* Washington, DC: Georgetown University Press.

Kass, Leon R. (1998). "The Wisdom of Repugnance," in Leon R. Kass and James Q. Wilson (eds.), *The Ethics of Human Cloning.* Washington, DC: American Enterprise Institute Press.

Kass, Leon (2003). "Beyond Therapy: Biotechnology and the Pursuit of Human Improvement," U.S. President's Council of Bioethics (January). http://www.bioethics.gov/.

Keller, Evelyn Fox (1992). "Nature, Nurture, and the Human Genome Project," in Daniel J. Kevles and Leroy Hood (eds.), *The Code of Codes: Scientific and Social Issues in the Human Genome Project*. Cambridge, MA: Harvard University Press.

Kevles, Daniel J. (1992). "Out of Eugenics: The Historical Politics of the Human Genome," in Daniel J. Kevles and Leroy Hood (eds.), *The Code of Codes: Scientific and Social Issues in the Human Genome Project*. Cambridge, MA: Harvard University Press.

King, Patricia A. (1992). "The Past as Prologue: Race, Class, and Gene Discrimination," in George G. Annas and Sherman Elias (eds.), *Gene Mapping: Using Law and Ethics as Guides*. New York: Oxford University Press.

Kniess, A., E. Ziegler, J. Kratzsch, D. Theime, and R.K. Muller (2003). "Potential Parameters for the Detection of hGH doping," *Analytical Bioanalytical Chemistry* 376(5), 696–700. Epub May 16, 2003.

Lebacqz, Karen (1997a). "Fair Shares: Is the Genome Project Just?," in Ted Peters (ed.), *Genetics: Issues of Social Justice*. Cleveland, OH: Pilgrim Press.

Lebacqz, Karen (1997b). "Genetic Privacy: No Deal for the Poor," in Ted Peters (ed.), *Genetics: Issues of Social Justice*. Cleveland, OH: Pilgrim Press.

Lissett, C.A., and S.M. Shalet (2003). "The Insulin-Like Growth Factor-I Generation Test: Peripheral Responsiveness to Growth Hormone is not Decreased with Ageing," *Clinical Endocrinology* 58(2), 238–245.

Lombardo, Paul (2003). "Taking Eugenics Seriously: Three Generations of ??? Are Enough," *Florida State University Law Review* 30(2), 191–218.

Mackler, Aaron L. (2003). *Introduction to Jewish and Catholic Bioethics: A Comparative Analysis*. Washington, DC: Georgetown University Press.

Mahowald, Mary Briody (1993). *Women and Children in Health Care: An Unequal Majority*. New York: Oxford University Press.

McBrien, Richard P. (1995). *The Harper Collins Encyclopedia of Catholicism*. New York: HarperCollins, 910.

McKenny, Gerald P. (2000). "Gene Therapy, Ethics, Religious Perspectives," in Thomas J. Murray and Maxwell J. Mehlman (eds.), *Encyclopedia of Ethical, Legal, and Policy Issues in Biotechnology*. New York: Wiley.

Minuto, F., A. Barreca, and G. Melioli (2003). "Indirect Evidence of Hormone Abuse. Proof of doping?," *Journal of Endocrinological Investigations* 26(9), 919–923.

Moltmann, Jürgen (2003). *Science and Wisdom*, Margaret Kohl (trans.). Minneapolis, MN: Fortress.

Moosa, Ebrahim (2002). "Interface of Science and Jurisprudence: Dissonant Gazes at the Body in Modern Muslim Ethics," in Ted Peters, Muzaffar Iqbal, and Syed Nomanul Haq (eds.), *God, Life, and the Cosmos: Christian and Islamic Perspectives*. Aldershot,: Ashgate.

Nasr, Seyyed Hossein (1996). *Religion and the Order of Nature*. New York/Oxford: Oxford University Press.

National Council of Churches in the U.S.A. (1980). *Human Life and the New Genetics: A Report of a Task Force*. New York: National Council of Churches.

Nie, Jing-Bao (2000). "'So Bitter That No Words Can Describe It': Mainland Chinese Women's Moral Experiences and Narratives of Abortion," in Rosemarie Tong (ed.), *Globalizing Feminist Bioethics*. Boulder, CO: Westview Press.

Novak, David (1995). "Natural Law, *Halakhah*, and the Covenant," in Elliot N. Dorff and Luis E. Newman (eds.), *Contemporary Jewish Ethics and Morality: A Reader*. Oxford/New York: Oxford University Press.

O'Mathuna, Donal P. (2002). "Genetic Technology, Enhancement, and Christian Values," *The National Catholic Bioethics Quarterly* 2(2), 227–295.

Peters, Ted (1997). *Playing God: Genetic Determinism and Human Freedom*. New York: Routledge.

Peters, Ted (2003). "Designer Children: The Market World of Reproductive Choice," in Ted Peters (ed.), *Science, Theology and Ethics*. Aldershot: Ashgate.

Peterson-Iyer, Karen (2004). *Designer Children: Reconciling Genetic Technology, Feminism, and Christian Faith*. Cleveland, OH: Pilgrim Press.

Ramsey, Paul (1972). "Genetic Therapy: A Theologian's Response," in Michael Hamilton (ed.), *The New Genetics and the Future of Man*. Grand Rapids: Wm B. Eerdmans.

Reichenbach, Bruce R. and Elving Anderson (1995). *On Behalf of God: A Christian Ethic for Biology*. Grand Rapids, MI: Wm. B. Eerdmans.

Rifkin, Jeremy (1983). *Algeny*. New York: Viking.

Roberts, Dorothy E. (1996). "Reconstructing the Patient: Starting with Women of Color," in Susan M. Wolf (ed.), *Feminism and Bioethics: Beyond Reproduction*. Oxford: Oxford University Press.

Robitscher, Jonas (ed.) (1973). "Eugenic Sterilization: A Biomedical Intervention," in *Eugenic Sterilization*. Springfield, IL: Charles C. Thomas.

Sachedina, Abdulaziz (1997). *Testimony to the National Bioethics Advisory Commission*, March 14. Washington, DC: Eberline Reporting Service.

Sachedina, Abdulaziz (2003). *Islamic Biomedical Ethics: Issues and Resources*. Islamabad: Comstech.

Sherwin, Susan (1992a). *No Longer Patient: Feminist Ethics and Health Care*. Temple University Press.

Sherwin, Susan (1992b). "No Longer Patient: Feminist Ethics and Health Care," in Helen Bequaert Holmes and Laura M. Purdy (eds.), *Feminist Perspectives in Medical Ethics*. Bloomington, IN: Indiana University Press.

Shinn, Roger Lincoln (1996). *The New Genetics: Challenges for Science, Faith, and Politics*. Wakefield, RI: Moyer Bell.

Sina, Ibn (1952). *Avicenna's Psychology: An English Translation of Kitab al-Najat*, Book II, Chapter IV, Fazlur Rahman (trans.). London: Oxford University Press.

Tong, Rosemarie (1997). *Feminist Approaches to Bioethics: Theoretical Reflections and Practical Applications*. Boulder, CO: Westview Press.

Tong, Rosemarie (ed.) (2000). *Globalizing Feminist Bioethics: Crosscultural Perspectives*. Boulder, CO: Westview Press.

Townes, Emilie M. (1998). *Breaking the Fine Rain of Death: African American Health Issues and a Womanist Ethic of Care*. New York: Continuum Publishing.

U.S. President's Council on Bioethics (2002). "Staff Background Paper: Human Genetic Enhancement" (December). http://www.bioethics.gov/.

Walters, LeRoy, and Julie Gage Palmer (1997). *The Ethics of Human Gene Therapy*. New York: Oxford University Press.

Wendell, Susan (1992). "Toward a Feminist Theory of Disability," in Helen Bequaert Holmes and Laura M. Purdy (eds.), *Feminist Perspectives in Medical Ethics*. Bloomington, IN: Indiana University Press, 63–81.

Wertz, Dorothy C., and John C. Fletcher (1992). "Sex Selection Through Prenatal Diagnosis: A Feminist Critique," in Helen Bequaert Holmes and Laura M. Purdy (eds.), *Feminist Perspectives in Medical Ethics*. Bloomington, IN: Indiana University Press, 240–253.

Wolf, Susan M. (ed.) (1996). *Feminism and Bioethics: Beyond Reproduction*. Oxford: Oxford University Press.

Zoloth, Laurie (2001). "The Ethics of the Eighth Day: Jewish Bioethics and Research on Human Embryonic Stem Cells," in Suzanne Holland, Karen Lebacqz, and Laurie Zoloth (eds.), *The Human Embryonic Stem Cell Debate*. Cambridge, MA: MIT.

Chapter 4
How Bioethics Can Inform Policy Decisions About Genetic Enhancement

Robert Cook-Deegan, Kathleen N. Lohr, and Julie Gage Palmer

Among its many functions, bioethics applies philosophy, law, history, social sciences, humanities, and religion to normative analyses of new biotechnologies. We show how explicit moral analysis, religious perspectives, and contributions from the humanities informed public policy decisions about the beginning of human DNA transfer experiments; we also examine the value that bioethics added to the policy-making process. We then turn to an emerging genetic technology that appears thorny through the bioethics lens: genetic memory enhancement. We describe current and potential contributions of bioethics to public policy in this arena. Finally, we contemplate how bioethics might contribute to similar policy-making for enhancement technologies in the future. We conclude that genetic interventions such as inserting or altering DNA to enhance memory or cognition—whether inherited or affecting only the person whose cells are genetically altered—will likely be introduced from the edges of medicine, and we call for broad bioethics conversations regarding genetic changes in memory and cognition.

The previous chapter in this volume ("Religious Traditions and Genetic Enhancement"; Chapter 3) addresses genetic intervention, focusing particularly on its links to eugenics, and religious and moral perspectives on its acceptability. That chapter is a review of normative analyses, and we do not plow that ground again here. Instead we focus on how normative analysis informs policy, and we specifically examine the roles bioethics and religion have played—or failed to play—in making policy decisions about genetic intervention.

4.1 Historical Backdrop to "Human Genetic Engineering"

In 1923, J.B.S. Haldane proposed genetic selection to breed humans. Haldane was one of the most prominent biologists of his day, and was thoroughly familiar with both experimental and quantitative genetics. Haldane's lecture revolved around the mythical figure Daedalus. In Haldane's hands, Daedalus was a quiet hero, a prototype for the 1920s biologist. Daedalus and the biologists of the Haldane era were devoted to their art and science, work fated to transform the world. Haldane lauded Daedalus for his freedom from religious constraint, for his ability to work without getting deeply caught up in the machinations of the Greek gods, and for his liberation from the intellectual shackles of "natural law." Daedalus forged ahead with his experiments without much regard for whether something was or was not "natural" (Haldane, 1923). A decade later, Aldous Huxley turned Haldane's essay on its head in his 1932 biotechnological dystopia, *Brave New World*. Haldane's positive image of intellectual progress gave way to Huxley's mordant vision of a world where reproductive and genetic technologies became tools to repress thought, hobble emotion, and make sex plentiful but shallow and impersonal. Today we might say Haldane was speculating about human genetic engineering, or "human inheritable genetic modification." He clearly anticipated genetic enhancement of humans, knowing well that religious groups would oppose it, a point he dwelt upon in the 1923 paper.

Genetic enhancement of humans was contemplated decades before DNA was discovered to be the material substrate for "genes," or units of inheritance. Avery, Macleod and McCarty identified DNA as the carrier of genetic information in 1944 (Avery et al., 1944[1]). More famously, James Watson and Francis Crick revealed DNA's double helical structure in 1953 (Watson and Crick, 1953a, b[2]). Within a year physicist George Gamov speculated that DNA was, in essence, an informational molecule storing a quaternary code in the order of its base pairs (Gamov, 1954[3]). That advance made possible the idea of exchanging biological instructions from one organism to another, or tweaking the genetic code in other ways—an idea that was realized with recombinant DNA techniques beginning in 1973 (Cohen et al., 1973). Haldane's speculation about the prospect of genetically altering humans predated the means to actually do it.

Public debate about genetic enhancement by deliberate alteration of a human genome dates back at least 80 years, although the terminology was different and technical details were vague. If we include the implicit promises of human breeding by conventional means that were so prominent with the eugenics movement—which dates back to the 1860s and Sir Francis Galton (Kevles, 1985)—then the debate is even older.

4.1.1 Policy History of Human DNA Transfer: The Interplay of Bioethics and Public Policy in the 1980s

Public bioethics committees studying DNA transfer reached two crucial conclusions reviewed in Chapter 3. First, there is nothing distinctive about a change in DNA as long as changes affect only body cells, not gametes (and hence changes

---

[1] Its significance is reviewed in Lederberg (1994).

[2] Available online with other seminal papers at http://www.nature.com/nature/dna50/archive.html (accessed 19 September 2007).

[3] Its significance is reviewed by Trifonov (2000).

cannot be transmitted to progeny); therefore, genetic changes in body cells should be treated no differently than other medical interventions. Second, if changes could be inherited, moral analysis of new issues raised by transmitting changes from parent to child (and possibly beyond) would be needed; however, no compelling reason exists to engage those questions now because the technical means to carry out germ line interventions in a way that might be applied to human reproduction safely and reliably do not exist.

As to the second point, transgenic animal experiments are not a model for human germ line intervention because they are currently too unreliable and would be associated with risks of deformity or functional disorders in resulting humans. A debilitated animal is an experimental result (and a moral problem for only some groups); a genetically debilitated human is a serious moral problem for every group that has considered the possibility. Human germ line changes will become morally acceptable only when technical means to carry them out safely and with predictable outcomes have been developed.

If techniques that were likely to be safe, such as the ability to "correct" disease-associated DNA mutations to a normal "wild type" genotype, were to emerge, then that prospect would signal the need to debate ethical issues and social implications of deliberate engineering of DNA in ways that could be inherited. Many commissions have considered questions surrounding deliberate genetic alterations in humans, and have recommended broad discussions that include but extend well beyond technical audiences (Parens and Knowles, 2003; Frankel and Chapman, 2000; Chapman and Frankel, 2003; President's Commission, 1982; Walters and Palmer, 1997). Bioethics commissions have concluded that no religious doctrine or moral theory opposes altering DNA to treat disease just because it entails genetic change. That consensus disappears once the focus shifts away from clear-cut disease, toward genetic enhancement. Consensus also disappears when the altered DNA might be passed on to progeny through inherited genetic modifications (Walters and Palmer, 1997; Frankel and Chapman, 2000; Chapman and Frankel, 2003; Parens and Knowles, 2003).

The foremost policy decision about human genetic modification in the 1980s was whether to allow medical research on DNA transfer to proceed in humans. The consensus embodied in various reports was sufficient to open the gate to DNA transfer experiments in humans under regulatory scrutiny: it was (and is) acceptable to pursue research, including clinical DNA transfer experiments, to treat diseases so long as the changes are not inherited. The policy decision was to allow gene transfer protocols to proceed. (Gene transfer is inserting DNA into cells to be infused into a person where they are expected to reproduce, giving rise to genetically altered cells; it is not inherited unless the cells to be treated are gametes—egg cells, sperm or the cells that produce them and are involved in reproduction.) By June 2007, 842 such protocols were listed by the National Institutes of Health.[4]

---

[4] Protocol list 8 June 2007, accessed 23 September 2007; http://www4.od.nih.gov/oba/rac/PROTOCOL.pdf

One distinctive feature of the policy process was the degree to which bioethics, in its many varieties, was explicitly incorporated into deciding the right course of action. Various groups agreed that genetic changes to address a disease were just another medical technology, to be introduced as other medical technologies have been. In the end, the policy framework was quite similar to other research—focusing on safety of research participants, informed consent, and independent ethical review of the protocols—but augmented by special mechanisms for broader public debate. The debate about gene transfer experiments notably did not directly address many of the questions that would follow if genetic alterations began to be widely used, such as social justice and equitable distribution of costs and benefits. The policy decisions about gene transfer experiments centered on the permissibility of doing the clinical studies in the first place, rather than on social justice.

If a genetic intervention were demonstrated to be successful, that is, shown safe and effective and therefore promising for clinical use, it would raise questions of adoption, coverage, payment, distribution of benefits, and priority relative to other technologies. These questions pertain to introducing new medical technologies more generally, and would mark a transition from research ethics to the ethics of health policy decisions. Many policies affect how a new technology is introduced, adopted, and paid for—issues of mainstream health policy. Those issues raise questions about how to be sure the technology actually works, how well it works compared with alternative approaches, whether it should be covered as a core health benefit, and how much to pay for it. The role of bioethics may be to broaden health policy debates to include explicit normative analysis and incorporate disparate perspectives much as it broadened research policy debate. In the context of health policy decisions, however, bioethics encounters a much more decentralized and complex set of stakeholders and political processes than in making a go/no-go decision about human gene transfer experiments.

The policy questions about introduction of genetic enhancements are likely to differ from the early debate about noninherited gene transfer experiments for the treatment or prevention of disease in two important respects:

- Debate will draw on experience with gene transfer experiments already under way, so the question of "whether to begin" research has already been addressed in principle for genetic intervention, and the remaining question is about whether a particular study of a particular genetic intervention is warranted for a particular purpose (in this case, enhancement).
- A genetic intervention intended to enhance function could well be posed as an expansion in use of an existing technology, and thus with some track record of safety and efficacy in disease treatment, rather than a decision *de novo*.

Both of these features suggest that the debate will be different in type, because it will take place in the context of experience with existing technologies, and will be about using existing technologies in new ways. The debate will be more about the relative priority of doing such research than about whether to do it at all. And it will entail weighing the anticipated benefits, risks, and costs, as well as deciding when, whether, and how to pay for such uses should they prove effective.

4.2 The Role of Bioethics in Regulation of Human DNA Transfer Research

Bioethics, including representatives from religious organizations (and at times surveys of religious perspectives), played a highly public role in the regulated introduction of human DNA transfer. Religious concerns led to explicit analysis of the ethical implications of advances in genetics. This set the stage for a regulatory process, and bioethicists and religious scholars interacted directly with those making policy decisions through institutions, such as bioethics commissions, created to gather broad public views.

Human DNA transfer was not treated as just another new medical technology subject to technical review. Instead, it became the focus of conspicuous public debate, drawing explicit attention to its ethics, religious implications, and possible social impact. Reviewing the specific role of bioethics in the process will illuminate our later comments on the future of genetic enhancements of human beings.

This history is both directly relevant and compelling in its own right. The history of human gene transfer has been dogged by scandal—early research conducted without IRB approval, the very public death of Jesse Gelsinger, and cancers in some children treated with gene transfer for immune disease—and engagement of complex and sometimes inconsistent government oversight a history that will affect future debate about genetic enhancement technologies.

4.3 From Recombinant DNA to Biotechnology

In the early 1970s, scientists discovered methods for splicing DNA from one organism into another, and growing the resulting recombinant DNA molecules in bacteria or cultured cells (Krimsky, 1982; Berg and Singer, 1995; Fredrickson, 2001). Fearing the possible consequences of unregulated recombinant DNA research, particularly the possibility that recombinant DNA could be incorporated into living organisms that would then acquire unpredictable characteristics with means to replicate and spread uncontrollably, scientists imposed a moratorium on recombinant DNA experiments. This self-imposed moratorium was lifted after the National Institutes of Health (NIH) established the Recombinant DNA Advisory Committee (RAC) to formulate safety guidelines for recombinant DNA research (Krimsky, 1982; Fredrickson, 2001).

Once the oversight mechanism was in place, recombinant DNA experiments proceeded. The guidelines were relaxed as experience accumulated, as genes such as those for human insulin and growth hormone were cloned and seemed to offer tangible medical benefit, and as public attention drifted to new concerns. Nonetheless, recombinant DNA experiments remained subject to review by Institutional Biosafety Committees at research institutions, with oversight from the national RAC (with analogous mechanisms in other countries).

In the early to mid-1980s, concern about the biohazard of recombinant DNA became less intense, and attention turned to introducing recombinant DNA in humans. The new debate about "human gene therapy" was not always about human genes, however, and the prospect of benefit implied by "therapy" was premature. The nascent debate about human gene transfer converged with growing attention to commercial uses of the promising new technologies of recombinant DNA and cell fusion. Wall Street analysts began to refer to "biotechnology," co-opting an academic term, expanding its use, and changing its meaning (OTA, 1984a). Biotechnology came to mean (more or less) the commercial application of molecular genetics and cell biology.

4.3.1 Splicing Life

In 1980, the general secretaries of the three largest religious denominations in the United States wrote a open letter to President Jimmy Carter expressing concern about advances in genetics.[5] This letter became a pivotal event in "public bioethics." It was sent just 4 days after the U.S. Supreme Court decision in *Diamond v. Chakrabarty* (447US303, 1980) that explicitly accepted patenting of engineered bacteria. Three points about the clerical letter are particularly relevant to the story.

First, the clerics were concerned that commercial uses of biotechnology would drive its application in the face of incomplete and unsettled federal regulation and without sufficient attention to ethical concerns. They addressed human DNA transfer as one of many issues. They feared that financial incentives would dictate the future direction of molecular biology in general. As the story unfolded, rules for introducing DNA transfer into humans became the main thrust of policy engagement, but the clerical letter raised many other points. Indeed, the issue the clerics worried most about—financial incentives and a profit-driven technological imperative—were addressed far less in subsequent policy documents than other points that required immediate attention, such as whether it was acceptable to conduct gene transfer research, how to ensure public oversight, and how to analyze clinical scenarios in which gene transfer might be plausible.

Second, the letter's authors included Jewish, Roman Catholic, and Protestant denominations through their respective national organizations. It did not include representatives of Islam, Buddhism, Hinduism, or other religious traditions whose perspectives are reviewed in both volumes of the current project, "Altering Nature."

Third, the letter called on the President to establish a panel with representation well beyond the technical considerations of biotechnology to consider ethical issues explicitly. This set in motion a cascade of events that have become a defining story of contemporary bioethics.

---

[5] Claire Randall, General Secretary, National Council of Churches; Rabbi Bernard Mandelbaum, General Secretary, Synagogue Council of America; Bishop Thomas Kelly, General Secretary United States Catholic Conference, letter of 20 April 1980, reprinted in President's Commission (1982, Appendix B, "Letter from the Three General Secretaries," 95–96).

4 How Bioethics Can Inform Policy Decisions About Genetic Enhancement

The letter was referred to the President's Commission for the Study of Ethical Problems in Medicine and Biomedical and Behavioral Research (hereafter President's Commission). The 1980 clerical letter became the impetus for the President's Commission 1982 report *Splicing Life* (President's Commission, 1982).

Scandal fueled further interest in human gene transfer. In July 1980, Dr. Martin Cline performed experiments on Israeli and Italian youths suffering from the hemoglobin disorder beta-thalassemia. The experiments took place the same month as the clerical letter, but the Cline experiments did not become public until months later, when word began to leak out in the European press, then spilled across the Atlantic (Thompson, 1994). Cline used recombinant DNA without approval of the Institutional Review Board or Institutional Biosafety Committee of the University of California at Los Angeles (UCLA) (Thompson, 1994). Following an investigation, Dr. Cline's federal grants were terminated, and he resigned as chair of a division in the department of medicine at UCLA (Sun, 1981; House of Representatives, 1982).

In *Splicing Life*, the President's Commission briefly reviewed this history, although it never mentioned Cline by name (President's Commission, 1982, 44–45). Cline's experiments nonetheless set the context for public debate. Among other things, Cline's ambition made clear that use of recombinant DNA in humans was likely to be attempted, and the UCLA experience highlighted the lack of clear authority for oversight among institutional biosafety committees, institutional review boards (IRBs), the RAC, and the Food and Drug Administration (FDA). The President's Commission focused on human gene transfer because it was immediate and concrete and because it highlighted the ambiguity of regulatory authority, a public policy issue that needed to be resolved.

The President's Commission took as its foremost task "clarifying the issues," the title of its first chapter. The report reviewed regulatory oversight of biotechnology and identified some gaps. The possibility of using recombinant DNA to treat human diseases emerged as one of the most concrete applications that might require a clear oversight mechanism. The second chapter opened with an explicit analysis of religious perspectives, leading into analysis of issues that should be of concern to government.

Splicing Life based this recommendation for clearer regulatory oversight on a key finding, that "issues raised by the projected human uses of gene splicing, which heretofore have not received attention, are as at least as complex and important as those addressed by RAC thus far."[6] The final paragraphs, under the subheading "Revising RAC," focused on how highly technical developments in diverse uses of biotechnology should be overseen by a body that included nontechnical expertise and representation, not just technical competence to review the science. The President's Commission called for such a body to address broader social, ethical, and legal issues explicitly.

---

[6] *Splicing Life* pp. 81–82.

4.4 National Bioethics Commissions

Prospects of human gene transfer received another dose of public attention at a November 1982 hearing before the Investigations and Oversight Subcommittee of the House Committee on Science and Technology, chaired by Rep. Albert Gore, Jr. (U.S. House of Representatives, 1982). The hearing had three significant outcomes. First, it became a highly public release event for *Splicing Life*, and it signaled Gore's interest in public bioethics. Second, it became the seed for establishing a new bioethics commission, a successor body to the President's Commission. Rep. Gore introduced a bill to that end in 1983. The bill, in amended form, ultimately became Section 11 of the Health Research Extension Act of 1985,[7] which gave rise to the congressional Biomedical Ethics Board and Biomedical Ethics Advisory Committee, which operated briefly in 1988–1989 before being defunded.[8] That short-lived committee, however, was not the end of the story of U.S. national bioethics commissions.

Two two-term Presidents, Bill Clinton and George W. Bush, later also established bioethics commissions. The National Bioethics Advisory Commission was chartered by President Clinton, and the President's Council on Bioethics was chartered by President George W. Bush. Rep. Gore's 1982 hearing was an early signal that U.S. national bioethics commissions had a future. It also indicated clearly that political leaders expected debates about genetic technologies to engage constituencies well beyond the technical communities directly involved in research and its application.

Third, Rep. Gore's hearing featured a high-profile airing of the Cline case, in which Cline defended his decisions in public and under oath, in response to pointed questioning. Finally, the hearing led Rep. Gore to request a report on human gene therapy, from the congressional Office of Technology Assessment (OTA). That OTA report was released at another Gore hearing in December 1984 (U.S. House of Representatives, 1982; OTA, 1984b).

4.4.1 Mandates for the Recombinant DNA Advisory Committee

The President's Commission recommendation that the RAC take up the question of gene therapy had a real effect. In April 1983, a RAC working group recommended that the full RAC declare its readiness to review human gene therapy proposals

---

[7] PL 99-158; 42 USC 275; accessed 23 September 2007; available online at http://history.nih.gov/01docs/historical/documents/PL99-158.pdf.

[8] The Biomedical Ethics Board and Advisory Committee were established in PL 99-158, enacted in 1985, but the process of appointing Advisory Committee members took until late 1988, by which time one member died. Replacing that member led to an irreconcilable conflict among the 12 members of the congressional Board. This "bioethics" commission for Congress officially operated for a year and held three meetings, but its funding was eliminated at the beginning of Fiscal Year 1990 and it ceased operations in September 1989.

(Walters and Palmer, 1997). OTA was writing its report even as the RAC subcommittee on human gene therapy was formulating its list of questions for those proposing to introduce genetic changes into humans (OTA, 1984b). The RAC subcommittee took a page from the book of FDA processes in listing "points to consider" when contemplating such experiments. Unlike the FDA review procedures, however, which were focused on collecting data about safety and efficacy while protecting proprietary information, the RAC process was public, entailed substantial nontechnical representation, and considered broader implications beyond safety and efficacy.

Both the OTA and the RAC processes were explicitly informed by religious perspectives and ethical analysis. The composition of RAC's human gene therapy subcommittee was diverse, as recommended by the President's Commission to address concerns beyond technical matters. The RAC subcommittee addressed both protecting human research participants and broader social and ethical concerns. Bioethicist LeRoy Walters chaired this group, which included members with legal, policy, religious studies, and philosophy backgrounds, as well as technical experts in molecular biology and clinical research. Walters also chaired the OTA panel, which included a Jewish rabbi, an Episcopal priest, and a Catholic theologian, along with several prominent philosophers and other bioethicists. OTA report drafts were shared with the RAC working group, and vice versa (Walters and Palmer, 1997). Bioethics, policy, and technical review were thus commingled in the formulation of policy guidance in both the executive branch and in congressional oversight of DNA transfer experiments in humans.

By 1985, the RAC had published and adopted the "Points to Consider in the Design and Submission of Human Somatic-Cell Gene Therapy Protocols," which became a part of the guidelines for recombinant DNA research as Appendix M ("Points to Consider"). The "Points to Consider" posed questions to investigators who submitted gene modification protocols to the RAC for approval. The RAC thus began direct oversight of human gene transfer research, implementing a response to the *Splicing Life* recommendation.

Both the FDA and the NIH had a role in oversight of human gene transfer research. The FDA applied its existing review procedures to gene therapy, focused on assessment of safety and efficacy of the clinical trials as assessed by FDA officials, but with relatively little opportunity for public discussion. The RAC took the lead in commencing a broad public conversation.

In 1987, W. French Anderson and colleagues submitted a prototype gene therapy protocol to the RAC to try out the new regulatory system. The RAC used the "Points to Consider" to review the mock protocol as if it were real. The first actual human DNA transfer clinical protocol was submitted by Steven Rosenberg in April 1988. Rosenberg's proposed study was a gene-marking study, introducing a nonhuman "marker" gene into immune cells to monitor cancer therapy. This was a human DNA transfer experiment, but not gene therapy. The experiment was opposed by activist Jeremy Rifkin. The protocol took a tortuous path to approval, but was approved on a second iteration by the RAC and its Human Gene Therapy subcommittee. In January 1989, NIH Director James Wyngaarden, who formally had authority (RAC and its subcommittee were advisory only), gave his approval. Rifkin promptly filed suit to block the experiment. The lawsuit was settled out of court in May 1989, and the first human gene transfer study commenced a few days later (Lyon and Gorner, 1995, 160–175; Thompson, 1994, 300–350; Rainsbury, 2000).

In March 1990, Michael Blaese, French Anderson, and colleagues submitted the first true gene therapy protocol. They proposed to treat children who suffered from adenosine deaminase (ADA) deficiency, a severe illness that compromises patients' immune systems. The RAC approved the study in July 1990; the investigators treated the first study subject in September 1990.

Some researchers found the dual RAC/FDA review unduly burdensome. In 1996, NIH Director Harold Varmus recommended that the FDA take over most responsibility for reviewing gene transfer protocols. Varmus reduced the size of the RAC and curtailed its functions. RAC retained its role as a mediator of public debate about novel uses of gene transfer methods. The primary effect of this change was a reduction in transparency of DNA transfer clinical protocols, because the FDA oversight process was far less public until the time of an FDA public advisory committee meeting, which was not convened for DNA transfer protocols until the cataclysmic events of 1999 noted below.

From 1995 through September 8, 2003, 485 DNA transfer protocols were submitted for review. Of these, 68 were submitted to the RAC under its new, more limited mandate.[9]

4.5 Outcomes of DNA Transfer Experiments

Despite the many DNA transfer experiments completed or under way in the United States and elsewhere, few reports of successful clinical outcomes have emerged. Some failures have been widely reported. The most catastrophic event was the death of Jesse Gelsinger in September 1999. Gelsinger was a University of Pennsylvania gene therapy trial participant who died as a direct result of a DNA transfer procedure. He reached age 18 with a form of a sometimes fatal disease, ornithine transcarbamylase deficiency. At the time of the experiment, he was clinically stable and in no immediate danger from his disease.[10] Researchers attempting

---

[9] The RAC Charter states that the purpose of the RAC is "to provide advice and recommendations to the Director, National Institutes of Health (NIH) on (1) the conduct and oversight of research involving recombinant DNA, including the content and implementation of the NIH Guidelines for Research Involving Recombinant DNA Molecules, as amended (NIH Guidelines), and (2) other NIH activities pertinent to recombinant DNA technology" (RAC (NIH) 2005).

[10] Paul Gelsinger, father of Jesse Gelsinger, wrote an account of Jesse's illness online at http://www.circare.org/submit/jintent.pdf (accessed 28 June 2006), and commented specifically about his son's health at the time of the experiment in a letter to *FDA Consumer* November–December 2000. His article was reprinted in Guinea Pig Zero, a newsletter for people who frequently volunteer for medical studies or have interest in those who do: http://www.guineapigzero.com/jesse.html (accessed 23 September 2007). It was also reprinted in a book anthology, Robert Helms (2002) (Ed.). *Guinea Pig Zero: An Anthology of the Journal for Human Research Subjects*. New Orleans, LA: Garrett Country Press, pp. 178–198.

to determine the appropriate dose of a viral vector that would carry the gene for the defective OTC protein infused the vector into blood vessels draining into his liver, as they had done before with five other OTC sufferers under the protocol. In Gelsinger's case, the vector caused a catastrophic, systemic inflammatory reaction, leading to his death.

The FDA cited the researchers for multiple regulatory violations, suspended all DNA transfer research at University of Pennsylvania for a time, and stipulated restrictions under which James Wilson and his colleague Steven E. Raper could conduct clinical research subject to FDA oversight. Those restrictions could be removed upon fulfillment of an agreement with the U.S. Department of Justice on behalf of FDA and NIH.[11]

Additional investigations revealed that fewer than 6% of the adverse events in DNA transfer research had been reported to the NIH as required (Woo, 2000). None of the adverse events reports received by the FDA had been made public; few had been reported to IRBs responsible for assessing potential risks and benefits of the experiments.

Following Gelsinger's death, both NIH and FDA changed regulation and oversight of human gene transfer research. The FDA notified gene transfer researchers that it intended to share adverse event reports with the NIH Office of Recombinant DNA Activities (ORDA). In March 2000, the FDA and NIH announced new initiatives to protect participants in DNA transfer trials, including closer monitoring, procedures to avoid conflicts of interest, and independent oversight and review mechanisms (FDA, 2000). The NIH part of the oversight was more open, so adverse event reporting and protocol review were also more public than FDA regulation of other drugs, biologics, and devices.

In 2001, the FDA published proposed rules that would provide for the public disclosure of safety data related to DNA transfer experiments (FDA, 2001). In addition, the NIH modified its adverse events reporting requirements and procedures, with the goal of, among other things, "enhancing knowledge about scientific and safety trends" (OBA (NIH) 2001). In June 2002, the FDA completed a Gene Therapy Patient Tracking System, which included systems for collecting information about long-term safety outcomes in DNA transfer research (FDA, 2002).

In the meantime, while the FDA and NIH were revising regulations in response to Jesse Gelsinger's death, gene transfer research began to show promise in a few trials. In early 2000, French and American researchers reported success in gene therapy trials designed to treat X-linked Severe Combined Immune Deficiency (SCID) and hemophilia, respectively. Excitement about these research triumphs was then tempered when two of the nine "cured" French children developed T-cell acute lymphoblastic leukemia (T-ALL cancers) 3 years after treatment. The T-ALL

---

[11] FDA letters and an agreement negotiated with the U.S. Dept. Justice are available online: http://www.fda.gov/ora/compliance_ref/bimo/restlist.html; restrictions themselves are listed in more detail at: http://www.circare.org/foia3/wilson_exhibit2.pdf; and the settlement agreement at: http://www.circare.org/foia3/wilson5_settlementagreement.pdf (accessed 28 June 2006).

appeared to be caused by the retroviral vectors used to transfer genes. The FDA determined that leukemia was an inherent risk of the study design and put 27 similar U.S. gene therapy trials on clinical hold.

In 2003, the FDA announced that it would still consider retroviral gene therapy trials involving fatal disorders for which no viable alternative treatments exist (FDA, 2003). A subsequent review of U.S. DNA transfer trials found "There are not sufficient data or reports of adverse events directly attributable to the use of retroviral vectors at this time to warrant cessation of other retroviral human gene transfer studies, including studies for non-X-linked SCID" (RAC, 2005b).

4.6 Summary and Lessons

This 20-year overview of human DNA transfer history illustrates several features about the role of religious leaders, social critics, and bioethicists in public policy. The story began when religious leaders wrote to a U.S. President. The President turned to his bioethics commission, which in turn made recommendations for government action. One role of the President's Commission was to filter a profusion of inchoate concerns into some concrete policy options for oversight and regulation and to issue recommendations for mediating a broad and open debate about emerging genetic technologies. Another role was to conduct a highly public debate by collecting facts, engaging in public discussion, and producing highly public reports. The Commission's findings were publicly aired in Congress, which then requested a more specific analysis of human gene therapy from OTA, a congressional support agency. That congressional report, in turn, helped inform an executive branch advisory group that prepared policy guidance for those proposing to commence human gene transfer experiments. Religious views and philosophical analysis were explicitly included in the process, at all three levels—in the bioethics commission, the congressional process, and in the executive branch.

The President's Commission report of 1982 explicitly addressed religious perspectives. The review committees established to oversee human gene transfer research drew on scholars in philosophy, humanities, and social sciences as well as religious leaders, not just technical experts. The NIH's RAC was chaired by a bioethicist and included philosophers, lawyers, and academics from religious studies, as well as scientific and clinical experts.

This direct involvement of "public bioethics" is not unique in the annals of U.S. research programs, but it is unusual in its breadth, its direct influence on policy, and its conscious focus on bringing issues beyond mere technical concerns for safety and efficacy into debate about introduction of a new medical technology. This framework for anticipating broad social implications is a partial model for other contexts, including introduction of genetic technologies to improve memory function.

We believe the introduction of DNA transfer into humans benefited from the engagement of philosophical and religious bioethical views in several ways, but three stand out:

1. **This scientific advance was far more public than it would otherwise have been**. Bioethics led to an open, transparent debate, in advance of experiments, which encouraged public discussion and resulted in thoughtful policy. The NIH review process, as recommended by the President's Commission, was very public. Later, from 1996 until 1999 when oversight was less public, a research death and attendant scandal ensued. Policy was then driven by analyzing mistakes—a retrospective rather than prospective model—and the public media had to penetrate shrouds of secrecy to find information that previously might have been more public.
2. **Policy debates incorporated questions of fairness and sensitivity to moral and religious pluralism**. Because the NIH review process was interdisciplinary, it could and did engage stakeholders beyond science, medicine, and academe. In particular, it incorporated law, ethics, and religious perspectives into formulating review procedures and public debate.
3. **Involvement of the bioethics community, defined broadly, forced questions about "where are we going in the long run" to the surface**. Questions about "playing God," and the long-term implications of going down a particular technical pathway, were addressed directly.

In short, this historical path brought new constituencies and new questions—many of them going beyond science and medicine—to the table. Often, such questions are ones that scientific and technical communities find uncomfortable and well beyond their expertise. Thus, despite the fact that they may be central concerns of the general public, they are apt to be set aside for later or for others to deal with.

This challenge—drawing attention to crucial issues that have bioethical as well as scientific features—is exemplified by the case study to which we now turn: the use of pharmaceuticals for treating disease and enhancing memory and cognition. Specifically, bioethics played a role in the introduction of DNA transfer experiments in humans. What might it play in other policy decisions that affect access to technologies that now or in the future might be used for enhancing human performance?

4.7 What Role for Bioethics in Addressing Technologies for Enhancing Memory and Cognition?

We propose to address the role of genetic enhancement by focusing on a set of plausible clinical scenarios. We focus on cognitive function and memory, because we want to address questions about enhancement that could arise in the real world in coming decades and to engage a case that addresses uncertainty about what is "natural," a major theme of this volume, when it comes to brain function. We specifically consider how treatment for Alzheimer's disease and other dementias involves drugs to improve memory or cognition as a possible precursor to genetic enhancement of memory, and how clinical treatment of dementia could morph into enhancement of memory or cognition in those not suffering from dementia. We propose that treatment for clear-cut disease is the indication driving first use of technology, which then gets applied to a prodrome or subclinical syndrome, and then for purposes of enhancement of normal function, a three-step cascade that has precedent in drugs affecting cognition.

In sports we have steroids and performance-enhancing drugs, which are based on drugs initially introduced for medical conditions. Some pharmaceuticals are used to enhance mental performance—such as wakefulness for pilots. This is the pathway by which some drugs have already raised questions of drug-induced "enhancement" that might pertain to future prospects of genetic enhancement technologies. We use this case to examine how ethical analysis could play a role, and indeed *should* play a role, in raising issues that purely technical analysis of risk and benefit might not address otherwise. We also aim to clarify whether and how religious perspectives and ethical analysis should influence the policies governing use of such technologies.

We build here on a literature in bioethics about the difficulty in distinguishing enhancement from treatment. That literature notes the distinction between treatment and enhancement cannot be treated as a "fire break" between one use and another (Buchanan et al., 2000). This conclusion that there can be no bright line between treatment and enhancement, however, is neither a counsel of despair nor a call to abandon any attempt to make the distinction. Cases that are clearly in one category or the other can be addressed even if there are difficult cases that cannot be so easily analyzed. Beyond the treatment-enhancement distinction, the literature supports two findings about the role of bioethics: (1) general moral principles are relevant, and become translated into policy when they confront real cases and real decisions; and (2) explicitly incorporating values beyond purely technical concerns about safety, efficacy, and individual risk invites both the engagement of bioethics and the involvement of constituencies beyond those doing science or introducing new technologies.

4.8 A Role for Bioethics in Health Policy Decisions?

We assume in this analysis that genetic enhancement technologies are likely to evolve from medical treatments that turn out to have enhancement potential. The policies most likely to affect such developments address incentives for research and development, safety and efficacy of treatments, and coverage and reimbursement of products or services when used beyond approved indications.

The debate about human gene transfer was about whether recombinant DNA should be introduced into humans, a distinct research decision; the debate about enhancement is instead likely to center on new uses of existing medical technologies. The context for examining "enhancement" uses is unlikely to be as analytically clean as the first introduction of gene transfer experimentation was. This is true for at least three reasons: first, many more considerations come into play and many more stakeholders are involved in health care than in biomedical research;

second, the decisions will require more sophistication about how coverage and reimbursement decisions are made for health care technologies, about fairness, and about setting priorities; and third, the introduction of enhancements may grow out of existing medical practices rather than distinct scientific "firsts."

Enhancement uses may be novel, but they will be more difficult to isolate from existing practices and norms. A debate assuming that genetic enhancement can be isolated and considered as its own case ignores the fact that the gates to enhancement are likely to opened and closed by policies for medical technology, rather than enhancement *per se*.

A debate about expansion of uses of an existing technology entails deciding whether that use is legitimate and, if legitimate, whether it is medically necessary and should thus be part of standard health benefits packages. Policies intended to stall or thwart enhancement could well spill over to other medical technologies, imposing unintended restrictions on them. Stopping research on memory and cognition for fear of "enhancement," for example, is hardly an attractive option if at the same time it thwarts developing better treatments for Alzheimer's disease. Yet some enhancement potential may indeed be inherent in research intended to reduce the suffering of Alzheimer's disease. The R&D priorities debate is likely to be about more than just enhancement, and likely to be entangled in nitty-gritty decisions about what research should proceed. That research will also often have multiple aims of which enhancement may be one, although in some cases enhancement may not even be initially recognized as a possibility, let alone a goal.

Just as it was wise explicitly to consider religious and moral views when deciding about the first uses of human gene transfer in a broad and public debate, so also will bioethics be relevant to the introduction, adoption, and distribution of genetic enhancement technologies, because they touch on issues well beyond technical considerations. Topics will include research ethics and the ethics of "innovative therapies" as they morph into enhancements, but they will also include social justice and conceptions of the "natural" that are sure to be publicly contested, and for which bioethics is suited to sift through arguments. Public deliberation will require considerable sophistication about how the health care system works, however, as well as grounding in theories of social justice. Engaging questions about genetic enhancement will require bioethicists to acquire knowledge of the health care system, the incentives that affect development of new medical technologies, and how fairly medical advances are distributed once created.

4.9 Human Genetic Modification Is Already Being Pursued as a Treatment for Alzheimer's Disease

Lest our clinical scenarios seem unduly speculative, we note that the University of California, San Diego, announced a gene therapy trial involving the gene for nerve growth factor for Alzheimer's disease in April 2004 (Philips, 2004), and results were reported widely as "promising" in April 2005 (Neergard, 2005). The website for the biotechnology firm Ceregene[12] reports that Phase I trials are near completion and a Phase II trial is in the planning stages. This trial tests implantation of cells that express neural growth factors, which could stimulate growth not only of cells dying as a consequence of Alzheimer's disease, but also other nerve cells. Thus, cell therapies that incorporate DNA changes fit plausibly into technologies with "enhancement potential" in the scenarios sketched below, in which a technology addressing a disease finds other uses down the line.

4.10 Precedents for Enhancement Technologies: Pharmaceuticals Affecting Cognitive Function

A debate is already under way about pharmaceuticals that can enhance cognition, wakefulness, and other mental functions. In Farah et al. (2004), a working group of neuroscientists and scholars in public policy and bioethics considered ethical, legal, and social implications of pharmacologic enhancement as beginning points for other forms of intervention. The group noted:

> Neurocognitive enhancement raises ethical issues for many different constituencies. These include academic and industry scientists who are developing enhancers, and physicians who will be the gatekeepers to them, at least initially. Also included are individuals who must choose to use or not to use neurocognitive enhancers themselves, and parents who must choose to give them or not to give them to their children. With the advent of widespread neurocognitive enhancement, employers and educators will also face new challenges in the management and evaluation of people who might be unenhanced or enhanced (for example, decisions to recommend enhancement, to prefer natural over enhanced performance or vice versa, and to request disclosure of enhancement). Regulatory agencies might find their responsibilities expanding into considerations of 'lifestyle' benefits and the definition of acceptable risk in exchange for such benefits. Finally, legislators and the public will need to decide whether current regulatory frameworks are adequate for the regulation of neurocognitive enhancement, or whether new laws must be written (Farah et al., 2004).

The issues to be confronted for genetic enhancements have some precedent in current debates about "cosmetic pharmacology" or "enhancing use" of drugs initially tested for specific diseases but then more broadly used for other purposes, including improved brain function. The drug modafinil, for example, was tested and approved for marketing as a treatment for narcolepsy (a rare disorder whose symptoms include falling asleep frequently and uncontrollably). According to the FDA's approved labeling text from 2004 for Provigil (the brand name in the United States), it is now approved as a drug "indicated to improve wakefulness in patients with excessive sleepiness associated with narcolepsy, obstructive sleep apnea/hypopnea syndrome, and shift work sleep disorder" (FDA, 2004). This was an expansion of the original 1998 indication for narcolepsy alone. Modafinil's main effect is to counter loss of alertness from sleep deprivation, but it has also been

---

[12] http://ceregene.com/pipeline.asp, accessed for this purpose 11 February 2007.

tested for improving cognition in schizophrenia and adult attention-deficit/hyperactivity disorder (Turner et al., 2003). The U.S. Air Force has tested modafinil use among pilots[13]; and its use for jet lag among frequent travelers and among students is common (O'Connor, 2004). The drug appears to have modest but consistent effects on normal volunteers who are not sleep-deprived (Randall et al., 2004), leading one group of investigators to conclude that "in healthy volunteers without sleep deprivation, modafinil has subtle stimulating effects" (Müller et al., 2004). On a website devoted to uses of modafinil (www.modafinil.org), only a few entries address narcolepsy or other labeled uses; most are about "off label" uses. Indeed, the problem of off-label use provoked an FDA warning letter to the president of the manufacturer, Cephalon, in February 2007.[14]

Is this enhancement? Modafinil was approved initially to treat symptoms of narcolepsy, and then for two other disorders associated with significant sleep deprivation. Now, however, it appears to be most often used to curb the effects of sleep deprivation among healthy people, arguably an enhancement rather than treatment of a medical condition.

Other drugs with more powerful effects may be found in the future—or not; we simply cannot know. Prudence suggests thinking through enhancements before they arrive on the doorstep, and several neurologists and psychiatrists are bracing for the issues that will come with more potentially enhancing drugs and technologies (Hauser, 2004; Dees, 2004; Chatterjee, 2004; Hyman, 2006; Gerlai, 2003; Rose, 2002). The pathway to the market for other pharmaceutical enhancements will likely follow the modafinil precedent: first efficacy (and acceptable safety) will be demonstrated for one or more specific disease conditions, followed by approval for marketing; then there will be reports in the literature of other medical uses or milder symptoms, and then the drug will be used for "nonmedical uses," that is, to enhance normal functioning. Once the drug is on the market, then either individuals may buy it out of pocket, or they may seek coverage of the drug by third-party payers (both private insurers and government programs such as, e.g., Medicare, Medicaid, Veterans Health Administration, or Indian Health Service). Payers can define when they will or will not cover the drug, and they will often cap their reimbursement or mandate levels of copayment.

For example, Cigna HealthCare modified its policy in August 2005 to cover modafinil when patients with narcolepsy failed to benefit from cheaper and longer-used first-line drugs. Cigna specifically excluded coverage for shift-work sleep disorder or as an adjunct to treatment of Alzheimer's disease, Parkinson's disease, or multiple sclerosis,[15] and by implication certainly

---

[13] Walz (2003); Woodring (2004); both quoting Dr. John Caldwell of Brooks City Base in San Antonio, TX.

[14] Thomas Abrams, Director of Drug Marketing, Advertising and Communications, U.S. Food and Drug Administration, letter to Frank Baldino, CEO of Cephalon, Inc., re NDA # 20-717, MACMIS # 14707, released by FDA 27 February 2007; posted 1 March 2007, available online at http://www.fda.gov/foi/warning_letters/archive/g6270d.pdf. Accessed 21 March 2008.

[15] Cigna HealthCare Coverage Position, Modafinil, Coverage Position Number 4031, 1 August 2005.

excluded jet lag or use by pilots. The degree to which Cigna can implement this coverage policy depends on the coding decisions of physicians and monitoring procedures of the Cigna health system. Clearly, many wanting to use the drug to stay alert on their trip to Athens could obtain a physician's prescription and pay for the drug out of pocket. Some could seek reimbursement through an employer-based reimbursement account (e.g., a flexible spending account) because such plans generally do not distinguish one kind of use from another, but cover all prescription drugs (and some plans also cover over-the-counter drugs).

These same general points have been made regarding other cases—from Prozac to Viagra and human growth hormone—in which drugs first developed as treatment for diseases have come to be used to improve normal functioning (Elliott and Kramer, 2003). A consideration of such examples dispels the notion that distinctions between medical and nonmedical uses are bright lines (Allen and Fost, 1990; Rose, 2002; Gerlai, 2003). The point is that enhancement is apt to grow from the outer margins of clinical practice. Indications expand not by a process of deliberate policy-making, but rather as a result of decisions made about individual products and services that become available and then disseminate into practice (Chatterjee, 2004; Dees, 2004). Stephen Hauser places great emphasis on the responsibilities facing the next generation, noting that neurologists: "must be absolutely honest brokers if we hope to be major players in the coming debates on the appropriate uses of biosciences" (Hauser, 2004, 949).

4.11 Clinical Scenarios for Genetic Enhancement of Memory

To illustrate the policy choices likely to emerge if genetic enhancement is considered for brain function, and building on the model of the incremental expansion of use of existing technologies, we examine three clinical scenarios:

1. Patients demonstrating the frank pathology of the Alzheimer's type.
2. Persons at risk for neurodegenerative disease of the Alzheimer's type but asymptomatic or with only mild symptoms.
3. Persons not apparently at risk (i.e., "normal") who desire enhancement of cognitive function.

The use of genetic modification to address these scenarios is briefly summarized below, followed by a more extensive review of the health policy considerations for treating Alzheimer's disease. This lays the groundwork for coming back to consider the role of bioethics in formulating policies that might guide the emergence of genetic enhancements of memory that develop via treatment of memory-impairing diseases.

4.12 Patients with Frank Symptoms of Alzheimer's Disease

Here the need for genetic intervention arises either if no alternative to somatic cell or germ line therapy exists or if genetic intervention promises to be more effective than the alternatives, has fewer adverse effects, or is lower cost in the long term (all other things equal).

The possibility of somatic cell transfer or germ line therapy raises two questions about the autonomy of decision-making and how dependent it is on level of impairment. One purpose of the intervention is to improve memory and capacity for reasoning and thus to promote self-determination. In frank dementia, most decisions will be made by others on behalf of the patient or research participant. Informed consent for clinical research is therefore more complex than when those being studied can give consent for themselves. This is clearly an area in which bioethics can make its contribution through research ethics.

If an intervention appears to be effective, questions about access, cost, quality of care, and fair distribution then arise. In the face of symptoms of Alzheimer's disease, the intervention would be intended to reverse or ameliorate symptoms or slow deterioration. This scenario thus sweeps in the fewest people and presents the most compelling case for using collective resources among our three scenarios. Another aspect of distributive justice relates to age, because most (although not all) of those with dementia are over 65. Issues of intergenerational transfer[16] of wealth to cover such treatments then arise, because most people of that age group in the United States will be enrolled in the Medicare program and thus have a variable portion their medical services financed by current workers.[17]

---

[16] The intergenerational transfer includes elderly people using their savings to pay for these drugs, leaving less in estates to pass along to children or grandchildren. It is also about whether adult children feel obligated to help cover the costs, which (see next footnote) may entail substantial out-of-pocket expenditures, rather than being free to spend on themselves or on their own children.

[17] The advent of the Medicare Part D benefit, covering outpatient pharmaceuticals, has changed the dynamics of drug coverage for those enrolled in this program (or who are also covered by Medicaid as "dually eligible" beneficiaries. At the risk of oversimplification, some Medicare beneficiaries are still covered for outpatient drugs through retirement plans or some types of Medigap (insurance to supplement Medicare) policies, but many who are not enroll in a Part D plan; the dually eligible were randomly assigned to Part D plans, with mixed effects so far (probably a net loss for those in states with "richer" Medicaid benefits). Of Part D enrollees, most pay a monthly premium (unless covered through premiums paid to a managed care organization that offers outpatient drug coverage). If, in 1 year, they hit a certain level of expenditure, they then face significant out-of-pocket costs (the so-called donut hole), before Part D coverage returns in that year. More information is available at http://www.medicare.gov/medicarereform/drugbenefit.asp, but knowing whether any individual plan might cover these particular medications is a difficult matter to learn (and can change each year).
With respect to antidementia drugs, all Medicare drug plans are required to cover and have at least two cholinesterase inhibitors and memantine on their formularies (see www.alz.org/national/documents/MedicareRX_Chooseplan.pdf). Each plan can pick the cholinesterase inhibitors to cover and decide on the co-payments (which can differ by drug).

Once short-term safety and efficacy for amelioration of dementia and improvement of short-term memory are generally established, questions of long-term safety (neurotoxic effects) and unknown long-term behavioral effects (memory, but also unknown cognitive, behavioral, or even personality effects) then emerge. The risk-benefit profile pits the potential harms of the intervention (e.g., the drugs) against the expected benefits of slowing the progressive deterioration of brain function. In the face of a severe dementing condition that is nearly certain to ensue without intervention, some risk is reasonable, even if others must make the call on behalf of a cognitively impaired person.

4.13 Patients Who Are at Risk but Asymptomatic

Here, the need for the intervention is less certain than in fully expressed disease. One theoretical reason for intervention is to slow deterioration of brain function so disease never appears. This is a plausible preventive rationale, although it is not "treatment" of frank symptoms. Mild cognitive impairment also raises questions about competence to make decisions, and informed consent to participate in research, although these are less prominent than for frank dementia.

The number of people with mild cognitive impairment who have a higher likelihood of developing Alzheimer's dementia is many times higher than the number who have an established diagnosis. For that reason, the market for such treatments is correspondingly larger, as are potential expenditures for prevention or symptom amelioration. These cost implications mean that the stakes for cost containment are high, but this scenario also starkly illustrates the potential balancing of benefits and costs, because many individuals might well choose a preventive measure with even marginal prospect of clinical benefit to avoid dementia. Issues of insurance coverage again arise. The burden here would fall less on Medicare and more on private health plans, employer-based health plans, Medicaid, and out-of-pocket expenditures because more people would be younger than the 65-year criterion for Medicare eligibility.

Should policy attempt to prevent individuals from seeking such interventions when the probability that they will develop symptoms is low? To what extent should such individual decisions, analogous to use of contraceptives or drugs for erectile dysfunction or baldness, draw on collective resources through insurance benefits, health plans, or government health programs? Policy will involve decisions about what technologies are available at all, as well as who pays for them. Treatments on the clinical margin, such as expensive drugs for subclinical syndromes including mild cognitive impairment, are likely to be difficult to resolve.

The intended medical outcome of genetic intervention in those with mild cognitive impairment would be reduction in number of Alzheimer's cases by preventing or slowing progression from mild loss of brain function to full-blown dementia. Many more people would be eligible if the criterion for use were mild cognitive impairment rather than Alzheimer's disease. Cost-effectiveness hinges on the likelihood of preventing progression to Alzheimer's disease. Coverage could prove costly because the number of people qualifying for the intervention is many times higher than the number with dementia, and the duration of treatment for an ongoing treatment would be longer. If the treatment were a one-time genetic intervention, however, cost per person might remain constant, and only the number eligible would expand.

4.14 Persons Who Have Normal Cognitive Functioning or Memory and Who Desire Improvements

This scenario is enhancement by definition. The issues raised with respect to persons at risk but asymptomatic apply here as well. However, sensitivity to risk is further heightened, and safety requirements are likely to be increased. The number of people who might use an intervention is very large, because it could include all those without contraindications, and any side effects would be balanced only by enhancement benefits, not amelioration of clear disease. Safety is therefore paramount. Those seeking the intervention would be, by definition, not impaired. Issues of substitute consent would thus not be relevant.

Fairness comes to the foreground with such a broad range of possible users. Wide use of genetic enhancements if paid for through collective means such as health insurance or government health programs would divert health resources away from uses for treating frank disease. The intervention could, however, enhance other capacities or nonmedical outcomes, such as ability to reason, capacity for attachments, and respect for self and others that are considered in some theories of social justice (Sen, 2000; Powers and Faden, 2006). If the intervention has such effects, however, then it raises general issues of social justice, because "haves" can purchase interventions that "have-nots" cannot afford (and are not likely to gain access to through public insurance programs).

Even in a broad framework of justice in health such as that advanced by Powers and Faden (2006), an enhancement would by definition be above the "sufficiency" threshold, so justifying use of public resources would be difficult. Under a modified Rawlsian framework as elucidated by Daniels (1985), such enhancement would not be an effort to restore normal functioning and would therefore not have standing to draw on public resources for purposes of achieving justice in health care.

If such use of genetic modification for enhancement purposes were introduced only after it were already used for disease treatment or to forestall progression from mild cognitive impairment to dementia, then the genetic intervention would have presumably been already tested for safety, and there would also be a base of experience in people treated for those purposes on which to base estimates of safety. That track record, albeit imperfect, would help assess safety and prospects of efficacy.

4.14.1 Adoption of Medical Technology: Where Enhancement Meets Health Policy

The scenarios above illustrate how genetic treatment for memory loss from dementia caused by disease could be linked to prospects for enhancement of normal human memory. Once such technologies are available, we will confront questions of coverage and reimbursement, distributive justice, and "how far should we go?" These are policy questions.

We proceed now to review the kinds of decisions that must be made about regulation, coverage, and reimbursement of technologies, focusing on dementia, and specifically Alzheimer's disease. Alzheimer's disease is at present incurable, so therapies are directed at symptoms and support. Both pharmacologic and nonpharmacologic therapies are used. Nonpharmacologic treatments include interventions to lessen disruptive behavioral disturbances, training caregivers to deal with symptoms, and social stimulation and cuing that may account for some of the cognitive impairment.[18] In addition, some pharmaceuticals are directed at symptoms or comorbid conditions (e.g., psychotropic drugs for accompanying depression) (OTA, 1990).

For the purposes of this chapter, however, we do not consider such treatments, but focus on interventions intended to remedy dementia itself, or the underlying death of nerve cells, because they are a closer analogy to genetic modification treatments that could subsequently morph into enhancements. We review the health policy considerations of Alzheimer's disease because they illustrate how many normative issues are embedded in them, even in the framework of conventional treatment for a known disease. Our hypothesis is that enhancements would add a further layer of normative issues to existing health policy issues, which are likely to be inextricably connected to health policy as the enhancement technologies evolve from medical treatments.

Pharmaceutical interventions specifically indicated for dementia—i.e., antidementia drugs—include cholinesterase inhibitors and N-methyl-D-aspartate (NMDA) receptor antagonists (Sauer et al., 2007).[19] Cholinesterase inhibitors increase the duration of action and concentration of a particular neurotransmitter; they include donepezil (Aricept), galantamine (originally Reminyl, later renamed Razadyne), rivastigmine (Exelon), and tacrine (Cognex) (which is rarely used today because of toxicity, particularly potential liver damage). NMDA is another neurotransmitter, whose action is blocked by memantine (Namenda; Axura), the only

---

[18] See, e.g., www.alz.org/alzheimers_disease_behavioral_symptoms_ad.asp#Non-drug_interventions

[19] See also www.alz.org/AboutAD/Treatment/Standard.asp. Cholinesterase inhibitors temporarily boost levels of acetylcholine, a messenger chemical that becomes deficient in the brains of Alzheimer's disease patients. NMDA regulates glutamate, one of the brain's specialized messenger chemicals involved in information processing, storage, and retrieval. Excess glutamate overstimulates NMDA receptors to allow too much calcium into nerve cells, leading to disruption and death of cells. Memantine may protect cells against excess glutamate by partially blocking NMDA receptors.

NMDA antagonist approved for use in Alzheimer's dementia (and only for those with moderate to severe disease, i.e., not mild Alzheimer's disease or mild cognitive impairment). Various placebo-controlled trials and systematic reviews of these trials provide fair evidence that they are efficacious (in clinical trials) and effective (in clinical practice) in slowing the rate of cognitive degeneration or stabilizing symptoms and behavior by slowing functional decline, at least in patients with mild to moderate Alzheimer's disease.

The rate of adverse events for these agents varies by drug. Harms relate chiefly to gastrointestinal side effects (e.g., nausea, vomiting, diarrhea) and related loss of body weight. Other problems may involve psychological effects, hepatotoxicity, and hematologic effects; more theoretical concerns involve respiratory failure and muscle weakness (Sauer et al., 2007). Effects of these medications do not seem to vary consistently by age, sex, or other demographic or social factors.

Overall, direct evidence (head-to-head trials) or indirect evidence (from placebo-controlled trials) is still insufficient to permit clinicians, patients, or policy-makers to draw robust conclusions about comparative efficacy, effectiveness, or safety of these pharmaceutical agents. There is ample room for improvement in drug treatments for Alzheimer's dementia.

Many lines of research are being pursued to improve treatment or prevent progression of Alzheimer's disease. The pathways that lead to deposition of fiber-like proteins, both within cells and between cells in the brain, are under intensive study. Treatments could include inhibiting the deposition process or breaking down and clearing the proteins once deposited. Many chemical-receptor systems in the brain are also being studied in hopes that symptomatic relief of memory loss and other symptoms might be achieved. Finally, as noted above, cell therapies to introduce growth factors into specific locations in the brain are under investigation, including DNA transfer into cells that are then engrafted into the brain.

4.15 Policy Issues

In considering health care issues of interventions that are already on the market, the chief concerns are with access, quality, and cost of these services. We address these first, and then turn to some cross-cutting issues. The discussion focuses chiefly on "antidementia drugs" introduced above (cholinesterase inhibitors and, more particularly, NMDA).

4.16 Access to Antidementia Drugs and Interventions

Access refers to the ability of individuals or populations to obtain appropriate and needed health care from appropriate clinicians in a timely way (Millman, 1993); implicit in this definition is the idea that patients do not face undue burdens in seeking or acquiring such care (President's Commission, 1983). It is typically considered in terms of factors such as income, race and ethnicity, age, and insurance status (and the nature of benefit plans). Access to care (and thus to the positive outcomes of care) is well understood to be highly variable in the United States, as documented in the National Health Disparities Reports by large variations and inequalities in use of services and common health status measures across population groups (AHRQ, 2004, 2005a).

Access questions raise particular alerts insofar as the U.S. population is aging. Perhaps more germane are the patterns that may arise as the baby-boomer generation reaches ages at which dementia incidence begins to rise precipitously and as their demands for health care place perhaps unsustainable pressure on Medicare, especially with the addition of the Part D drug benefit. Added to this specific development are other factors such as direct-to-consumer (DTC) advertising on the part of the pharmaceutical industry and the rapidly rising amount of information available on the Internet—both of which can be expected to increase demand for health care, and pharmaceuticals in particular. DTC advertising presents special concerns for antidementia drugs because of possible bias in favor of information about putative benefits (often unclear as to how long they will be present) and against information about harms. Apart from influences on retirees, these same factors affect information seeking and action on the part of adult children who may need information on options for their parents.

As use of these medications moves toward enhancement, and especially if such use were not covered by insurance, some problems of access would arise. Assuming one sees enhancement uses as discretionary, questions such as timeliness, burden posed by geographical distance to providers, and similar barriers may fade. In addition, in principle autonomy is not breached if the decision is voluntary, informed, and a matter of preference. Whether justice is served if some have access to enhancements while others do not has long been a core question in the literature on the ethics of enhancement and raises complex issues that cannot be pursued here.[20]

4.17 Quality of Care

Quality of care is "the degree to which health services for individuals and populations increase the likelihood of desired health outcomes and are consistent with current professional knowledge" (Lohr, 1990, 4). Quality concerns the structural components of health care delivery (e.g., characteristics of facilities and qualifications of professionals), processes of care (e.g., what is done to and on behalf of patients) and outcomes or end results of that care. Issues of the benefits and harms—and their relative balance—thus come into play in considering quality of

---

[20] For an extensive examination of issues of distributive justice raised by the prospects of genetic interventions in humans, see Buchanan et al., (2000).

care and patient safety. "The right care, at the right time" is another hallmark of high-quality care.

Exactly how Medicare, state Medicaid agencies, or private sector health plans or institutions might track and evaluate the quality of care relating to the use of these drugs is not clear (as of mid-2007). More than 30 authoritative guidelines for managing patients with Alzheimer's disease (or dementia, often related to other conditions) can be found.[21] Only three deal specifically with prescription medications for dementia (cholinesterase inhibitors); none apparently covers memantine. Two authoritative sets of quality measures are available.[22] For Medicare beneficiaries, some review of medication use may fall to the Quality Improvement Organizations (QIOs) charged with oversight of quality under Medicare. At some point, QIO responsibilities may extend to the new Part D drug plan. Given that Part D has not reached steady state, relying on QIOs to monitor and oversee the quality of care for Alzheimer's disease patients, at least as regards medication use, lies well in the future.

A significant quality issue may arise as to who can make the best judgments about the balance of benefits and risks of antidementia drugs. Many clinicians and patients (and their families) subscribe to varying degrees of shared or informed decision making—initiatives that reflect principles of beneficence and respect and take patient autonomy (e.g., self-determination) at least indirectly into account. In such an approach, health professionals can provide comprehensive information on both expected outcomes and risks, and patients or families can express their preferences about those outcomes and risks, so that a mutually satisfactory decision can be made across all feasible options.

These pharmaceuticals may present especially knotty challenges to this desirable approach because of the scanty evidence base (especially for a broad range of patients that extends beyond those included in the initial trials) and because of the fear and other emotions that accompany the diagnosis of Alzheimer's disease (indeed, of any incurable ailment). For those reasons, the quality of dementia care may be fraught with more than usual problems of measurement and improvement.

Moreover, to the extent that what *can* be done in monitoring and improving quality of care relates to patients with established diagnoses, the problems multiply as use expands to other patient populations and, especially, to off-label use. Few if any authoritative guidelines exist for treatment of mild cognitive impairment *per se*. Even the label or diagnosis may vary depending on who is making it, using what signs, symptoms, or valid instruments. When uses move beyond disease or at-risk groups to those who are otherwise healthy and using these medications for memory or cognition enhancement, the challenges to ensuring high quality care become extraordinarily thorny. Presumably, along this spectrum, bioethical concerns related directly to quality of care would shift toward nonmaleficence (i.e., avoiding the harms of adverse events).

---

[21] See those documented in the National Guideline Clearinghouse at http://www.guideline.gov/.

[22] One concerns process of care and one, relating to registries, concerns the structural aspects of care; www.qualitymeasures.ahrq.gov/, accessed for this purpose 12 February 2007. See also AHRQ (2005b).

4.18 Costs of Care

Costs can be both direct medical costs and indirect costs (e.g., loss of income for patients or caregivers). Value—i.e., quality in the context of costs—is increasingly invoked as a core goal for health care, especially relevant for purchasers and policymakers.

According to Consumer's Union, as of late 2006, the monthly costs were about $160 to $180 for the main cholinesterase drugs and about $150 for memantine. They recommend (as "best buys") donepezil and galantamine for patients with early Alzheimer's disease and memantine for those with middle-to-late stage disease (Consumer Reports, 2006).

To such costs would need to be added blood and laboratory tests that might be needed to monitor possible adverse effects (especially liver toxicity if tacrine is used). Set against direct costs, of course, would be any delays in institutionalization and those attendant offsets. Indirect costs—such as those incurred by family and caregivers who lose time from work or must hire additional help—might be presumed to be lower with access to and coverage for such drugs, at least in the short run, to the degree clinical stability or even improvement lessens demands for care.

Who bears these costs differs across patient populations. Those with good insurance coverage may face only copayments (coinsurance, deductibles); those with no or poor insurance may incur substantially higher medication costs, unless they are covered by Medicaid in a state that includes these types of medications for these indications. For the elderly (or dually eligible), the Medicare Part D has an average premium of about $32 per month for basic benefits and an annual $250 deductible that the beneficiary is responsible for paying. Between $251 and the initial coverage limit of $2,250, the Part D plan is responsible for 75% of costs and the beneficiary pays a 25% coinsurance. Part D provides no coverage between $2,251 ($3,600 in true out-of-pocket costs) and $5,100 (i.e., the "donut hole"). Once beneficiaries hit the $3,600 threshold, coverage is split in three ways, and they pay a 5% coinsurance, or copayments of $2 for generic drugs and $5 for nongeneric drugs (Greenwald, 2007).

Two other issues arise with respect to costs. The first is drug pricing. Most observers believe that costs of medications will drop, and access rise, once they go off patent. For all these drugs, various formulations have patents that extend to as near-term as 2008 and as far term as 2020.[23] Given that pharmaceutical firms often can extend their patents to cover new uses or new formulations (often either "me-too" or combination drugs), the impact that losing patent protection for antidementia drugs might have lies well in the future. Moreover, new drugs in these classes may come on line, *with* patent coverage, and if they appear to offer better (longer) benefits or fewer risks of harms, then physicians may shift their patients to these newer (presumably more costly) drugs. Of course, pharmaceutical firms have little reason to reduce prices directly. In addition, although some large health insurers or federal programs (e.g., the Veterans Health

---

[23] For detailed information, visit the "Electronic Orange Book" at www.fda.gov/cder/ob/default.htm, accessed 12 February 2007.

4 How Bioethics Can Inform Policy Decisions About Genetic Enhancement   187

Administration) can and do negotiate meaningful discounts, Medicare is currently precluded from negotiating such discounts. One might posit, therefore, that pricing will stay stable (i.e., flat) for some years, unless price competition occurs, driven by either introduction of new products or changes in the market that create incentives for reduced prices (such as competition to be listed on formularies of large buyers).

The second cost issue—the cost-effectiveness of these specific antidementia medications (against each other)—will come into play in the policy arena. (We are not here considering the cost-effectiveness of alternative drug therapies, such as these Alzheimer's disease agents vis-à-vis psychotropic medications or cognitive and behavioral counseling and interventions.) To examine these issues will require better effectiveness data in all types of patients than are available today as well as accurate cost data for comparable dosages (levels of activity or potency). Those data all remain to be amassed.

4.19   Cross-Cutting Policy Topics

4.19.1   *Efficacy and Effectiveness of Treatments*

Cutting across these topics are questions about the efficacy (in rigorous clinical research studies), effectiveness (in clinical practice), appropriateness, and necessity of services[24] For pharmaceuticals in particular, efficacy as documented in controlled trials or effectiveness as observed in demonstrations or studies of actual medical practice are critical elements in decisions about clinical practice, insurance coverage, and reimbursement. When multiple interventions have been shown to be effective—drugs in particular—then issues of costs and cost-effectiveness arise, such that policymakers, providers and others want to know whether medications have equivalent benefits and equivalent (presumably low) side-effect profiles, taking dosage (equipotency) into account. Similarly, they need to know whether benefits outweigh harms. If equivalency on these parameters can be assumed, then ceteris paribus, lower costs are better. When such information is available (which is relatively rarely, especially for comparatively new drugs), policymakers, health plan administrators, insurers, and others may take it strongly into account in establishing formularies.

---

[24] Efficacy refers to the outcomes of services for highly selected groups of patients provided in "ideal" settings from the "best" clinicians; effectiveness refers to outcomes in a broad range of patients provide in "average" or everyday settings from "average" clinicians. The studies to demonstrate these characteristics are, respectively, referred to as explanatory or pragmatic studies. An example of an explanatory study is a randomized clinical trial; an example of a pragmatic study is a population survey of routine clinical practice. Pragmatic studies are done in primary care or other ambulatory settings and permit enrollment of patients representing a broad set of population characteristics, severity of diseases, and comorbidity; they have longer follow-up periods; and they usually attempt to measure health outcomes of interest to patients and their families (including preferences), not simply surrogate or intermediate outcomes such as results of laboratory tests or rates of use of other health services. For more details see Gartlehner et al. (2006).

4.20 Relevance to Genetic Enhancement

As noted in the information reviewed above, little is known with certainty about Alzheimer's drugs. Genetic interventions would start from an even lower base of evidence. Genetic therapies might be attractive in requiring less frequent administration, but their unit costs might well be higher. The prices of cellular therapies (and the likely vehicle for genetic modifications would be transgenic human cells) cannot be known in advance, but such interventions would likely entail substantial costs for the initial remedy, which would involve placing cells into a patient's body under close medical supervision. Moreover, monitoring of treatment afterward will almost surely be required, at least until experience accumulates that suggests the treatments are both safe and effective.

Current drugs show very modest benefits in terms of improving memory or slowing cognitive decline, comparative effectiveness information is sparse, the quality of trials and systematic reviews about these pharmacotherapies is generally only fair, and costs (all still being on patent) are high. In short, much remains to be learned about pharmacological treatment for conditions such as dementia (Alzheimer's in particular), and there is little to no evidence about their impact on patients with less severe problems (e.g., mild cognitive impairment) or for persons who "simply" want to improve memory and cognition for reasons not necessarily related to health *per se*. Uncertainty over any new genetic interventions would layer on top of this already-high mountain of uncertainty concerning drug treatments.

4.21 Regulatory Issues

In the face of today's limited evidence base, many policy hurdles and questions remain ahead for introduction of any genetic interventions for dementia. The basic steps of bringing genetic therapies to market include moving through the FDA approval process for biologic and cellular treatments. Pharmaceutical firms must demonstrate that their drugs are efficacious (but not necessarily effective in actual clinical practice) and that they are safe; the latter idea is sometimes characterized as tolerability (measured in terms of reports of overall and serious adverse effects and withdrawals or attrition from trials because of adverse events). If approved for the indications for which they are studied, firms may then move the drugs to market and study them for new indications, to broaden the market.

In the case of dementia, the logical market expansion leads to testing efficacy in patients with mild cognitive impairment, because it expands the potential patient pool many-fold. Moving from mild cognitive impairment to testing humans for enhancement purposes would be yet another huge step, broadening the market to almost all people.

Regulation of brain enhancement technologies would face issues that accompany all technologies, but would be more intense with very broad use in healthy individuals. The first is the emergence of unanticipated adverse events or higher-than-expected known adverse events. Apart from straightforward physiologic effects are interactions with other drugs, with herbal and other alternative compounds, and with food. The occurrence of such events is expected to be monitored through post-marketing surveillance and adverse drug event reporting, but none of these steps is certain protection against direct harms or less benefit than originally claimed. The removal of some drugs from the market (and voluntary withdrawals do occur), and "black box" warnings on labels or bold letters warnings are continually being added to existing drugs. These steps, in theory, affect all users, whether for those with frank disease, those with milder forms, or those who simply want to enhance performance. Such individuals can then move to other therapeutic modalities or relinquish their quest for enhanced performance. The need for stringent safety monitoring would be particularly acute for enhancement genetic interventions used in healthy individuals.

The second issue is rising demand for off-label use. Once pharmaceuticals are approved for market with a given indication (even with complete, accurate, and clear statements of contraindications and possible adverse effects), physicians (or dentists, or others with prescription-writing privileges) are free to prescribe the medication for any purpose and for any patients. In the case of antidementia drugs, the possibility of expanded use for mild cognitive impairment, other cognitive problems, or even enhancement of memory or cognition in individuals with average (or even better than average) memory is clear. Some clinicians may balk at satisfying consumer demand in the most extreme "enhancement cases," a clear prerogative grounded in provider autonomy (if not also some element of paternalism), but in this country, somebody *will* write that prescription. Whether bioethics has anything to say *at this point* about genetic modifications to improve cognition and memory matters may be open for debate, but our point is that these challenges can be anticipated, and bioethicists can begin now to draw out and analyze the implications for the health sector and society at large.

4.22 Back to Bioethics and Religion

Our chapter has taken us far from where it started, when the heads of three large religious denominations sounded the alarm to a U.S. President about how commercial forces could propel genetic technologies into use before adequate public debate and assessment of the long-term consequences were taken into account. That concern was channeled into a focus on introducing DNA transfer experiments in humans, which became embodied in oversight at NIH and FDA.

In making decisions about introducing DNA transfer into humans, some other, much broader questions of "ethics," about which philosophy and religion have much to say, such as fairness in access, did not have to be addressed squarely because they were premature. In a debate about expansion of existing technologies into enhancement applications, however, fairness and social justice will be the battleground along with safety, efficacy, and informed consent; they will not be just in the background.

Engaging those questions is not likely to take shape as simple "either-or" decisions about genetic enhancement technologies *per se*. Rather, if enhancement technologies do grow organically from the outer margins of medical treatments as we predict, then there will be a role for broad public debate about the move from treatment to enhancement. There should be a process for including religious perspectives and explicit attention to ethical analysis, building on top of health policy decisions analogous to those for other medical technologies. This will pose several issues for bioethics.

For one thing, there will not be a sharp technological departure from current practice to signal the broader social stakes in the move from treatment to enhancement. The trigger for engaging bioethics commissions has often been a technological shock such as recombinant DNA, cloning a sheep, or developing human embryonic stem cell lines, or the prospect of new kinds of research such as inserting recombinant DNA into humans. Bioethicists are of course writing scholarly articles about enhancement prospects for their peers, but the jump discontinuity that often invokes a call for "public bioethics" in the form of a commission or special proceeding may not occur. Either the ongoing scholarship will inform policy, or some constituency or constituencies, perhaps bioethicists among them, will have to flag the need for normative analysis, and ensure the broader public debate.

We must consider whether the give and take (some might say push and shove) of ordinary democratic politics is adequate for responding to the challenges that the prospects of enhancement will present. One feature of health policy decision-making is that it is highly decentralized, takes place in many different kinds of institutions, and therefore is even more apt than many other kinds of public policy to drift without an explicit strategy or real "national policy," let alone international consensus. The fact that bioethics commissions have contributed to biotechnology policies in the past will be of little comfort if the political leaders who can authorize their creation mistakenly believe that they are useful only for new technologies and fail to see that they may play a valuable role in responding to the problem of new uses for existing technologies. A debate about genetic enhancements should involve constituencies beyond the direct stakeholders, entail explicit considerations of social policy well beyond safety and efficacy and technical feasibility, and engage difficult long-term questions about where enhancement applications are likely to lead us collectively.

Bioethics has successfully grappled with health policy in the past. For instance, the President's Commission issued *Securing Access to Health Care* more than 20 years ago. Some prominent moral philosophers have directly engaged questions of health policy. Norman Daniels, for example, has a body of work on justice in health care. His work is noteworthy in being grounded in moral theory and thoroughly immersed in empirical health services research. His 1985 book, *Just Health Care*, laid a theoretical foundation still in use, and his subsequent work has included analysis of how medical decisions could be justified in a system of accountability that entails transparency and reasonable, explicit standards and procedures (Daniels,

1985, 1996; Daniels and Sabin, 1998). Daniels has also directly addressed the question of enhancement very much along the lines followed in this chapter (Daniels, 1992).

The Department of Clinical Bioethics is the current name of a group at the National Institutes of Health Clinical Center that has been growing over the past decade, and many scholars and investigators there have wrestled with the intricacies of health policy.[25] Scholars have returned to the foundations of moral theory concerning health and health care. Madison Powers and Ruth Faden, for example, stipulate six factors that need to be sufficiently present to enable health (and elaborate how needs and capacities may change over a life span), a promising approach for addressing the inevitable questions of justice that will arise with enhancement technologies (Powers and Faden, 2006).

Other work in bioethics addresses the priorities of the research enterprise and contemplates enhancement genetics. Much of this work delves less deeply into the mechanics of the health care system, and thus it represents the more theoretical style of bioethics. Norman Daniels, Eric Juengst, Erik Parens, and Daniel Brock, among others, have addressed genetic interventions and enhancement (Daniels, 1992; Juengst, 1997; Parens, 1998; Juengst and Parens, 2003; Juengst et al., 2003). In addition, Walters and Palmer devoted a chapter of *The Ethics of Human Gene Therapy* (1997) to prospects for enhancement that combines history and theory.

The President's Council on Bioethics produced a report that directly engages the question of genetic enhancement at a particularly high level of abstraction from the nitty-gritty of health policy (President's Council, 2003). *Beyond Therapy: Biotechnology and the Pursuit of Happiness* gives us an opportunity to examine the value of bioethics in public discourse and in informing policy decisions. We focus in particular on policy, the central theme of this chapter. A contrast between *Beyond Therapy* and the President's Commission report of 1982, *Splicing Life*, is particularly instructive. It enables comparison, assessing a "public bioethics" commission with particular attention to policy impact.

Beyond Therapy took as its task "to clarify the relevant scientific possibilities and, especially, to explore the ethical and social implications of using biotechnical powers for purposes beyond therapy" (President's Council on Bioethics, 2003, p. xx). It addressed three "case studies," which were actually domains in which uses of biotechnologies might be used for enhancement: having "better babies," improving human performance, and reducing the ravages of aging. The analysis proceeds through the arguments for and against many uses, including some of the examples cited in this chapter. The report returns repeatedly to a few basic arguments: that technology is valuable, that it is limited in what it can do (particularly if it is expected to fulfill deep human needs by technological means), and that achieving technological ends does not necessarily bring happiness, fulfillment, or meaning. These points are illustrated through quotes ranging from Homer to Shakespeare to Jane Austen to Greek philosophers and playwrights. This aspect of the analysis—use of literature

---

[25] See http://www.bioethics.nih.gov (accessed 11 February 2006).

and philosophy and occasional reference to religious traditions—was also used by the President's Commission (1982), which referred to the myth of the Golem, Mary Shelley's *Frankenstein*, and a work that recurs frequently as a leitmotif in the President's Council's arguments, Aldous Huxley's dystopic *Brave New World*.

In contrast to the President's Council in 2003, The President's Commission in 1982 went beyond literary analysis to map government jurisdiction, stakeholders, and policy decisions bearing on genetic intervention. Although some general policy directions are implicit in the arguments, *Beyond Therapy* (2005) gives little attention to policy decisions; the President's Commission (1982) directly addressed them. The health policy concerns about cost, access, and quality are noted in a final chapter of *Beyond Therapy*, but only in passing and without a map of jurisdictions, stakeholders, a decision tree, or any arguments about how the President's Council's normative analysis might inform policy decisions.

As our previous section argues, the chief arbiters of which technologies might become available for enhancement purposes by expansion from medical use might well be health care payers, particularly government health programs (e.g., Medicare, Medicaid, the Veterans Health Administration, Indian Health Service, and Federal Employee Health Benefits Program). Research on uses would likely be funded by the National Institutes of Health or the Agency for Healthcare Research and Quality. One principal gatekeeper for which technologies are available for use will likely be the Food and Drug Administration. These are all units of government under the President's jurisdiction. The President's Council gives only a hint of direction based on literary analysis, leavened with a final chapter that addresses American political philosophy. The Council's main client, the U.S. President, might want a better analysis of the decisions he faces.

The President's Council report is primarily a hortatory document that cautions against overly optimistic adoption of technologies, but it lacks principles for putting on the brakes should those making policy decisions choose to do so. The line of argument consistently suggests that American society has a lead foot when it comes to technology, and that the role of careful deliberation is to keep one foot close to the brake pedal. That is, *Beyond Therapy* is an exhortation to think through long-term consequences, collective impacts of individual decisions, and the downside of the technological imperative. The "cashing out" in the final chapter relies extensively on theories of "the natural" echoed throughout the present volume, with all the strengths and weaknesses of "natural law" approaches. The purpose of that final chapter seems to be an effort to meld natural law cautions into American political theory, rather than to map political choices.

As a policy document, the contrast between the 2003 President's Council and the 1982 President's Commission is stark. *Beyond Therapy* did not purport, however, to address policy concerns, only to air publicly the arguments about enhancement. *Beyond Therapy* concludes that the most likely pathway to enhancement is by extension from medical practice, but the argument ends only in rhetorical cautions, rather than policy options.

Is it fair to criticize the absence of policy content in a report from the President's Council on Bioethics? We believe it is. Because the pathway to enhancement will

4 How Bioethics Can Inform Policy Decisions About Genetic Enhancement

be through health policy, it is incumbent on a deliberative body advising the President of the United States to do homework on the pathway by which enhancement might become policy, and to identify the policy decisions that could influence that process.[26] The health policy decisions that could lead to enhancement uses are foreseeable, and the decisions and criteria to guide them could be specified in a document of more than 300 pages.

Although the examples and technical reviews in *Beyond Therapy* are recent, the basic arguments are the natural law arguments that J.B.S. Haldane rejected 80 years ago, and that he would certainly reject again today were he alive. Haldane celebrated Daedalus's commitment to experiment despite the concerns of the gods. This is where the President's Council also leaves off, with a discussion of the concepts of humility and hubris, arguments that most scientists and technologists will reject outright. That is, the arguments in *Beyond Therapy* are persuasive only to those who regard "the natural" as a normative guide.

For those whose careers and belief systems are focused on technological innovation that deviates from the past, or devoted to science that expands the frontiers of knowledge, those arguments are unpersuasive. The coda section of *Beyond Therapy* interprets the principles of "life, liberty and the pursuit of happiness" as American values. It presents arguments for caution against wholesale adoption of enhancing biotechnologies, and asserts that caution in adoption of technology could comport with American political philosophy. The connections between American values and technological caution, however, are not easy to follow. U.S. history suggests that one salient feature of American culture is low barriers to new technologies and high enthusiasm about using them. The technological caution of *Beyond Therapy* is therefore swimming upstream—a point the report concedes, but fails to argue for persuasively.

As a policy document, the 1982 President's Commission report *Splicing Life* had obvious impact, and thus demonstrable utility, by locating a place where decisions were being made and offering some ideas about what decisions might be contemplated by the Recombinant DNA Advisory Committee and the Food and Drug Administration. *Beyond Therapy* is obviously not intended to have such direct policy impact, but instead to articulate a set of arguments about limits of biotechnology. As previous sections clearly indicate, we believe this is selling bioethics short, leaving it in the role of making natural law arguments devoid of policy context. A presidential bioethics commission advises the head of government. It would seem logical to address governance. We believe bioethics would serve a much more useful service if it had higher aspirations and aimed to inform policy decisions with sufficient precision to enable policy change, rather than recycling natural law arguments without a policy map.

The center of mass in bioethics, however, has remained far from mainstream health policy. With some exceptions noted above, bioethics has generally been content

---

[26] Note that these policy decisions need not be designed to halt or slow new technologies or their uses. In fact, a President's Council could make policy recommendations that are designed to minimize bureaucratic incursions where decisions would be better left to doctors and their patients.

with airy arguments about oughts and shoulds, and this is a fine way to start. In our view, it is not a good place to stop. *Beyond Therapy* is hardly alone in its style and substance. Our main point is that policies that influence the introduction of genetic interventions for enhancement are likely to grow as offshoots of medical care. Ethical and religious perspectives on the technology are likely to influence policy—the actual decisions about specific uses of technology—in part to the degree they demonstrate knowledge of how medical technologies are introduced, adopted, and diffused. If bioethics truly wishes to engage questions about the ethics of genetic enhancement, the most effective means to do so will of course include theoretical treatments of oughts and shoulds. However, that analysis will make little difference if it floats free of policy analysis. Rather, the community of scholars will need hands-on engagement with the policy levers that influence what technologies are approved for use, covered by health plans, and used in clinical practice, in addition to abstract normative concerns that are already a staple of bioethics.

Ours is a call to engagement in policy analysis, based on optimism that bioethics has much to contribute. It is not a rejection of the "big picture" concerns. Recall that a letter from religious leaders that was highly diffuse and raised inchoate concerns rather than precisely formulated problems nonetheless led to a process for introduction of DNA transfer into humans that successfully drew on diverse constituencies and raised questions that would otherwise not have been part of the policy process. But those "airy" concerns raised by religious leaders were translated into policy content through a presidential commission. Bioethics served three purposes: to raise concerns, to articulate arguments, and also to lay out a policy map. The third goal is as important as the first two, and normative analysis is relevant to both public discourse and making policy choices. Normative questions matter, and the role of bioethics can be both to broaden the debate into constituencies that might otherwise be left out, and to ensure that normative issues are explicitly addressed when policy decisions are being made.

Acknowledgments The authors wish to thank Matthew DeCamp and Allen Buchanan of Duke University, as well as Judy Illes of Stanford University, for their helpful comments. The work has been supported in part by a grant from the Ford Foundation to Rice University and Davidson College, by a P50 Center grant to Duke University from the National Human Genome Research Institute and U.S. Department of Energy, and by RTI International. Andrew Lustig and Baruch Brody convened the group and organized the overall efforts. Lisa Rasmussen edited this chapter.

References

Agency for Healthcare Research and Quality (AHRQ) (2004). *National Healthcare Disparities Report: Summary*. Rockville, MD: Agency for Healthcare Research and Quality. Available at http://www.ahrq.gov/qual/nhdr03/nhdrsum03.htm (accessed for this purpose 4 September 2005).

Agency for Healthcare Research and Quality (AHRQ) (2005a). *National Healthcare Disparities Report*. Rockville, MD: Agency for Healthcare Research and Quality. Available at http://www.ahrq.gov/qual/nhdr05/nhdr05.htm (accessed 12 February 2007).

Agency for Healthcare Research and Quality (AHRQ) (2005b). *National Guidelines Clearinghouse.* Available at: http://www.guideline.gov/search/searchresults.aspx?Type = 3andtxtSearch = alzheimer%27s + diseaseandnum = 20 (accessed 4 September 2005).

Allen, David B., and Norman C. Fost (1990). "Growth Hormone for Short Stature: Panacea or Pandora's box?" *Journal of Pediatrics* 117, 16–21.

Avery, Oswald T., Colin M. McCleod, and Maclyn McCarty (1944). "Studies on the Chemical Nature of the Substance Inducing Transformation of Pneumococcal Types," *Journal of Experimental Medicine* 79, 137–158.

Berg, Paul, and Maxine Singer (1995). "The Recombinant DNA Controversy 20 Years Later," *Proceedings of the National Academy of Sciences (USA)* 92, 9011–9013.

Buchanan, Allen, Dan W. Brock, Norman Daniels, and Daniel Wikler (2000). *From Chance to Choice: Genetics and Justice.* New York: Cambridge University Press.

Chatterjee, Anjan (2004). "Cosmetic Neurology: The Controversy over Enhancing Movement, Mentation and Mood," *Neurology* 63, 968–974.

Chapman, Audrey R., and Mark S. Frankel (eds.) (2003). *Designing our Descendants: Promises and Perils of Genetic Modification.* Baltimore, MD: The Johns Hopkins University Press.

Cohen, Stanley N., Annie C.Y. Chang, Herbert W. Boyer, and Robert B. Helling (1973). "Construction of Biologically Functional Bacterial Plasmids *In Vitro*," *Proceedings of the National Academy of Sciences (USA)* 70(November), 3240–3244.

Consumer Reports (2006). *Evaluating Prescription Drugs Used to Treat: Alzheimer's Disease—Comparing Effectiveness, Safety and Price.* Consumer Report Best Buy Drugs, www.crbestbuydrugs.org/PDFs/costupdates/AlzfheimersDrugs-CostUpdate-Nov2006.pdf (accessed 12 February 2007).

Daniels, Norman (1985). *Just Health Care.* Cambridge: Cambridge University Press.

Daniels, Norman (1992). "Growth Hormone Therapy for Short Stature: Can We Support the Treatment/Enhancement Distinction?" *Growth: Genetics and Hormones* 8(Supplement 1), 46–48.

Daniels, Norman (1996). *Justice and Justification: Reflective Equilibrium in Theory and Practice.* Cambridge: Cambridge University Press.

Daniels, Norman, and James Sabin (1998). "The Ethics of Accountability in Managed Care," *Health Affairs* 17(5), 50–64.

Dees, Robert H. (2004). "Slippery Slopes, Wonder Drugs and Cosmetic Neurology," *Neurology* 63, 951–952.

Elliott, Carl, and Peter Kramer (2003). *Better than Well: American Medicine Meets the American Dream.* New York: W.W. Norton.

Farah, Martha J., Judy Illes, Robert Cook-Deegan, Howard Gardner, Eric Kandel, Patricia King, Eric Parens, Barbara Sahakian, and Paul Root Wolpe (2004). "Neurocognitive Enhancement: What Can We Do and What Should We Do?" *Nature Reviews Neuroscience* 5(May), 421–425.

Food and Drug Administration (FDA) (2000). "U.S. Department of Health and Human Services, New Initiatives to Protect Participants in Gene Therapy Trials." Press Release of 7 March 2000. Available at http://grants2.nih.gov/grants/policy/gene_therapy_20000307.htm (accessed 23 September 2007).

Food and Drug Administration (FDA) (2001). "U.S. Department of Health and Human Services, Availability for Public Disclosure and Submission to FDA for Public Disclosure of Certain Data and Information Related to Human Gene Therapy or Xenotransplantation," *Federal Register* 66(12): 4688–4706. Available at http://www.fda.gov/cber/rules/frgene011801.pdf (accessed 23 September 2007).

Food and Drug Administration (FDA) (2002). "Gene Therapy Patient Tracking System, Final Document." U.S. Department of Health and Human Services. Available at http://www.fda.gov/cber/genetherapy/gttrack.pdf (accessed 23 September 2007).

Food and Drug Administration (FDA) (2003). "FDA Advisory Committee Discusses Steps for Potentially Continuing Certain Gene Therapy Trials That Were Recently Placed on Hold," FDA Talk Paper T03-16, 28 February 2003. Available at http://www.fda.gov/bbs/topics/ANSWERS/2003/ANS01202.html (accessed 23 September 2007).

Food and Drug Administration (FDA) (2004). "FDA Approved Labeling Text for NDA 20-717/S-005 andS-008 (Approved-23-JAN-2004)," PROVIGIL (modafinil) Tablets (C-IV). Available at http://www.fda.gov/cder/foi/label/2004/20717se1-008_provigil_lbl.pdf (accessed 23 September 2007).

Frankel, Mark S., and Audrey R. Chapman (eds.) (2000). *Human Inheritable Genetic Modifications: Assessing Scientific, Ethical, Religious and Policy Issues.* Washington, DC: American Association for the Advancement of Science.

Fredrickson, Donald S. (2001). *The Recombinant DNA Controversy: A Memoir, Science, Politics, and the Public Interest 1974–1981.* Washington, DC: ASM (American Society for Microbiology).

Gamov, George (1954). "Possible Relation between Deoxyribonucleic Acid and Protein," *Nature* 173, 318.

Gartlehner, G., Hansen, R. A., Nissman, D., Lohr, H.N., and Carey, T. S. (2006). "A Simple and Valid Tool Distinguished Efficacy from Effectiveness Studies." *Journal of Clinical Epidemiology* 59(10), 1040–1048.

Gerlai, Robert. (2003). "Memory Enhancement: The Progress and Our Fears," *Genes, Brain, and Behavior* 2, 187–190.

Greenwald, Leslie M. (2007). "Medicare Part D Data: Major Changes on the Horizon," *Medical Care* (Supplement 2), S9–S12.

Haldane, John Burdon Sanderson Haldane (1923). *Daedalus, or Science and the Future*, an essay read to the Society of Heretics, Cambridge, England, 4 February 1923; reprinted in Krishna R. Dronamraju (ed.), *Haldane's Daedalus Revisited.* Oxford: Oxford University Press, 1995, 23–51. Also available online at http://www.cscs.umich.edu/~crshalizi/Daedalus.html (accessed 19 September 2007).

Hansen, Richard A., Gerald Gartlehner, Daniel Kaufer, Kathleen Lohr, and Timothy Carey (2006). "Drug Class Review of Alzheimer's Drugs." Final Report Update 1. Portland, OR: Oregon Health and Science University 2006. Available at www.ohsu.edu/drugeffectiveness/reports/documents/Alzheimer%20Final%20Report%20Update%201.pdf. Accessed for this purpose 12 February 2007.

Hauser, Stephen (2004). "The Shape of Things to Come," *Neurology* 63, 948–950.

House of Representatives, U.S. Congress (1982). Hearings on Human Genetic Engineering Before the Subcommittee on Investigations and Oversight, Committee on Science and Technology, 97th Congress, Committee Report No. 170. Washington, DC: Government Printing Office.

Hyman, Steven E. (2006). "Improving our Brains." *BioSocieties* 1, 103–111.

Juengst, Eric, Robert Binstock, Maxwell Mehlman, Stephen Post, and Peter Whitehouse (2003). "Biogerontology, 'Anti-aging Medicine' and the Challenges of Human Enhancement," *The Hastings Center Report* 33(4), 21–30.

Juengst, Eric, and Erik Parens (2003). "Germ-line Dancing: Definitional Considerations for Policymakers," in A. Chapman and M. Frankel (eds.), *Designing Our Descendants: The Promises and Perils of Genetic Modifications.* Baltimore, MD: The Johns Hopkins University Press, 20–39.

Juengst, Eric (1997). "Can Enhancement Be Distinguished from Prevention in Genetic Medicine?" *Journal of Medicine and Philosophy* 22, 125–142.

Kevles, Daniel J. (1985). *In the Name of Eugenics: Genetics and the Uses of Human Heredity.* New York: Knopf.

Krimsky, Sheldon (1982). *Genetic Alchemy: The Social History of the Recombinant DNA Controversy.* Cambridge, MA: MIT.

Lederberg, Joshua (1994). "The Transformation of Genetics by DNA: An Anniversary Celebration of Avery, MacLeod and Mccarty (1944)," *Genetics* 136 (February), 423–426.

Lohr, Kathleen N. (ed.) (1990). *Medicare: A Strategy for Quality Assurance.* Volume I. Institute of Medicine. Washington, DC: National Academy Press.

Lyon, Jeff, and Peter Gorner (1995). *Altered Fates: Gene Therapy and the Retooling of Human Life.* New York: W.W. Norton.

4 How Bioethics Can Inform Policy Decisions About Genetic Enhancement

Millman, Michael (ed.) (1993). *Access to Health Care in America*. Institute of Medicine. Washington, DC: National Academy Press.

Müller, Ulrich, Nikolai Steffenhagen, Ralf Regenthal, and Peter Bublak (2004). "Effects of Modafinil on Working Memory Processes in Humans," *Psychopharmacology* 177, 161–169.

Neergard, Lauran (AP) (2005). "Alzheimer's Gene Therapy Success," as reported on CBS Newswatch, 25 April.

O'Connor, Anahad (2004). "Wakefulness Finds a Powerful Ally," *The New York Times*, 29 June.

Office of Biotechnology Activities (OBA) (2001). National Institutes of Health, U.S. Department of Health and Human Services, Recombinant DNA Research: Actions Under the NIH Guidelines, *Federal Register* 66(223) (19 November 2001), 57970–57977. Available at http://www4.od.nih.gov/oba/rac/frproactions/11–01pro.pdf (accessed 23 September 2007).

Office of Technology Assessment (OTA) (1984a). *Commercial Biotechnology: An International Analysis*. Washington, DC: Government Printing Office. U.S. Congress, Office of Technology Assessment, OTA-BA-218. Available at http://www.princeton.edu/ ota/disk3/1984/8407/840701. PDF (accessed 23 September 2007).

Office of Technology Assessment (OTA) (1984b). *Human Gene Therapy—A Background Paper*. Washington, DC: Government Printing Office, U.S. Congress, Office of Technology Assessment, OTA-BP-BA-32.

Office of Technology Assessment (OTA) (1990). *Confused Minds, Burdened Families: Finding Help for People with Alzheimer's and Other Dementias*. Washington, DC: U.S. Government Printing Office, U.S. Congress, Office of Technology Assessment, OTA-13A-403, July 1990.

Parens, Erik (1998). *Enhancing Human Traits: Ethical and Social Implications*. Washington, DC: Georgetown University Press.

Parens, Erik, and Lori P. Knowles (2003). *Reprogenetics and Public Policy: Reflections and Recommendations*. Special Supplement to the *Hastings Center Report*, July–August 2003.

Philips, Helen (2004). "Alzheimer's Gene Therapy Trials Shows Early Promise," *NewScientist.com News Service*, 28 April.

Points to Consider in the Design and Submission of Protocols for the Transfer of Recombinant DNA Molecules into One or More Human Research Participants (Points To Consider), Appendix M of the Recombinant DNA Guidelines, Office of Biotechnology Activities, National Institutes of Health. Available online at http://www4.od.nih.gov/oba/rac/guidelines_02/Appendix_M.htm (accessed 23 September 2007).

Powers, Madison, and Ruth Faden (2006). *Social Justice: The Moral Foundations of Public Health and Health Policy*. New York: Oxford University Press.

President's Commission for the Study of Ethical Problems in Medicine and Biomedical and Behavioral Research (President's Commission) (1982). *Splicing Life: A Report on the Social and Ethical Issues of Genetic Engineering with Human Beings*. Washington, DC: President's Commission for the Study of Ethical Problems in Medicine and Biomedical and Behavioral Research. Available online at http://www.bioethics.gov/reports/past_commissions/splicinglife.pdf (accessed 23 September 2007).

President's Commission for the Study of Ethical Problems in Medicine and Biomedical and Behavioral Research (President's Commission) (1983). *Securing Access to Health Care: A Report on the Ethical Implications of Differences in the Availability of Health Services*. Washington, DC: Government Printing Office. Available online at http://www.bioethics.gov/reports/past_commissions/securing_access.pdf (accessed 23 September 2007).

President's Council on Bioethics (2003). *Beyond Therapy: Biotechnology and the Pursuit of Happiness*. Washington, DC: Government Printing Office. Available online at http://www.bioethics.gov/reports/beyondtherapy/ (accessed 23 November 2007).

Rainsbury, Joseph M. (2000). "Biotechnology on the RAC-FDA/NIH Regulation of Human Gene Therapy," *Food and Drug Law Journal* 55(4), 575–600.

Randall, Delia C., Nicola L. Fleck, John M. Shneerson, and Sandra E. File (2004). "The Cognitive-enhancing Properties of Modafinil are Limited in Non-sleep-deprived Middle-aged Volunteers," *Pharmacology, Biochemistry and Behavior* 77, 547–555.

Recombinant DNA Advisory Committee (RAC) (2005a). Charter of the Recombinant DNA Advisory Committee, Office of Biotechnology Activities, National Institutes of Health, U. S. Department of Health and Human Services, modification of charter approved 13 April 2005. Available at http://www4.od.nih.gov/oba/rac/RACCharter2005.pdf (accessed 23 September 2007).

Recombinant DNA Advisory Committee (RAC), National Institutes of Health (2005b). Conclusions and Recommendations of the NIH Recombinant DNA Advisory Committee Gene Transfer Safety Symposium: Current Perspectives on Gene Transfer for X-SCID, March 15, 2005. Available at http://www4.od.nih.gov/oba/rac/SSMar05/pdf/X-SCID_RAC_rcmdtns_06_2005.pdf (accessed 23 September 2007).

Rose, Steven P.R. (2002). "'Smart Drugs:' Do They Work? Are they Ethical? Will They be Legal?" *Nature Reviews Neuroscience* 3(December), 975–979.

Sauer, Brian D., Judy A. Shinogle, Wu Xu, Matthew Samore, Jonathan Nebeker, Zhiwei Liu, Randall Rupper, Linda Lux, Jacqueline Amozegar, and Kathleen N. Lohr (2007). *Improving Patient Safety and Pharmacovigilance: Methods using Observational Data and Cohort Studies*. Final Report. Rockville, MD: Agency for Healthcare Research and Quality.

Sen, Amartja (2000). *Development as Freedom*. New York: Anchor.

Sun, Marjorie (1981). "Cline Loses Two NIH Grants," *Science*, 214 (December 11), 1220.

Thompson, Larry (1994). *Correcting the Code: Inventing the Genetic Cure for the Human Body*. New York: Simon & Schuster.

Trifonov, Edward N. (2000). "Earliest Pages of Bioinformatics," *Bioinformatics* 16, 5–9.

Turner, Danielle C., Trevor W. Robbins, Luke Clark, Adam R. Aron, Jonathan Dowson, and Barbara J. Sahakian (2003). "Cognitive Enhancing Effects of Modafinil in Healthy Volunteers," *Psychopharmacology* 165, 260–269.

U.S. House of Representatives (1982). *Genetic Engineering*. Subcommittee on Investigations and Oversight, Committee on Science and Technology, U.S. House of Representatives, November 16–18.

Walters, LeRoy, and Julie Gage Palmer (1997). *The Ethics of Human Gene Therapy*. New York: Oxford University Press.

Walz, Chris (2003). "Air Force Testing New Fatigue-Combating Drug," *Pentagram* 14.

Watson, James D., and Francis H.C. Crick (1953a). "A Structure for Deoxyribonucleic Acid," *Nature* 171(April 25), 737–738.

Watson, James D., and Francis H.C. Crick (1953b). "Genetical Implications of the Structure of Deoxyribonucleic Acid," *Nature* 171(May 30), 964–967.

Woodring, Tech Sgt. J.C. (2004). "Air Force Scientists Battle Aviator Fatigue," *Air Force Link* 30(April).

Woo, Savio (2000). "The Last Word: Researchers React to Gene Therapy's Pitfalls and Promises," *FDA Consumer* (September–October). Available at http://www.fda.gov/fdac/departs/2000/500_word.html (accessed 23 September 2007).

Chapter 5
The Machine in the Body: Ethical and Religious Issues in the Bodily Incorporation of Mechanical Devices

Courtney S. Campbell, James F. Keenan, David R. Loy, Kathleen Matthews, Terry Winograd, and Laurie Zoloth

A substantial portion of the developed world's population is increasingly dependent upon machines to make their way in the everyday world. For certain privileged groups, computers, cell phones and PDAs, all permitting the faster processing of information, are commonplace. In these populations, even exercise can be automated as persons try to achieve good physical fitness by riding stationary bikes, running on treadmills, and working out on cross-trainers that send information about performance and heart rate.

Still, these examples of everyday human interaction *with* a mechanized world presuppose an ability to differentiate between our selves in our organic and bounded embodiedness and the "other" we encounter as an external mechanical artifact of technology. For the most part, we do not yet have a composite bodily nature comprised of flesh and machine. This boundary, which in actuality has been permeable for several centuries, may be dissolving further as new mechanical devices are introduced in biomedicine and incorporated in the body. The terrain of contemporary medicine is, in fact, permeated with innovative technologies to restore, repair, rehabilitate, and, in rare cases, enhance our physical and psychological capacities. Consider the following illustration from a recent article in *The New York Times*, which reported on soldiers who have lost limbs in the war in Iraq and are recipients of "high-tech limbs."

> Sgt. David Sterling, 23, of Placerville, Calif., lost his right hand and forearm to a rocket-propelled grenade near Fallujah. He was on a mission to rescue a group of Marines when his mission was ambushed. He now wears an $85,000 myoelectric forearm, powered by a lithium battery that approximates hand movements through electrical impulses when he flexes the remaining muscles in his arm.
>
> For routine tasks, like shaking hands and holding a glass, he snaps an artificial hand onto the end of the device. For other challenges, he removes the hand and snaps in the hook or the pliers-like grip—'It's great for changing an oil filter'–that he carries in his backpack. At home, he has snap-on kitchen devices, work tools, separate hands that help him write, play golf, shoot pool, even cast a fishing rod (Janofsky, 2004, A13).

The article was accompanied by a front-page photo of Sgt. Luke Wilson, who having lost his leg in Iraq, was walking with a device known as a C-leg. The picture portrays him walking with a rapid stride, wearing only gym shorts and sneakers. The C-leg has no appearance of being a human leg except that it is attached to Wilson. It is made out of several pieces of metal that together are strapped to his thigh. It has no wax or other substance to hide its mechanical structure, yet by the way it is incorporated into his body and in his walking, it is clearly one of Sgt. Wilson's two legs.

For most persons, including Sgts. Sterling and Wilson, such prosthetic devices are a necessary compensation for loss of normal organic functioning and are not a desired aspect of the self. Conventionally, we have come to understand the use of such mechanical devices as a complex development in rehabilitative medicine and an advance toward simple mobility. At some point, however, the "other" that is machine becomes increasingly difficult to differentiate from the "self" that is embodied. Indeed, as suggested by Rodney A. Brooks, director of the Artificial Intelligence Laboratory at MIT, a comprehensive convergence of human and machine can be imagined in the not-too distant future: "While we have come to *rely* on our machines over the last fifty years, we are about to *become* our machines during the first part of this millennium" (Brooks, 2002, 212).

Is the use of such machines, and the prospects for our converging with or "becoming" one with our machines, in some way a violation of natural orders and boundaries? Alternately, is it part of our nature as tool-making creatures to pursue incorporation of various forms of the mechanical within the body? To what extent does the incorporation of such devices require new understandings of our identity as embodied beings? What difference does it make, scientifically, morally, and existentially, as to whether a mechanical device affects, facilitates, or replaces the *organic processes of body functioning* (as in making artificial limbs to enhance mobility) in contrast to devices that are proposed to interact with and influence *higher-order brain* functioning, such as memory and consciousness?

The evolution of these technologies and their current application, imaginative visions of future development, and the comparatively minimal normative commentary within religious traditions make it difficult to present definitive responses to such questions. We will propose three patterns of normative response from the ethics of the world's classic religions—appropriation, ambivalence, and resistance. Generally speaking, mechanical interventions on bodily processes, such as restoring mobility to impaired soldiers, are ethically uncontroversial and follow the pattern of appropriation in religious bioethics. However, certain circumstances (military applications) or forms (neurocognitive interventions) of bodily incorporation of mechanical devices may invite not wonder but caution or ambivalence, if not outright resistance. Researchers have begun preliminary studies and scholars have begun to discuss, for example, the prospect of a computer chip implanted in the brain that might expand our capacity to assimilate and transfer information, and repair and augment such faculties as memory and reasoning powers (Huang, 2003; Maguire and McGee, 1999, 7, 10). These interventions are more ethically controversial because they present improvements over our natural bodily endowments or pose implications for human nature, identity, and destiny. Nonetheless, we recognize that much of our analysis, while suggesting diverse ways of ethical and religious reflection on mechanical devices in the body, must necessarily be tentative and inferential rather than conclusive.

We will begin our analysis with an overview (Sections 5.1, 5.2) of the fields of scientific research that open up the possibilities of bodily incorporation of mechanical devices. We present a taxonomy to help differentiate methods of research intervention and application, both current and foreseen. This conceptual overlay must be understood as fluid rather than rigid, but the categories we present do provide a basis for distinguishing between devices that are available, those currently in research and development, and those that belong (for now) to the realm of scientific fantasy.

We then turn to a discussion of the religious meaning and implications of incorporation of mechanical devices. We consider arguments (Section 5.3) that such technological advances have no religious meaning because the conceptual and methodological resources of the major traditions of East and West are too bound to their origins in non-technological cultures and the sacred writings that emerged from such contexts. Our analysis will suggest that metaphorical imagination and casuistic deliberation are intrinsic to the nature of ongoing and vibrant religious communities and provide resources for these communities to address questions posed by advances in biomedical technology.

In subsequent sections, we further develop the three patterns of appropriation, ambivalence, and resistance as themes of religious response, contending it is important to differentiate between those mechanical devices that primarily impinge on the organic processes and functioning of the body (Section 5.4) and those devices that are directed at interventions in cognitive states or in the nature of human consciousness (Section 5.5). The somatic interventions tend to elicit patterns of appropriation and acceptance, although ambivalence is sometimes generated either because of the prospects of employing the device for purposes of enhancing human capacities or because these types of intervention may appear to offer technological solutions to what are primarily questions of religious meaning. By contrast, incorporated devices that impinge on various cognitive states tend to generate patterns of ambivalence and even resistance because they appear to challenge conceptions of human nature and identity.

While the benefits of pursuing biomedical and bioengineering research on incorporated devices for individuals seem relatively straightforward and morally uncomplicated in many cases, such as artificial limbs to provide mobility, we find that it is important for religious traditions to raise issues of the common good and social justice (Section 5.6). This concern, although not uniquely religious, has deep roots in various traditions, and is manifested both in concerns about equitable distribution of beneficial resources, and the priorities such research and its clinical diffusion should have when many basic health care needs in some societies, such as the United States, continue to be unmet. The concerns raised by the issue of the common good lead us to some brief reflections on the significance of this discussion by religious communities for questions of public policy (Section 5.7).

In writing about these technologies, and their ethical and religious implications, we will use the language of "incorporation" technologies or devices. Incorporation, with its etymological roots stemming from Latin words for the body, seems to best convey a broad understanding of the technological-human phenomena we seek to describe and assess. That is, "incorporation" refers to the union or combination or inclusion of a mechanical device within something already formed, namely, the human body, so as to form a united whole. The meaning we intend through the language of "incorporation" is expressed well in the words of one researcher describing research success with the movement of an external robotic arm through a neural-computer interface in a rhesus monkey: "...the brain is so amazingly adaptable that it can incorporate an external device ... as a natural extension of the body" (Naam, 2005, 180).

5.1 Scientific Dreams and Research Realities

"Sophisticated microelectronics for signal processing are bringing the dream of merging man and machine closer to reality." So observes an article in *Science* that provides an historical review and discloses general future prospects in medicine related to "bionics" (Lavine et al., 2002, 995). It is not clear for whom such a merger is a "dream" or how widely this dream is shared. Nevertheless, a wide spectrum of technologies can be interpreted as holding some form of mechanical implications for the body. These range from the basic addition of clothing to the body to the full transembodiment of an individual human consciousness into an artificially created device (Moravec, 1988; Kurzweil, 1999). Neither of these ends of the spectrum, it should be noted, reflect the core concern of this paper, the bodily incorporation of mechanical devices. In the first instance, clothing or other ways of adorning the body are largely external to the body, and while the device (a shirt, a piercing) is made by a machine, it does not *function as* a machine in relation to the body. In the second instance, consciousness is disembodied and incorporated into a machine; while this is an example of humans literally "becoming" their machines, this phenomenon currently resides primarily in the realm of speculative fantasy and thus has simply not yet generated much scholarly reflection.

The umbrella of bodily incorporation of mechanical devices covers a broad and motley array of possibilities, which differ along a number of conceptual dimensions. It will be useful to delineate these dimensions clearly as a way of scientifically situating our analysis. A primary goal in setting out this technological background is to enable differentiation of scientific reality from hyperbole generated by science fiction, entertainment or news media, industry and entrepreneurs, and researchers. These stakeholders often present incorporation technologies as more foreboding, more imminent, or more promising of a panacea than they actually are.

We can present a scientific context for incorporation technologies in a framework of four primary dimensions: *System*, *Method*, *Impingement*, and *Objective*. We will use a series of tables to clarify conceptual possibilities.

5.1.1 System

Some conventional scientific understandings of the body provide one set of distinctions for the diverse ways in which mechanical devices may be incorporated within the body.

Mechanical: The body can be scientifically understood as a physical mechanism of moving parts. Mechanical devices can function to repair these parts or provide different parts that may augment standard human capacities. These devices can range from a hip-joint replacement to a powered exoskeleton that gives superhuman strength (BBC 2004a).

Organic Function: The body can be scientifically understood as a collection of organs that perform various involuntary functions to maintain bodily processes. Mechanical device technologies may be incorporated in the body to provide assistance or regulation of these processes. Kidney dialysis units, heart valves, insulin pumps, assisted reproductive techniques (ART) and in general most drugs are interventions in the internal functioning of the body's organs and tissues; the bodily incorporation of pharmacological agents is a common example of a device made by a machine that does not retain its mechanized form once assimilated by bodily systems.

Sensory: The body can be understood as a collection of sensory channels by which we perceive and experience the world. For nearly a half-century, mechanical devices have been implanted in the body to aid or improve sensory perception. Cochlear implants, for example, have had a long history of use, and more recent developments in auditory and retinal implants, as well as sensory prostheses, also intervene in the processes of hearing, sight, and touch.

Mental: The body can be understood as the corporeal home for mind and consciousness. Although mind-body dualism has been subject to centuries of debate in philosophy, theology, and science, a distinction between mind and body underlies the way many people experience the world most of the time. Currently, pharmaceuticals that alter perception or mood, enhance performance, or affect our sense of self and identity remain subject to critical scrutiny. It comes as no surprise then that incorporation technologies that affect the mental realm of human beings are among the most problematic both in terms of scientific technique and ethical and religious acceptability.

5.1.2 Method of Intervention

There are three distinct modes of technological intervention into the human body: chemical, mechanical, and electrical. Of course, a specific kind of intervention may involve a mixture of these basic intervention methods.

Chemical: The most pervasive kind of incorporation of a machined device in a broad sense is the introduction of chemical substances into the body (or parts of it) to alter its functioning. For example, an underlying assumption of both testing for "performance-enhancing" substances at athletic competitions and the recreational use of "mind-altering drugs" is that the combination of drug-plus-person comprises a different being from the person without such substances.

Ethical issues related to chemical or pharmacological alterations, such as enhancements or identity changes, have correlates in non-chemical forms of incorporation.

In addition, we note parallels with technologies discussed more extensively in other chapters of this volume. For example, ARTs rely on chemical modes of intervention, while research into genetic modification, which offers extensive and longer-term possibilities for incorporation of machines or machine-made devices, is also an extension of the chemical mode (Mehlman, 2003).

Mechanical: Many interventions into bodily function occur through machines in the most straightforward sense: they operate through moving parts and physical forces. Although the "dream" of human-machine merger may be a relatively new phenomenon, machines have long been at the center of historical discussions of bodily incorporation of mechanical technologies because mechanical devices were exploited for centuries before other methods of intervention became available (Science, 2002).

There are several active areas of new research in mechanical devices that address the scale on which the devices are built and the ways in which they are controlled. At the "micro" level, technologies that were developed for computer chips can be used to create Micro-Electro-Mechanical Systems (MEMS), which integrate mechanical elements, sensors, actuators, and electronics on a common silicon substrate through microfabrication technology. A device the size of a grain of rice can contain a complex mixture of physical and computational elements. MEMS research is a leading edge to a much wider range of devices that can be embedded in the body because of their small size.

On the much smaller scale of nanotechnology, researchers are proposing to create molecular-sized physical machines. The scale of nanotechnology has been defined as follows: "Nanomachines are mechanical devices so small that the parts are single molecules. While nanomachines abound in nature, nanotechnology seeks to understand how they work and build them synthetically for applications in medicine, computer science, [and] space exploration, ..." (Nanomachine).

The phrasing that "nanomachines abound in nature" already presumes that nature is mechanized, so introducing nanomachines into nature is portrayed as only a change in scale, not kind. The definition thereby claims that most of the molecular systems in living beings are operating as nanomachines. In this account, we could say, for example, that the cellular elements that replicate, copy, and regulate DNA comprise the best known nanomachine in popular culture; as Ramez Naam puts it, "each protein (encoded by DNA) is a tiny molecular machine" (Naam, 2005, 11). There is no sharp boundary for many researchers between nanotechnology, chemistry, and genetic engineering.

Electrical: Whereas mechanical and chemical devices tend to be limited to simple fixed functions, electrical and computational devices have the ability to provide complex situation-dependent control. Electrical devices function through the flow of currents through switches, wires, and a multitude of electronic devices that can produce light, heat, and motion, sense the environment, and perform computations. The use of electrical devices in biomedicine occurred long before the advent of the computer, but in today's world almost any electrical mode of bodily intervention will also be computational.

5 The Machine in the Body: Ethical and Religious Issues

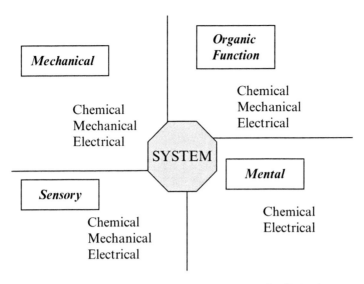

Fig. 5.1 Possible Systemic and Intervention Methods in Incorporation Technology

Table 5.1 Examples of Incorporation Technology: Systemic and Intervention Methods

METHOD	SYSTEM Mechanical	Organic Function	Sensory	Mental
Chemical	Steroids	Pharmacologic agents Assisted reproduction	Pharmacologic agents that treat vision problems	Pharmacologic agents such as psychedelics, anti-depressants, etc.
Mechanical	Hip replacement	Dialysis machines, Heart valves	Glasses	
Electrical	Electro-muscular stimulation devices	Heart pacemaker	Cochlear/retinal implants	Neural chip

Figure 5.1 and Table 5.1 summarize the array of possibilities in incorporation technologies as suggested by their systemic and intervention method dimensions.

5.1.3 Impingement

Our initial response to different kinds of embodied incorporations of mechanical devices is often strongly shaped by the way in which the technology impinges on the body. A device we don as part of our clothing is experienced as much less problematic than something embedded within the body, even if it is the same device (e.g., a radio frequency identification chip (RFID)), which is put into clothing (McCullagh, 2003a) or embedded in persons' bodies (Weissert, 2004). The degree of impingement varies with the device, but for our purposes, incorporation requires a more-or-less permanent and irreversible level of impingement.

Worn: The least intrusive technology can be donned and doffed at will. Eyeglasses, hearing aids, false teeth, limb braces, and other assistive technologies possess a quality of easy reversibility. This quality supports a perception of voluntary use of such technologies, though of course such uses are not immune from coercive contexts, including societal expectations. For example, condoms are regulated as a medical device, but the voluntariness of their use may be socially controlled by expectations about contraception or about preventing the sexual transmission of infectious disease.

Attached: Some technologies are external to the body but are attached in a more permanent or invasive way. Technologies of attachment may be imposed, as with the electronic tracking anklet (BBC, 2004b), or voluntary, as with the attachment of performance artist Stelarc's third arm (Stelarc, 2004). Although external to the body, mechanical devices that are attachments seem more invasive of the body and of privacy than an article of clothing. While attached devices still offer voluntary removability, the freedom to do so may be constrained by public safety. In a well-publicized case in 2002, an experimenter who had been wearing a computer-video helmet for several years had it removed at an airport security station. The person claimed bodily injury due to the disorientation he experienced by having to go without the computer-video helmet (McCullagh, 2003b).

Penetrating: Some technologies, such as feeding tubes, have components both inside and outside the skin, requiring some kind of ongoing bodily penetration. ART methods such as in-vitro fertilization require penetration of the skin and body for both oocyte retrieval and blastocyst implantation. External insulin pumps for diabetics have chemical tubes through the skin, while many devices, including experimenter Kevin Warwick's surgically implanted transponder, require electrical connectivity (Warwick, 2004). A current development with MEMS technologies and new electronics is the ability to have computational control of embedded devices without risking the potential infections of ongoing skin penetration.

Many people experience a psychological difference between a device attached externally to the body and a device that penetrates the skin. The skin has always been a boundary for the self, and penetrating this boundary can constitute a profound invasive violation of not only the body but the person (Lappe, 1996). Penetrating technologies thereby raise different experiential and ethical issues than worn or attached technologies.

Embedded: A further level of impingement involves embedding the device fully within the body. The device now becomes fully incorporated, a part of the body, rather than something worn or attached with relative ease of removability and reversibility. One estimate puts the number of persons worldwide who live with some form of mechanical implant at approximately three million (Maguire and McGee, 1999, 7). This estimate of what has been referred to as the "cyborg continuum" (Fry, 2000) includes cosmetic implants (breast, penile, pectoral, chin, calf, hair), prosthetic devices (testicular, hormonal, dental), as well as artificial limbs and pacemakers. An embedded technology generally implies permanence and irreversibility

5 The Machine in the Body: Ethical and Religious Issues   207

Table 5.2 Examples of Incorporation Technologies: Method and Impingement Intersections

METHOD	IMPINGEMENT Worn	Attached	Penetrating	Embedded	Transem bodied
Chemical		Nicotine or hormone patches	Insulin pumps	Contraceptive implants	
Mechanical	Eyeglasses	Artificial limbs	Feeding tubes, dialysis, Insulin pumps	Cosmetic implants	
Electrical	Hearing aids	TENS nerve stimulation	Direct brain stimulation	Pacemakers	Computer substrate for mind "transplant"

(or at least difficulty in removal) and may generate further reconceptualization of the boundary of self, body, and machine.

Transembodied: The final (and still purely fantasy) degree of impingement would involve the creation of a new physical substrate into which a person's memories or mind would be transferred. In some scenarios, the self-as-mind is disembodied or transembodied and with the aid of electrical, digital, and computational technology is transported into a mechanized home, such as a computer. This speculative scenario is compelling for some for whom self-identity is comprised entirely of mind and who find appealing the prospect of a technologically-based immortality, as the unique individual mind evolves through a never-ending sequence of "machine upgrades" (Bostrom, 2004b).

Table 5.2 summarizes the possibilities of incorporation technologies at the intersection of the method and impingement dimensions.

5.1.4 Objectives

A fourth dimension of bodily incorporation of mechanical devices concerns the intended purpose of the intervention, particularly as distinguished between therapeutic or enhancement purposes. This issue, a central consideration germane to ethical issues in reproductive and genetic technologies, is of no less importance in assessing incorporation technologies.

Therapeutic: The great bulk of mechanical devices incorporated in the body are intended to restore some aspect of normal functioning that a person has not developed or has lost through an acute incident or through chronic degeneration. Devices ranging from eyeglasses to artificial hearts to retinal implants find their justification through bringing a person "back" to some putative state of normalcy. A problematic aspect, however, resides in the concept of "normalcy." Prosthetic arms or legs for injured soldiers aim to restore aspects of everyday mobility and function, whereas some cosmetic implants or life-extension technologies may render the concept of normalcy vacuous.

Enhancement: The boundary between use of medical devices for therapeutic and enhancement objectives is porous. Olympic runners are allowed to take anti-inflammatory drugs for asthmatic conditions, but not stimulants or corticosteroids that increase muscle mass. Drugs that have restorative purposes for some persons (e.g., Viagra) may be used for enhancement by others who already possess "normal" function.

The same blurring of conceptual boundaries applies to technologies such as sensory implants: research advances have made feasible a cochlear implant that allows a person to hear in sonic ranges that are inaudible to people with a normal range of hearing; in addition, a visual sensory implant can provide infrared night vision (Goldblatt, 2002). The technological trend is moving society toward the super-sensory enhancement devices envisioned in the science fiction of William Gibson (Gibson, 1984). Not to be overlooked, the prospects of enhancements through implanted medical devices provide lucrative commercial opportunities.

The boundary between therapeutic and enhancement objectives is blurry not simply because certain enhancement interventions have already been adopted, but also because the possibility of enhancement in many circumstances is parasitic on the knowledge gained through therapeutic interventions. Scientist Theodore Friedmann has observed that gene transfer research for purposes of therapy, such as for muscular dystrophy or anemia, inevitably opens the scientific pathway to the manipulation and transfer of genes to stimulate and enhance muscle growth or oxygen delivery (Friedmann, 2005). Naam has made a similar point in his recent book, *More Than Human*: "Scientists cannot draw a clear line between healing and enhancing, for they're integrally related. Promising research on curing Alzheimer's disease, on reducing the incidence of heart disease and cancer, on restoring sight to the blind and motion to the paralyzed is the very same research that could lead to keeping us young, improving our memories, wiring our minds together, or enhancing ourselves in other ways" (Naam, 2005, 6).

Table 5.3 Examples of Incorporation Technologies: Impingement and Objective Intersections

OBJECTIVE	IMPINGEMENT Worn	Attached	Penetrating	Embedded	Transembodied
Therapeutic	Eyeglasses	Prosthetic Limbs	Cochlear, retinal implants	Artificial heart	
Enhancing	Hearing or visual augmentations	Super strength "exoskeleton"	Neural implants (Warwick, e.g.)	Cosmetic implants	
Hedonistic		Super strength "exoskeleton"	Body piercings	Embedded neural stimulation	Computer substrate for mind "transplant"

Hedonistic: Drugs again give us a window into how technologies can be applied in search of pleasure. Other forms of interventions can provide hedonic physical and even neural stimulation. Early research on pleasure centers in the brain (Olds and Milner, 1954) demonstrated that direct electrical stimulation could create motivation in mammals that was stronger than the stimuli of food or sex.

Table 5.3 summarizes the possibilities of incorporation technologies at the intersection of the impingement and objective dimensions.

5.2 Research Applications: Representative Technologies and Trends

Relying on the framework of this background account of the conceptual array of incorporation technologies, we now discuss some representative examples in more detail. We include both current and proposed technologies, their applications within the next decade to generation, and common ethical concerns they raise. The sequence of this discussion is arranged from current uses to speculative possibilities.

5.2.1 Implanted Medical Electronics

Implanted electronics have been used for medical purposes for almost half a century, beginning with cochlear implants and the implantable heart pacemaker in the late 1950s. In general, these devices are penetrating or embedded mechanical devices that facilitate a therapeutic objective, namely, bringing functionality that is lost or missing through mechanical and electrical modes of intervention. The functionality sought can be physical (e.g., heart pacemakers and ventricular assist devices, artificial limbs), chemical (e.g., insulin pumps), or sensory (e.g., cochlear and retinal implants).

The wave of micro-electronics development since the 1970s—especially through the use of sophisticated signal processing and computational techniques implemented in tiny low-power devices—has greatly increased the therapeutic potential for implanted devices. These devices reflect a wide range of configurations, covering nearly all of the categories of impingement (worn, attached, penetrating, embedded) delineated above. Since progress in electronics (such as embedded power sources and receivers) often leads to new ways of accomplishing a task, the differentiation of therapeutic and enhancement objectives becomes very fluid.

Hearing: The most developed sensory device is the cochlear implant, which was introduced in 1957 (Science, 2002). The cochlear implant uses an external microphone to gather sound waves, which are converted into electrical impulses and transmitted to a microelectrode array permanently implanted within the cochlea. The electrodes stimulate the auditory nerve: Patients are able to hear, although not with complete normalcy, through the body's incorporation of a mechanical device

(Rauschecker and Shannon, 2002). Studies estimate that some 70,000 profoundly deaf persons worldwide have received cochlear implants, and of these 40,000 can hear well enough with their implant to engage in conversation (Stallard, 2004; Naam, 2005).

Society can expect a steady increase over the next decade in the capacity and convenience of these devices. Research has recently been directed to developing a new generation of electrode implants for brainstem auditory centers (Lavine et al., 2002). If the auditory nerve is damaged, an auditory brain implant (ABI) can directly stimulate the auditory brainstem. The most successful method of stimulation at present is a penetrating ABI, which is inserted into the ventral cochlear nucleus. Two human subjects have received penetrating ABIs and clinical trials are underway (Business Wire, 2004). Although further research is needed to technically refine the capabilities of both standard and ABI implants, as one researcher puts it, "there really seems to be little in the way of a moral argument that could say that using technology in this way is bad" (Brooks, 2002, 217).

Despite their pervasive use and regulation as a legitimate medical device, some arguments claim that even therapeutic uses of cochlear implants can be disruptive to the integrity of sub-cultures (Crouch, 1997). Moreover, ethical questions for the broader culture will arise when such interventions are used for enhancement rather than restorative purposes. The design of hearing implants to pick up sounds that would not be audible in the normal human range is increasingly possible. Since the implant is likely (with foreseeable technology) to interfere with normal hearing, the recipient would be taking a risk in choosing (or being persuaded) to add enhanced hearing ability for recreational, professional, or military uses.

Vision: The use of contact lenses to correct vision was first reported in 1888, and hard plastic contact lenses were introduced in 1939. Current research now focuses on restoring some vision to persons who are blind through retinal implants. Retinal degeneration leading to blindness afflicts more than six million persons in the U.S.

While researchers are far from reproducing normal vision in a blind person, functional sight that enables people to navigate daily life situations can be restored. In September 2004, the FDA approved the first implantable lens for extreme nearsightedness (Henderson, 2004). One research group is working to develop an epiretinal implant for FDA approval by 2006 (Zrenner, 2002). A second research program has implanted an intraocular device (the Implantable Miniature Telescope, or IMT) and hopes to seek FDA approval for the device within three to five years (Daviss, 2004; Sandhana, 2003).

Visual implants are not as advanced as cochlear implants yet, since the retinal-neuronal system is far more complex than the auditory system. As well, experimental approaches differ on the optimal location at which the sensory information is directed into the nervous system, such as the retina, the optic nerve, or the visual cortex in the brain. Some approaches are invasive (e.g., retinal implants or direct stimulation of the optic nerve or visual cortex) and others are wearable and non-invasive (e.g., a vibrating pin array worn on the skin that receives input from a camera) (Zrenner, 2002). Given the plasticity of the nervous system, those receiving implants for vision may develop sight-like capabilities (though not as sensitive as normal vision) through a variety of modalities. Despite the organic complexities and differences in experimental methods, the research expectation is that retinal implants can be a customary clinical reality within the next decade to generation (Cohen, 2002; Loomis, 2002).

The therapeutic uses of retinal implants are ethically uncontroversial. However, ethical questions will be more pronounced when visual technologies reflect enhancement objectives. For example, cameras are able to respond to a much wider range of signals from different sources than the normal human visual system and the output of those cameras (infra-red vision, magnified vision, low-light vision) can produce a signal that enhances vision. Such improvements on the natural endowment of vision can create ethical questions both intrinsic to technically-enhanced capabilities of sight as well as about justice and access.

Internal Organs: Heart. There are many uses of embedded electronic technologies for internal organs. In some cases they provide direct electrical stimulation (e.g., pacemakers), in others they control a physical device (an artificial heart) or regulate a biological process (heart valves), and in others they control chemical flow (such as insulin for persons with diabetes). Here we provide a brief overview of developments in heart regulation or replacement.

It is relevant to our concerns that research on artificial pacemakers dates back over two centuries, and that an attempt to develop what became known as the "Hyman Pacemaker" in the 1930s was criticized on the grounds that such a device would supplant determinations of divine providence in the continuation or ending of life (Baxi, 2003). Nonetheless, a sustained programmatic effort was stimulated by a research challenge issued in 1964 by the U.S. National Heart, Lung, and Blood Institutes to develop an artificial heart. This research challenge evolved in parallel with discussion at prominent research conferences on scientific control of human nature (Wolstenholme, 1963). However, in the view of Al Jonsen, a leading historian of bioethics and a participant in these conferences, the prospects for heart replacement at this time did not generate much ethical discussion, nor controversy over whether such research would replace providential with technological control of life (Jonsen, 1998, 15–16; 2004).

Initial attempts to regulate heartbeat with an external heart pacemaker occurred in 1958, and the first totally implantable pacemaker was embedded in 1960. The pacemaker emits a small periodic electrical signal that stimulates the heart muscle, enabling it to beat regularly. Subsequent research on mechanical heart devices for persons with heart disease has seen some progress and noticeable setbacks. Still, as articulated by researchers writing in *Science*, "the ultimate goal is to develop a machine that is as good or even better than (a) donor heart, avoiding the need for a transplant altogether" (McCarthy and Smith, 2002, 998). A total artificial heart was first implanted in a human being in 1969, but only as a temporary measure until a human heart became available for transplantation. In 1982, Barney Clark became the first person to receive the Jarvik-7 permanently implanted total artificial heart. This event did generate substantial ethical debate (Annas, 1988, 391–396; Fox and Swazey, 1992).

In 1993, the FDA approved the use of the left ventricular assist device (LVAD) as a bridge to a heart transplant. The LVAD supports the left ventricle, enabling damaged heart muscle to repair itself and contract more vigorously. Researchers claim approximately 90% of patients with end-stage heart disease could benefit from implantation of the LVAD. In some cases, muscle repair has been substantial enough for the person to no longer need a transplant upon removal of the device. McCarthy and Smith contend that "the goal is for wider application of LVAD support so that heart failure patients are supported for weeks or months until cardiac damage has been repaired sufficiently such that a transplant is not necessary" (McCarthy and Smith, 2002, 998).

In the summer of 2001, AbioCor introduced the public to Bob Tools, the first patient to receive a self-contained total artificial heart (TAH) (AbioCor, 2004). The initial announcement generated much media fanfare but little ethical or religious debate (Morreim, 2003). Unlike the Jarvik heart, the AbioCor TAH is intended to be a destination device. "A TAH implant is a dramatic event that removes the patient's heart and results in complete dependence on the most sophisticated technology ever implanted in humans" (McCarthy and Smith, 2002, 998). More precisely, the device is a totally implantable biventricular replacement that largely leaves the native atria in place. Currently, AbioMed has completed 14 of the initial 15 scheduled implants.

The AbioCor heart is designed for 90% reliability after five years of operation compared to a five-year survival rate of 70% for a current heart transplant patient. However, because of machine malfunction, blood clots, the size of the TAH, and higher risks of infection, the AbioCor TAH has yet to meet its design expectations. The company has applied to the FDA for an HDE (Humanitarian Device Exemption), which would allow a very limited market use of the TAH. HDEs are permitted for a population of no more than 4,000 persons, so the criteria for patient eligibility even with FDA approval remain very narrow: only persons with end-stage heart disease and no other alternatives (including human heart transplant) meet the selection criteria (Morreim, 2004).

The heart has long had special cultural, religious, psychological, and symbolic status, and thus its replacement with a mechanical device would seem to pose questions about the moral implications of incorporation technologies more sharply than the embedding of mechanical devices to assist or replace other internal organs (Fox and Swazey, 1992). While heart pacemakers are commonplace—and in fact have barely caused any ethical ripples—whole-heart mechanical replacements continue to generate many serious medical complications. However, widespread public acceptance of cardiac assist devices suggests there is a range of moral flexibility with integration of machines and the heart so long as the technology is shown to be safe and effective.

5.2.2 *Neural Control of External Devices*

Incorporation technologies that do not merely address the therapeutic needs of the physical body but also engage the mind-body continuum create a fantastic (in both senses) set of possibilities, and raise social, psychological, and ethical questions about personal identity and selfhood. As with embedded microelectronic technologies, most of the neural cognitive research conducted so far is therapeutic. This research has been directed in particular to providing control over artificial limbs, as illustrated by the prosthetic devices available to injured soldiers, or providing motor function to people with neuro-muscular dysfunction. But science fiction presents a long history of speculations on how people can "jack in" (Gibson, 1984) to computers and use the mind to interact directly with a world beyond their immediate bodily reach. An equally engaging body of literature deals with possibilities for "mind control" in which implanted or external devices can reveal and/or exert influence on a person's thoughts. The basic neurophysiology of neuronal pathways suggests that these possibilities are not entirely fictional.

One branch of research uses externally received neurophysiological signals, such as EEG (brain waves), EMG (muscle nerve activity), and metabolism (PET scans) (Sabbatini, 1997). This research, including work on Biofeedback, dates back decades. As new technologies increase the ability to receive neurophysiological signals and process them in more sophisticated ways, researchers are gradually developing more effective ways to couple brain activities with action. The most developed method involves the use of neuromuscular signals to control devices that correspond to movement of the muscles. The clearest application of this research is in providing control of an artificial limb, where signals from the corresponding muscles (or other muscle groups, such as the back), can control the device.

A more invasive approach taps directly into neural signals through some form of implanted array. The obvious risks (infection and inadvertent destruction of neurons) mitigate against this approach, but such research has been conducted on patients with diseases such as ALS, a degenerative condition that makes almost all other forms of interaction impossible (Wickelgren, 2003; Kennedy et al., 2000). Studies on animals are gradually moving towards more precise forms of direct neural control over the outside environment, such as controlling a cursor on a video screen (Carmena et al., 2003) or movement of a robotic arm from neuronal signals generated by electrodes implanted in the brain.

Direct neural stimulation, in which electrical signals are transmitted to specific areas of the brain, can have a wide variety of physiological and psychological effects. This approach has been effective in the treatment of central nervous diseases, such as Parkinson's disease and seizures (Cyberionics, 2003). For example, Medtronic, a corporate leader in the heart pacemaker field, has developed Activa Therapy for "deep brain stimulation," the functional equivalent of a brain pacemaker to treat patients with symptoms of Parkinson's disease. This technology received FDA approval in 2002. Surgeons implant a thin, insulated, coiled wire with four electrodes, then thread an extension of the wire under the skin from the head down through the neck, and into the upper chest. The wire is connected to an implanted neurostimulator, which sends electrical pulses under patient control to stimulate the brain. Some patients have reported significant lifestyle functions restored, such as drinking a cup of coffee or walking (Medtronic, 2003). Thirty thousand patients have received the electrode implant worldwide, and in the U.S., Medicare pays the $30–$40,000 cost for the procedure (Naam, 2005, 190).

As researchers map more neural pathways in the next few years, more treatments of this type will become practical. One area of research focus is the stimulation of "pleasure centers" (Olds and Milner, 1954), which raises issues surrounding both voluntary (hedonistic) and imposed (for behavior control) stimulation. Anticipated applications will raise many of the same ethical problems as other implanted electronics, especially when extended beyond therapeutic uses to enhancement objectives. Observes Naam: "As we learn how to repair damaged brains, we'll discover an immense amount about how the brain works. That in turn will lead to devices that can improve our mental abilities.... These abilities will pose serious questions to our sense of identity and individuality. They will blur the line between man and machine" (Naam, 2005, 175, 176).

5.2.3 The Cyber Soldier

For millennia, one of the greatest incentives for developing technologies that enhance capacities has been warfare. Not surprisingly, several research projects on incorporation technologies have been sponsored by the U.S. Department of Defense, primarily through the Defense Advanced Research Projects Agency (DARPA). DARPA has long had a reputation for studying purported science-fiction goals in current research, with a mixed record of success. DARPA was one of the primary funders of Artificial Intelligence during the decades when hopes were high for this technology (and later turned out to be unrealistic). DARPA has sponsored work on robotics, command and control systems, and human-machine interfaces, and has funded research projects at Duke, the University of Southern California, SUNY, and Vanderbilt on a variety of incorporation technologies.

An increasing perception of threats to security, and a change in thinking about the types of weapons and warfare that conventional soldiers will experience, has culminated in a view that innovative biotechnologies, especially defensive and protective technologies, will need to be developed for 21st century warfare. These research projects endeavor to augment the capacities of soldiers or "warfighters" through more sophisticated weaponry, armor, and sensory enhancement. Michael Goldblatt, a researcher with DARPA, writes that the agency is exploring "augmenting human performance to increase the lethality and effectiveness of the warfighter by providing for super physiological and cognitive capabilities" (Goldblatt, 2002, 337).

One current DARPA project seeks to augment the soldier's perception and memory with "augmented reality," which amplifies natural vision and brings additional information to the soldier in real time (ASSIST, 2004; Etter, 2002, 333). This proposal is fairly near-term, as contrasted with the Moldice proposal (Moldice, 2004), a longer-term research project to generate real-time conversion of biomolecular signals into electrical signals, which could be used for neural control of external devices such as vehicles or weapons (Naam, 2005, 178). Other projects, such as the Peak Soldier Performance Program, carry forward the historical use of

5 The Machine in the Body: Ethical and Religious Issues

chemical and other biological augmentations to keep soldiers awake, alert, and strong. The "ultimate goal" of PSP is to "deploy the warfighter at peak physical condition and maintain that level of performance throughout the mission, providing the warfighter with the ability for prolonged activity without the loss of strength, endurance or mental acuity during mission critical periods" (DARPA). There are disputes among researchers and bioethicists over the extent to which such research considers ethical issues on an equal footing with technical enhancements.

5.2.4 *Nanomedicine*

The term "nanotechnology" applies broadly to any technology that takes advantage of mechanical properties on a molecular scale. This application has been advocated for many years. A famous speech in 1959 by Richard P. Feynman (Feynman was awarded the Nobel Prize in Physics in 1965) entitled "There's Plenty of Room at the Bottom" includes speculations such as: ... it would be interesting in surgery if you could swallow the surgeon. You put the mechanical surgeon inside the blood vessel, and it goes into the heart and 'looks' around. (Of course the information has to be fed out.) It finds out which valve is the faulty one and takes a little knife and slices it out. Other small machines might be permanently incorporated in the body to assist some inadequately-functioning organ (Feynman, 2004).

The range of possible applications of nanotechnology is as hard to envision now as it would have been for someone in the time of Faraday and Ampere to predict today's uses of electricity (including all of modern communications and computers). Within the next decade or two, however, we will see increasing use of nano-created substances, including biological applications. These will not be the tiny but capable robots celebrated by Robert Freitas's *Nanomedicine* (Freitas, 1999, 2003) or viewed with trepidation in Bill Joy's "grey goo" problem (Joy, 2000). Research will instead provide artificial materials designed to have unique chemical and physical properties that can be of use in many fields, including medicine. They will have the look and feel of chemical substances, not of sentient robots.

Some impressive applications of nanotechnology with biomedical implications have already occurred (Meyyappan, 2004). An interesting example is the use of gold nanoshells that can be "tuned" to specific wavelengths outside the visible range. Nanoshells that absorb near infra-red light have been shown to accumulate in tumors, which can be eradicated by shining near infra-red light on the tumor site. Normal cells are not affected, and the tumor cells are killed by the heat generated upon light absorption by the nanoshells (just as food is kept warm by infra-red light) (Hewitt, 1994).

Both the NIH and researchers in the field have ambitious aspirations for nanotechnology in the realm of regenerative medicine. As outlined by Samuel Stupp of Northwestern University, nanomedicine holds out the potential for the reversal or prevention of paralysis and blindness through spinal cord regeneration; heart and blood vessel regeneration; therapies for Alzheimer's, Parkinson's, and diabetes;

minimization of stroke impairments; repair of all bone fractures; and access to new cartilage. Organic nanostructures, containing 10^{15} signals of information per square centimeter, have been successfully shown to initiate spinal cord repair and bone growth in mice and rats (Stupp, 2005).

Other researchers have turned to nanotechnology as a mode of treatment for brain cancers. Medical advances in this area have been hindered by the "blood-brain barrier." This barrier prevents 98% of small molecule drugs and 100% of large molecule drugs from entering the brain, thus precluding effective treatment for the estimated 80 million Americans who experience some form of brain disease. Researchers have recently developed nano-containers to deliver drug treatment across the blood-brain barrier. In short, the nano-container functions, according to UCLA researcher William Pardridge, as a "molecular Trojan horse" to move drugs into the brain. This method has been successfully been carried out in mice, rats, and rhesus monkeys, as well as in cells in the lab, and the question now is "whether it is possible to move from the petri dish to persons" (Pardridge, 2005).

Despite its relative newness, some scholars have already decried the "gap" between the scientific research on nanotechnology and the sparse analysis of its ethical and social implications (Mnyusiwalla et al., 2003). They have identified several ethical considerations generated by nanotechnology—among them matters of equity, privacy and security, environmental risk, and metaphysical questions about the status of human-machine interactions—and maintained that neglecting ethical analysis of nanotechnology risks "derailing" scientific research on nanotechnology in a manner analogous to research on genetically modified foods. For his part, Joy has encouraged a self-imposed moratorium for researchers in nanotechnology. Other scholars have observed that ethical issues associated with nanotechnology and regenerative medicine will raise issues that are similar to those addressed in moral assessments of medicine's use of chemical and genetic engineering.

5.2.5 Consciousness and Disembodied Immortality

A speculative form of human-machine integration, one in which the person is disembodied and does become a machine, envisions that the essence of a person (the mind) can be transported to a completely different hardware substrate than the body, even while retaining the identity and consciousness of the individual (Moravec, 1988; Kurzweil, 1999). Relying on a strong mind/body dualism, the entire premise of this proposal requires an ability to dissociate mind and body that is fundamentally open to question. There is no current technological basis to believe this dissociation and portability of mind could be achieved. The hope for success is based on an analogy with digital computers: We can easily run the same computer program on different machines, so if our mind is a kind of program (as it is viewed in much of the literature on artificial intelligence), then the same transference could be possible for human minds.

This example of mechanistic incorporation of the mind-self does not really belong in a serious analysis of prospects for the next few decades, but because disembodied immortality has had such a strong grip on the popular imagination, it stands as a kind of far-point in anticipating potential developments in the dreams of human-machine mergers that possess unique ethical and metaphysical implications. Moreover, this is the one example that directly impinges on mental processes and consciousness, so it has prompted discussion among proponents and religious critics, as we illustrate in a later section of the paper.

5.3 Normative Religious Resources: Appropriation, Ambivalence, and Resistance

Our analysis of representative trends and technologies of incorporation indicates that the scientific dream of a merger of human beings with machines is yet to become a research reality. The mechanical devices described may directly impinge on the embodied nature of persons, but not to such an extent that human identity is deemed to be compromised. Mechanical impingements on human consciousness await development and refinement: The development of neural networks to control external objects, the military appropriation of mechanical enhancements and nanomedicine are deemed to be at such a speculative developmental stage that they have yet to stimulate religious reflection. In short, it can be claimed that, with the exception of the very speculative proposals for disembodied consciousness through a machine or computer substrate (an issue we address later), religious traditions have yet to discern in incorporation technologies a radical challenge or compromise to human nature or identity.

Given this differentiation between impingements on the body as distinct from prospective impingements on consciousness, it must be acknowledged that there are relatively few areas in which the normative teachings of religious traditions have directly addressed ethical questions posed by a specific incorporation technology. Nevertheless, mechanical devices implanted in the body do present implications for broader themes in religious thought and experience, including the nature and destiny of the human person, the significance of a person's embodied experience, including the experiences of pain and suffering, the person's relationship to ultimate reality, the divine or the sacred, and the vocation of medicine. Community-constituting convictions and narratives inform the method and content of reasoning about such conceptual questions as whether a moral line should be drawn between therapeutic or enhancement interventions, and/or between somatic and neural/cognitive interventions.

By attending to these broader community-forming concepts, it is possible to identify three general orienting themes in religious perspectives on mechanical devices that intervene in the body and/or mind, what we shall designate as perspectives of "appropriation," "ambivalence," and "resistance." Within any given tradition, of course, diverse values, stories, and norms may mean resources for support, caution, and opposition may be found internal to the tradition.

Appropriation. Certainly, religious traditions and values are not irrevocably obstacles to scientific inquiry and its application. Indeed, some religious world views can support, directly or indirectly, scientific research on mechanical devices upon which humans are almost completely reliant (Fox and Swazey, 1992). For some traditions, advances in basic or clinical technology are celebrated and encouraged (Feldman, 1986). Specific examples of "appropriation" may be found in the approach of Jewish and Islamic traditions to porcine heart valves. While both traditions have strict prohibitions on consumption of pork rooted in scriptural foundations, the compelling imperative in each tradition to save or preserve human life overrides the dietary or "polluting" proscription and permits use of porcine heart valves for Jewish (Rosner, 1999) or Muslim patients. This appropriation of a medical device is justified by the tradition's method of moral reasoning and its prioritization of values, both of which are embedded in broader, more comprehensive world views about the self, body, and the vocation of medicine.

Ambivalence. An approach of tentativeness and caution towards mechanical devices is most prominently displayed in Eastern traditions such as Buddhism and Hinduism. The incorporation of mechanical devices does not constitute any intrinsic violation because, in these traditions, our authentic Self is differentiated from our physical self. However, neither tradition affirms an overriding imperative to preserve biological life that grounds appropriation in some Western traditions. The core of these traditions, rather, is the cultivation of spiritual knowledge, knowledge that will advance the quest for liberation from the cycles for rebirth. Thus, medical devices, while promoting a good of life prolongation, may train a person's attention to a focus on the body, and divert it from the need to advance spiritual welfare detached from the body. In some circumstances, then, incorporation technologies may present an impediment to liberation.

Resistance. The pattern of "resistance" is most prominent in conservative and evangelical Protestant Christian thought regarding those devices that either offer possibilities of enhancement over our natural body endowment, or seek to alter our mental processes or identity of consciousness. In the view of this tradition, the vocation of medicine, as illustrated in the ministry of Jesus, is oriented by a commitment to healing of malady. This reasoning both justifies appropriation of those incorporation technologies that pertain to restoration of organic functioning or processes, while limiting or grounding resistance in relation to enhancement of capacities, or prospective technologies to alter conscious identity.

As indicated above, these general patterns are seldom explicitly articulated in religious teaching; indeed, in the view of one scholar, it would be surprising if the classical religious traditions had much of substantive moral significance to say on the incorporation of mechanical devices in the body. In this section, we will examine the legitimacy of this critique and suggest some meaning resources embedded in these traditions that provide a *context* for understanding the ethical question presented, if not a definitive moral conclusion. We will then proceed to examine the implications for religious perspectives of, in turn, mechanical impingements on the body, and prospective mechanical impingements on consciousness.

5.3.1 *Religious Resources: Metaphors and Casuistry*

In the view of sociologist Brenda E. Brasher, we should not expect meaningful dialogue regarding machines in the body to be forthcoming from religious traditions. Brasher contends that religious or theological questions about incorporation technologies may not be methodologically possible. The world's major religions, she observes, achieve internal coherence through reliance on foundational texts that have arisen from oral traditions of pastoralist and agrarian peoples. The questions of the implications of technology for culture, and for religious experience and doctrine, simply do not arise in these texts.

Brasher thereby claims the encounter with contemporary technologies, including biomedical devices and machines, generates problematic questions, even theological crises, for religious traditions regarding normative human identity, the integrity of the faith community, and their social ethics. The traditions are simply ill-prepared to address the instrumental rationality of contemporary science and technology, or the ethical ramifications of technical applications that penetrate or are embedded within the body. Thus, she argues, "For agrarian and pastoralist-linked traditional religions to be able to address these concerns, changes may need to occur within their symbol systems, changes that may be beyond their capacity to make" (Brasher, 1996, 818).

As one example of this anachronistic dissonance, Brasher contends that the textual religions assume embodied existence as a given. However, incorporation technologies challenge the foundational assumption of embodiment by mechanical intrusions, connections, or extensions. The boundaries of the body-self in the traditional religions are blurred by technology, and the experience of a body extended in space or even dis-embodiedness through a mechanistic representation (for example, a camera cell phone) may become experientially paradigmatic.

This theological crisis is particularly acute, Brasher contends, for Christianity. The doctrine of the Incarnation, that God was embodied as human person, is a defining claim of the Christian tradition. However, Brasher argues, "this notion is problematized by the coupling now underway of human and machine" (Brasher, 1996, 818). At the very least, the image of person-as-machine, rather than person-as-image of God, demands attention to foundational claims of the tradition about the self, God, the body, and nature. Unless the traditions take on these questions, Brasher believes that they will be unable to articulate "the meaning resources necessary to rejoin the ethical issues involved" (Brasher, 1996, 817) in human-machine incorporation or identity-convergence.

Brasher's critique is not without foundation. With few exceptions, religious traditions have not fully engaged the specific technologies or applications described above as representative of current research trends. Therapeutic interventions designed to improve sensory apprehension of the world, such as hearing and sight, have either not occasioned substantive religious reflection, or have been considered compatible with religious commitments to physical healing. For example, physician and ethicist C. Christopher Hook observes that the "dependence" of many persons on "technologies such as filled or false teeth, glasses or contact lenses, hearing aids, pacemakers, dialysis, hairpieces, and even vaccinations" is consistent with Christian understandings of the healing ministry of Jesus and the vocation of medicine (Hook, 2002, 53).

While Brasher critiques historical religious traditions as inadequate for the "coupling" of human and machine presented by incorporation technologies in part because of conceptions of embodiment, the main target of her criticism is that religious communities are so bound to their historical context that they simply do not have the intellectual resources to respond coherently to social changes, especially those brought about by science and technology. In short, her argument assumes a kind of conceptual fundamentalism as integral to the dynamics of religious communities. However, in claiming that the traditional religions are constricted by their textual origins, Brasher assumes that scripture is the exclusive source of moral guidance and overlooks communal interpretations of scripture embedded and enacted in tradition. Moreover, Brasher's appeal to an agrarian heritage as a way of discounting religious perspectives in a technological era assumes a static understanding of "the natural" that is no less misdirected.

The question of how faith communities respond (or adapt, or assimilate, or reject) to social changes stimulated by technological progress is very legitimate. It is not the case, however, that religious traditions are methodologically empty in addressing such changes in the way Brasher suggests. The general question for consideration is how such traditions understand and give meaning to a phenomenon that is unfamiliar—a technology, a machine, a mechanical device meant for bodily implantation—within the realm of more common experience. We suggest there are two methods integral to religious experience that enable a bridging between the innovative and the traditional worlds: *metaphorical interpretation* and *casuistical reasoning*.

Metaphorical interpretation is integral to religious experience simply because the realm of the ultimate, the sacred, the divine, the gods, etc., about which religious language speaks and which define religious communities, is itself a realm largely unfamiliar in human experience, even if disclosed in some fashion through sacred texts, persons, or in the natural world. Thus, religious experience and language is necessarily metaphorical insofar as it seeks to explicate and give meaning to a realm of experience and existence beyond the capacity of human language to describe effectively.

Casuistry is integral to religious experience as a mode of ethical deliberation and reasoning that relies on precedent and tradition in the experience of the faith community as one way to envision and guide moral thinking about seemingly innovative choices. That is, casuistry takes the unfamiliar circumstance requiring a moral decision and gives it a context of religious familiarity.

We here wish to provide a brief illustration of the significance of both methods as a way to counter Brasher's contention that religious traditions lack the intellectual resources for an adequate response to biomedical technologies.

5.3.1.1 Metaphorical Interpretation

In a very thoughtful exposition of the professional, cultural, and existential questions presented by the incorporation technology of mechanical ventilators, physician

Arthur Kohrman observes: "Much of the history of all cultures and their views and values is embedded and transmitted in stories and myths, the metaphorical repositories of wisdom and experience, and it is often through and in those old metaphors that we can find some ways to begin thinking about the new and unfamiliar" (Kohrman, 1995, 61). Kohrman's own appeal to story, narrative, metaphor, and myth indicates that such a resource for thought and interpretation of the novel is not an exclusive religious method. Science and medicine offer their own metaphors, mythic constructions and narrated meanings. The notion of "progress," which is so intertwined with the discourse of contemporary scientific research and biomedicine, is a prominent example of telling a story about science. Indeed, using the "race" to map the human genome, which researchers portrayed as "the holy grail" of the biological sciences (Gilbert, 1992) as an example, Protestant theologian Philip Hefner comments: "Stories are inseparable from the conception of our technology, but also from the uses we imagine for these technologies" (Hefner, 2003, 61).

While metaphorical interpretation is perhaps an inescapable human phenomenon (Lakoff and Johnson, 1980), it is necessarily an integral religious resource for envisioning the world because of the limits of our language in describing and understanding religious experience of the unfamiliar realm of the sacred. Moreover, if it's the case, as Kohrman suggests, that the cultural repositories of the wisdom of metaphorical understanding about the unfamiliar reside in myth and story, then we have ventured onto terrain that is deeply religious in its texture. With this as background, we offer brief examples of metaphors and myths that bear on the resources religious communities can bring to the discussion of incorporation technologies. This analysis will make evident that part of what makes moral reflection on incorporation technologies both so intriguing and so complex is the concept of "liminality" or in-between status that emerges about the relation of humans, machines, and nature.

5.3.1.2 Chimeras

While traditionally the image of a chimera refers to a human-animal hybrid, an entity that has generated substantial moral controversy of its own in recent scientific research (Mott, 2005), the salience of the chimera for incorporated mechanical devices is, as noted, thoughtfully discussed by Kohrman (Kohrman, 1995). The specific situations for which Kohrman believes the chimera metaphor and mythic narrative are especially applicable are children who are technologically dependent through systems of life-support. The resort to metaphor and mythic narrative provides a new way of seeing what is unfamiliar and perplexing to us, and therefore may illuminate "our senses of the identity and value of these human machine hybrids [ventilator-dependent children]" (Kohrman, 1995, 53).

In the classical Greek narrative, the chimera—a fire-breathing monster with a lion's head, a goat's body, and a serpent's tail—ravages the kingdom of Iobates. Iobates sends his houseguest Bellerophon to slay the chimera as part of a series of heroic but anticipated fatal tasks. Bellerophon is given a golden bridle by the goddess Minerva that he uses to harness the winged horse Pegasus. With the help of

Pegasus, Bellerophon slays the chimera. After completing this and many other dangerous tests, Iobates concedes that Bellerophon must be a favorite of the gods and therefore makes him successor to his kingdom. However, Bellerophon's arrogance provokes him to ride Pegasus from his earthly throne to the throne of the gods on Mt. Olympus. Displeased by this display of hubris, Zeus sends a gadfly to sting Pegasus, causing him to throw Bellerophon, who is maimed and blinded by the fall. Bellerophon is left to roam the earth as a solitary wanderer because of his offense against the gods.

Kohrman intentionally uses the narrative as a stimulus for our moral imagination: "Visualize for a moment a child: a quadriplegic, ventilator-dependent child... head of a lion (partially or completely useless) body of a goat, and—the tail of a dragon—the real technological tether, the tubes connecting the child to the ventilator and its alarm devices." Kohrman quotes from Alexander Pope's rendition of Odysseus' encounter with the chimera—"Her pithy nostrils flaky flames expire; Her gaping throat emits infernal fire"—to convey the travails of both child and caregiver as a tracheotomy tube is inserted to enable respiration for the child (Kohrman, 1995, 54).

What meaning and moral sense can we make of this terrifying situation? Kohrman contends such children can only appear as aberrations within a socially constructed world view and culture that affirms as its most celebrated values individualism, physical beauty, physical and social mobility, the ability to compete, and independence. Within this framing, the "child-machine creature" possesses liminal status, accompanied by both physical and moral vulnerability.

Profound questions of the nature of self and identity are posed for what Kohrman designates as "the relatively new chimeric species" of children who are technologically dependent (Kohrman, 1995, 58). The boundaries of self for a child who is connected to a machine are radically problematized. Moreover, it is unclear whether the child sees the machine as part of self, as an extension of body, or as an alien and overpowering object. As illustrated by experiences of patients with other penetrating, invasive, or embedded mechanical devices, these "boundary" concerns are applicable to adults as well; mechanical interventions that are well integrated into anatomy and physiology are typically experienced as part of one's self, and thus experientially different from invasive interventions that are independent of one's bodily organism. A pacemaker embedded in the body is unlikely to be seen as alien to the self in contrast to an invasive, penetrating device such as an endo-tracheal tube (McCullough, 2004).

The context of a mechanical chimera can also render professional status and caregiving norms problematic. Lacking clear conceptualizations for this "novel category of persons," society and professionals find themselves deprived of a common moral structure to direct or constrain actions (Kohrman, 1995, 56). While the modern chimera is a metaphor for a ventilator-dependent child, Kohrman also suggests Bellerophon may be a metaphor for the physician, that is, a rescuer commended for nobility but potentially "done in by his own arrogance and hubris" (Kohrman, 1995, 55). The punishment meted out to Bellerophon for overreaching presents a warning to medical professionals and to a society that enthusiastically

5 The Machine in the Body: Ethical and Religious Issues 223

employs and uncritically advocates for technology as the means of rescuing humanity from its maladies; indeed technological "immoderation leads to unfulfillable expectations and our own downfall" (Kohrman, 1995, 55).

While many people in the developed countries have constant personal interactions with technology such as computers and cell phones, Kohrman draws a distinction between those with chosen or voluntary dependence (the majority of persons) and those with an involuntary dependence (for example, quadriplegic children). Lacking clear social categories and moral direction for appropriate care for the modern chimeras amongst us, Kohrman asks rhetorically, "what is it worth (morally) to our society to keep (technology dependent people) alive and optimally supported?" (Kohrman, 1995, 60). Ultimately, he contends, we must resort to metaphor and mythic narrative to begin to make sense of such questions as we inevitably encounter them through the lives and bodies of persons for whom dependence upon machines is an experiential reality: "We cannot abandon, however, our efforts to know better what these people mean—for us, for themselves, for society—as if their situations had no parallels in our collective experience from which we might draw guidance or, at least, reassurance" (Kohrman, 1995, 61). Thus the metaphorical imagination is a prelude to casuistical deliberation on liminal status that draws on parallels from prior experience.

5.3.1.3 Golems

A persistent, if marginal, narrative in classic Jewish texts of an artificially created being, *the golem*, has long been noted as an interesting source for communal reflection on human creation of less-than-human life. A golem—either a calf or a man and only rarely a woman—is a being created as an adult, always with language, and often from the elements of clay: dust and water. Human golems in the earliest texts are created along with edible cows and exemplify how the tradition acknowledges the permission and possibility of practical magic under careful circumstances.

For prominent interpreters, such as Moshe Idel (1990) and Gershom Scholem (1965), the golem legend draws its potency precisely because it suggests the power of rabbinic scholars both to create and to control their use of the entities they create. Clearly not human—for not born of a human woman—golems are nonetheless useful for human interests, especially in Jewish communities of resistance who despair of embodied armies for revenge against an oppressive dominant community. The golem tradition resurfaces additionally in moments of tension between the body and the word—golems are created by placing the names of God into the body of a dead child, for example.

A careful account of the golem story can be instructive for the human encounter with the non-human. In his magisterial, scholarly analysis, Idel traces each appearance of such a being in rabbinic writings, while drawing on the earlier work of Scholem, whose attention to golem narratives was shaped by his scholarship about both the mystical and the messianic in Jewish tradition. Scholem identified the core elements of the golem tales, which begin in the Talmudic period, gain influence in the medieval period, and continue throughout the early modern era (Chabon, 2005).

Most scholars trace the origins of narratives about creation of golems to the intricate discussions on the *Sefer Yezirah*, or Book of Creation. Rabbinic responsum describe golems who are created by human beings, with marginal disputes occurring over whether the golem has the status in Jewish law of a human or an animal, or rather is unique. In most texts, as Idel notes, however, the line between creation of a golem and creation of a human is clearly established.

The golem stories themselves draw on a broader and distinctive Jewish worldview: the world as created is incomplete, and improvement, including healing, is a divinely entrusted responsibility for human beings. The world is not fallen, as in Christian tradition, but instead elicits from humans innovation, knowledge, and technical skill to achieve completeness. The warrant for *tikkum olam*—healing and repairing the unfinished world—in Jewish law is linked to the call for *pekkuah nefesh*—the requirement to set aside the entire Torah, if necessary, to save life. Normative Jewish responses to technology can always be understood in light of these two central responsibilities. The status of golems is that of a tool by which these requirements can be fulfilled; hence human use of a golem is necessarily instrumental.

Human self-identity emphasizes especially the capacities for creativity, innovation, and responsiveness to the mandate for healing. A golem may be a living being and have human form, but it lacks characteristics essential to personhood and utterly lacks core attributes of moral status, most tellingly, that of being born from a woman's body. The rabbinic disputes about the legal status of a golem in Jewish law raise yet again themes of liminality. To have one's bodily functions sustained by machines does not transform a human into a human-machine chimera along the lines suggested by Kohrman or into a golem; importantly, machines cannot have the experience of a human body. As stated by Jewish thinker Byron Sherwin "It is human embodiment that provides the basis for the uniquely human way of being-in-the world" (Sherwin, 2004, 145).

In Jewish commentary, those entities that are created by human beings, including new life forms, must remain under human control. In this regard, Sherwin contends, "the golem legend has important meaning for the unfolding of certain current technologies, especially robotics and nanotechnology" (Sherwin, 2004, 149). The golem therefore is a community resource of metaphorical imagination that can provide understanding, meaning, and perhaps constructive guidance on addressing aspects of incorporation technologies.

In response to Brasher's contention that the classical religions lack the conceptual resources to address the social and ethical challenges of incorporation technologies, we argue that the appropriation and use of metaphor to render familiar the unfamiliar is both a human and a necessarily religious enterprise. It is of course the case that metaphors both reveal and conceal, illuminate and hide, and this is of particular importance in the contexts of the chimera and golem metaphors we have discussed. While offering a resource of meaning for professional and religious reflection, we have indicated that what unites these two diverse narratives is the phenomenon of liminality, the status of being "in-between" two forms familiar in human experience. The chimera and golem provide grounds for familiarity, but perhaps conceal a compromised human identity. A ventilator-dependent child is ultimately a child, not a fire-breathing new species. The artificially-created being, the golem, sustains our attention, rather than its human creator.

Thus, our claim is first that metaphors train our vision, helping us to *see* moral issues against a backdrop of familiar moral horizons. They draw on and can expand our imaginative capacities, but do not themselves provide moral conclusions or solutions. They are the context within which both problem-seeing and problem-solving are situated. Secondly, metaphor and mythic narrative has a dual aspect of revealing and concealing: Metaphors can both expand our moral vision and also at times inculcate a kind of moral myopia that must be guarded against.

5.3.1.4 Casuistry

The concept and content of metaphor and mythic narrative provide a way for religious communities, as well as secular society, to situate the technologies of bodily incorporation within larger narratives about the world, the self, and community. Should relevant similarities or even precedents be identified from past experience, reasoning from similar cases, or casuistry, can then be employed. It is thus important to examine the extent to which mechanical devices incorporated within the body are portrayed or experienced as relevantly similar to societal experience with other technologies in biomedicine.

The application of such devices may be perceived as continuous with prior human experience with accepted mechanical, chemical, or genetic technology, as illustrated previously by the metaphor of DNA as a form of nanotechnology. If these other methods of altering nature and intervening in the body are apt ways of seeing incorporation technologies, we can then establish a continuum of technological permissibility. According to McCarthy and Smith (2002), for example, contemporary technological refinements are part of a continuum of basic scientific research into restoring bodily functionality that dates back five hundred years. Others situate incorporation technologies as one method among many—genomics, tissue engineering, and stem cells—in the domain of regenerative medicine (Haseltine, 2003).

The question thus seems to be not "whether" society should approve of permanently implantable mechanical devices; that matter of moral acceptability arguably appears to have been settled some time ago. Rather, the issues now concern the circumstances under which their use should be approved, and what restrictions should limit their use. According to this perspective, our moral cues can be derived from professional regulatory approaches already adopted to address ethical and social questions in genetics or tissue transplants. What is not required is any innovative ethical or religious thinking; this account of a continuum of technological intervention in the body may indeed offer one explanation for the relative absence of sustained religious discussion about incorporation of mechanical devices.

Alternately, incorporation technologies may be construed as different not just in scale, but in kind, and thereby as a novel human experience with technology. The use of diverse mechanical devices may be viewed as an opportunity for exploration and liberation from aspects of human suffering, and/or as radically dangerous, perhaps exploitative, and thus engender ethical resistance. Indeed, incorporation technologies may be portrayed as so radically novel that a new ethical approach is required. An example of the argument from novelty is expressed by Roco and Bainbridge: "Perhaps wholly new ethical principles will govern in areas of radical technological advance, such as the acceptance of brain implants, the role of robots in human society, and the ambiguity of death in an era of increasing experimentation with cloning" (Roco and Bainbridge, 2002, 22). Notwithstanding their speculation on the need for a "wholly new" ethics, Roco and Bainbridge do refer to a core cluster of values, and maintain that even given the uniqueness of technologies that combine nanotechnology and communication and information systems, "human identity and dignity must be preserved." However, the authors seem oblivious to the ways that utilizing incorporation technologies, as well as other biomedical and communication technologies, can call into radical question precisely those conceptions of "humanness" and "identity" upon which "human dignity" depends.

In either the perspective of a technological continuum, or of technological novelty, casuistic deliberation must be preceded by an interpretative situatedness of the technologies in social experience. Situational interpretation, relying on the metaphorical imagination, and practical decision-making about incorporation technologies, relying partly on casuistic methods, are thus integrally related. While the metaphorical imagination relies on bringing past stories and experience into the present, in advancing into the future, Catholics and Jews especially, but other communities like Muslims and Puritans as well, have turned to casuistic reasoning so as to both apply and revise the moral norms of the religious community. As illustrated in the Islamic reasoning method of *qiya*, if a religious community engages issues for which there is no direct guidance in its foundational texts, such as the *Qur'an* or the *hadith*, then the tradition's scholarly community relies on "making an analogy from precedents found in the authoritative texts" (Kelsay, 1993, 32).

In the recent re-appreciation of casuistry (Jonsen and Toulmin, 1988; Leites, 1988; Keenan and Shannon, 1995; Keenan, 1996), a number of bioethicists, Jewish scholars, and Roman Catholics have reflected on the dynamism intrinsic to the method of casuistry (Kopfensteiner, 1995; Noonan, 1995; Gallagher, 2000). They have shown repeatedly that through casuistry a variety of communities have not only entertained novel circumstances by applying their moral norms to new cases, but communities have come, in the applications themselves, to newer and finer understandings of the claims and relevance of their norms.

Through casuistry, understandings of ourselves and our nature inform the way we use casuistry and the way we understand moral responsibilities. Our understanding of these responsibilities keeps pace, then, with our own self-understanding and eventually religious communities begin to amend the moral norms. Through casuistic reasoning, traditions are constantly moving, erroneously, fraudulently, correctly, unselfishly, luckily, intentionally forward to newer perspectives on ourselves, our moral requirements, and nature. Casuistry then provides religious communities and ethicists with a mode of not only articulating moral claims, but more importantly of prudentially advancing across the newer and unfamiliar frontiers of our grasp of ourselves and nature. It appropriates the wisdom of the past to both the similar and different circumstances of the present.

Not surprisingly, Brasher's critique of the inadequacy of the classical religions to address technological innovations overlooks entirely the indispensable role of casuistry in numerous faith traditions. Patterns of casuistic reasoning offer a bridge from matters of metaphorical meaning and problem-seeing to normative ethics as current and prospective technologies are appraised in light of relevant community-forming narratives generated from past experience. Hence, religious traditions are not mired in the muck of a pastoral past, as Brasher contends, but can evolve within a technological context. Moral deliberation may not be immediately determinable but can be constructed out of resources currently embodied by the particular tradition.

Our argument then is that faith communities can draw on both imaginative and deliberative resources in their tradition. Yet, Brasher also contends that human-machine "coupling" creates insurmountable problems for religious conceptions of embodiment. We now wish to turn to the religious implications of incorporated medical devices that impinge on the body.

5.4 Body, Machine, and Self

Among the anticipated benefits of incorporation technologies in the next two decades is that "the human body will be more durable, healthier, more energetic, easier to repair, and more resistant to many kinds of stress, biological threats, and aging processes" (Roco and Bainbridge, 2000, 5). This observation is striking for many reasons, among which is the implicit metaphor of the body as a machine that will become more readily repairable. Indeed, some researchers deny that trends toward human-machine integration are actually occurring because human beings are, after all, just complex machines. Our research overview provided glimpses of this view, such as the understanding of the molecular composition of DNA as a form of nanotechnology, or more generally, the scientific representation of the human bodily organism as a complicated mechanism.

This position is explicitly advocated by MIT researcher Rodney Brooks: "If we accept evolution as the mechanism that gave rise to us, we understand that we are nothing more than a highly ordered collection of biomolecules (which) interact with each other according to well-defined laws." Brooks follows this logic of mechanistic reductionism to its conclusion: "This body, this mass of biomolecules, is a machine that acts according to a set of specifiable rules. ... This body is a machine, with perhaps billions of billions of parts, parts that are well ordered in the way they operate and interact. We are machines, as are our spouses, our children, and our dogs" (Brooks, 2002, 172–173). In such an account, incorporation of machines in the human body is not a unique phenomenon but simply extends those mechanistic processes and laws that govern evolving life. Concerns that such interventions will culminate in a transformed and compromised humanity fundamentally misunderstand human identity.

Existential distress over the prospect of becoming identified in body and self with our machines is attributed by Brooks to a deep-rooted belief that human beings, or their essential nature, are "special." Such a view has been at the core of religious and philosophical thinking for centuries, but in a new era that dares to dream of human-machine mergers, we must inevitably acknowledge that anthropocentrism is an anachronistic anthropological assumption that can now be discarded. We should, Brooks maintains, already have learned the error of human distinctiveness from the two "major blows" such a worldview received from the Copernican and Darwinian revolutions. This diminished claim of human uniqueness has been reinforced in the 20th century by advances in physics and genetics (Brooks, 2002, 163–164). On Brooks' view, then, it is much too late in the discussion to voice ethical questions about *bodily* incorporation of mechanical devices.

There are profoundly problematic assumptions in the reductionist account offered by Brooks (or the parallel "coupling" problematic referred to by Brasher). Brooks' perspective builds from the conception that biomolecules are machines, to the view that body parts, and then the body as a whole is a machine, and ultimately to a characterization of persons as machines. It is one thing to acknowledge that the human body can incorporate machined or mechanical devices into its organic processes in ways that restore functioning. It simply does not follow from this that the body in its organic totality *is* a machine, as though the sum were simply explicable through its parts. Secondly, attributing mechanical characteristics to the body as Brooks does hardly means that *a person* is a machine.

Indeed, a real contradiction between the paradigm and practice emerges when the reductionistic position of human as operating machine is enacted in everyday experience. Brooks acknowledges, "when I look at my children, ... I can see that they are machines interacting with the world. But this is not how I treat them." Instead, he accords his children (and presumably, many other persons) "special" treatment that is rooted in a version of unconditional love. The question is, if humans are no different in kind than a heart valve or an insulin pump (or a microscope), why such differential treatment is warranted. To this query, Brooks compares his position to that of a "religious scientist," conceding he maintains "two sets of inconsistent beliefs and act on each of them in different circumstances" (Brooks, 2002, 174).

What we should not conclude from this 21st century version of circumstantial or situation-based ethics is that there are *any* circumstances that justify treating a person just as we would treat a machine. Instead, the inconsistency Brooks acknowledges should instead reveal that something is very mistaken with his mechanistic reductionist account of the body and the person. Having used reductionism to discard a mind-body dualism, Brooks is unable to avoid his own pragmatic dichotomous compartmentalization. Actions are partitioned from beliefs, and love (and emotion in general) is portrayed as incompatible with rational analysis. Religion and science are no less presented in a simplistic oppositional relationship. This

5 The Machine in the Body: Ethical and Religious Issues   229

failure of intellectual coherence renders the claim of an already mechanized humanity for which incorporation interventions are immune from moral scrutiny deeply suspect.

Having divested nature and the body entirely of ends, meaning and purpose, Brooks is faced squarely with an ethical question about why he doesn't treat others (or himself) merely as a machine. By contrast, faith communities locate the core issue at a quite different juncture precisely because the body has a status and meaning distinct from mechanical functioning. The prospect of merging mechanical and human identity should necessitate profound reflection, and because the person and their body cannot be reduced to a machine, incorporation interventions require justification. Here we briefly discuss religious perspectives on embodiment and its implications for incorporating machines into the body.

5.4.1 Asian Traditions

While traditional Eastern cultures have not engaged in much direct reflection on matters such as retinal implants, nanomedicine or other manifestations of incorporation technologies, Asian attitudes about the body offer directions for a constructive discussion. Interesting agreements between South Asia and East Asia (otherwise so different philosophically) include an ambivalence towards the body that carries over into ambivalence towards engineering or mechanized models of the body. Asian attitudes on the human body also provide a sharp contrast with modern secular Western attitudes. For example, none of the three Asian traditions de-sacralizes the body in the Cartesian fashion that Brooks accepts without question. Rather than understanding the body as a machine open to re-engineering any way we like to augment its functional efficiency, Asian traditions affirm that improved psycho-physical integration is to be attained through spiritual practice. This view does not deny the value of technological interventions if they can reduce suffering, e.g., such as correcting disabilities. However, mechanical devices cannot address our deepest *dukkha* or suffering (see below for further discussion), which requires our own efforts to transform our ways of thinking, feeling and acting. Insofar as technologies of incorporation interfere with our ability or inclination to engage in spiritual practice, they become problematic.

5.4.1.1 India

Hindu emphasis on the spiritual identity of *atman* (Self) and *Brahman* (Being) de-emphasizes the body and its sense organs. The important contrast is not mind versus body, but human propensity for self-identification with one's "mind-body" instead of the impersonal and transcendental *atman*—one's true Self. What the West designates as "mind" is in Hindu thought a more subtle version of the material body, which includes all of one's karma and mental tendencies. At death, the functions of the physical body cease, but these "mental bodies" transmigrate to another physical body. Only spiritual ignorance allows us to think that we are or have any of these bodies: one's true Self is transcendent and indifferent to all of them. Spiritual practice involves not transforming the Self (which cannot be improved or damaged), but strengthening and controlling the psycho-physical integration of our mental and physical bodies. This comprises the authentic integration that enables alienation from those bodies and realization of the true Self.

What does this understanding of body, Self, and Being imply for incorporation technologies? A significant ambivalence. The basic problem of Hindu ethics is not sin but *avidya* or ignorance/delusion. The material world is not evil but displays the divine play of *maya* (illusion). The complete transcendence and indifference of our true Self suggests that there is no necessary religious objection to incorporating mechanical devices to modify either the physical body or its cognitive functions. However, in addition to customary medical concerns about safety and effectiveness of a device, Hindu thought would require consideration of how incorporation would affect (or damage or assist) the spiritual practice of psycho-physical integration. Moreover, an everyday concern of Hinduism with the body is the danger of *pollution*, e.g., physical contact with an "untouchable," which requires ritual purification. Would silicon chips, or nano-engineered bodily tissues, be polluting in a similar way? The acceptance of heart pacemakers, for example, among the Indian middle class suggests not.

Yet the spiritual *context* of the Hindu view about our body-mind complicates such a straightforward evaluation. If each *atman* is reincarnated many times according to our karma, too much concern about this particular body, even if we experience physical impairments, may reflect spiritual immaturity; indeed, if disability is due to karmic retribution for one's past actions, a person would do better to improve karma for a better rebirth. Ideally, a person should be singularly devoted to spiritual practice to realize the true Self, a liberation that will bring freedom from attachment to this fleeting world of appearances. The important struggle, therefore, is not outside oneself but within oneself; the most important goal is spiritual wisdom, not improving our physical, material or social environments.

5.4.1.2 China

As Chinese civilization is quite different from India in many ways, it is significant that the same ambivalence regarding the body emerges. This ambivalence is perhaps most obvious in the striking contrast between philosophical Daoism and popular Daoism. The Daoism of Laozi and Zhuangzi emphasizes accepting the alternation of life and death, joy and woe, for they are all part of the Dao (Way). Physical disabilities and mental limitations also originate in the Dao.

Popular Daoism, by contrast, is preoccupied with various methods—alchemy, breathing exercises, herbal drugs, magic, sexual hygiene/gymnastics, and special diets—to extend life, gain special powers, and attain immortality. The wide variety of practices employed implies appropriation of new technological devices that would assist in achieving such goals, provided they do not harm one's *qi*. The popular Daoist tradition offers medical precedents of impingement: early Chinese practice (along with some Buddhist hospitals) pioneered the insertion of penetrating metal objects (e.g., surgery, needles) in the body as ways to improve health.

Qi is our vital force or spiritual energy; the concept points to a hylozoistic unity of matter, spirit and energy. Chinese spiritual practice seeks to remove obstacles in one's own life-force and unify it with the *qi* that pervades the whole universe. Strictly speaking, there is nothing outside of *qi*, but the *qi* of silicon chips and other incorporated implants might be incompatible with the organic *qi* of human bodies. This question could, however, be open to empirical resolution since the practitioners of traditional medicine and martial arts have methods, such as acupuncture, to measure *qi*. Significantly, new technologies have led to a divide within acupuncturists: reformers embrace new techniques such as electro-acupuncture, while traditionalists reject such methods as too aggressive.

Traditional Chinese religion emphasizes ancestor veneration: an important aspect of filial piety is not defacing "the gift from the ancestors," i.e., one's own body, which should be carefully preserved. This is true even after death: both China and Japan continue to have strong taboos against dissection, limiting the availability of organ donation. These beliefs suggest possible resistance to incorporation technologies, but if the body is not defaced—a concept susceptible to different interpretations—there seems little basis in Confucianism or popular religion for opposition.

Philosophically, *qi* refers to a third ontic category that is neither mind nor body but is another dimension of being that underlies both. The non-duality between our own *qi* and the circulation of the Dao's universal *qi* implies that the human body is a microcosm of the macrocosm. Chinese texts emphasize the responsibility this entails: we are to "follow the Heavenly Principle without yielding to selfish desires"—i.e., without being self-centered in ways that diminish the human capacity to participate in the ceaseless transformative processes of heaven and earth.

Chinese religion is not concerned with a Reality transcendent beyond this world, but instead emphasizes the ebb and flow, the *yin* and *yang*, of flux. A concern with social harmony leads to primary focus on "person-making," by which the rational, emotional, physical, moral and aesthetic aspects of our being are developed and integrated.

5.4.1.3 Buddhism

What Lambert Schmithausen concludes about the Buddhist attitude towards nature also applies to that part of the natural world with which we are most familiar. Early Buddhist texts are impressed not so much by the beauty of nature as by its more somber aspects—the struggle for life, the prevalence of greed and suffering, and especially the universality of decay and impermanence (Schmithausen, 1991). However, the Buddhist response is not violent subjugation but rather effort towards transcending nature—and the physical body—spiritually. Within Indian Buddhism the body is unclean, even humiliating; we are strongly encouraged to cultivate non-attachment towards the body. A preoccupation with the body is the problem, not the solution.

According to traditional Buddhist teachings, we experience illness either because our four elements are unbalanced (i.e., our immune systems are weak) or because of karmic retribution. Neither source of illness implies a "hands off" medical or existential passivity. Bad karma is not punishment that must be patiently endured as fate; it is just another source of *dukkha* that should be relieved when it can be relieved. The same holds true for alleviation of physical disability.

With regard to incorporation of medical devices, then, the Buddhist questions become: how much do such interventions contribute to reducing physical and mental *dukkha*? And, how much do they aggravate *dukkha*, by encouraging attachment to the body, which remains subject to impermanence, illness, and death? If new technologies genuinely relieve suffering, the Buddhist response is straightforward: why not? Our bodies, like all other life forms, have no essence ("self-being") that could be violated by invasive procedures, mechanical devices, or artificial grafts. The more important issue is whether such interventions would render the recipient unable (or less likely to pursue) the Buddhist path of self-transformation—a way of life that emphasizes disciplining one's mind-body in order to realize a freedom from being preoccupied with its inevitable decay and death.

5.4.2 Abrahamic Traditions

5.4.2.1 Roman Catholicism

The understanding of the human body, as retrieved from the Scriptures and from the practices and theology of the early, Medieval and Renaissance Church, is discovered never as an object, but always as a person or as a subject. Were the body an object, Christian tradition might concur with classical Greek philosophers such as Plato that our body is incidental to who we are. A turn to Christian revelation and history shows that the task of the Christian tradition is to preclude the tendency to isolate and objectify the body (McGuire, 1990), and in the current context, to resist the scientific paradigm of mechanistic reduction of the body. Regarding this task, Cardinal Walter Kasper offers this reminder:

> The body is God's creation and it always describes the whole of the human and not just a part. But this whole person is not conceived as a figure enclosed in itself, as in classical Greece, nor as a fleshy substance, as in materialism, nor as a person and personality, as in idealism. The body is the whole human in relationship to God and humanity. It is human's place of meeting with God and humanity. The body is the possibility and the reality of communication (Kasper, 1976, 150).

For Roman Catholics, understanding the person as embodied provides a complement to the virtues in stabilizing the moral journeys of communities of culture and faith. Despite the commonplace belief that Christianity has maintained a negative stance toward the human body (Nelson, 1992, 9), historians, biblical scholars and theologians argue that the Christian tradition has continuously appraised the human body as integral to the Christian's identity (Newell, 2000). Basic liturgical practices (baptism, the Eucharist), beliefs (Incarnation, resurrection), and community (the church as the ongoing body of Christ in history) presuppose a profound Christian commitment to an embodied self-identity.

Reflecting on the Greek word for body, *soma*, Protestant scripture scholar Rudolf Bultmann resonates with Catholic insights when he argues that, for St. Paul, "*soma* belongs inseparably, constitutively, to human existence... The only human existence there is—even in the sphere of the Spirit—is somatic existence" (Bultmann, 1952, 192; Jewett, 1971). Moreover, the body is so integrated into human existence that the human does not *have* a *soma*, but rather *is soma*.

In this light Antoine Vergote contends that the human "is not someone who has a body but whose existence is corporal." The scriptural understanding of "the resurrection event does not imply the thesis of an immortal soul; on the contrary, it suggests the idea that the body is the whole man" (Vergote, 1991, 96–97). The human body provides then the stuff for resurrection. Through our corporeality, we are related and, thus, can be caught up in Christ, who transforms that corporeality. Biblical scholar Wayne Meeks makes a similar point in quoting St. Paul: "Christ will be magnified *in my body*, either by life or by death" (Phil. 1:20) (Meeks, 1993, 134). This scriptural interpretation is carried forward in the patristic age, Brian Daley finds, in which human destiny—as defined in the Risen Christ—is the opportunity and the demand for all people to find in their own bodies the fullness of the Spirit of Christ (Daley, 1990, 32).

It is not primarily belief in the goodness of creation that compels Roman Catholics to confess that the body is a temple of the Holy Spirit or to believe in the resurrection of the body. Rather Catholics understand the theology of the body through the lens of the incarnation of Jesus Christ. That God became human flesh, that God was incarnated in Jesus Christ, means that the human body is the suitable way of trying to understand ourselves as God understands us (Keenan, 1994).

Roman Catholic teachings on the body thereby provide theological-ethical resources to address the moral challenges of incorporating mechanical devices. How comfortable are Catholics with the fact that we are literally making ourselves more technological, more machine-like? Should the purposes of these technologies be restricted solely to healing or may they be used also to enhance? Catholic reflection on these questions is enriched by a consideration of ascetic practices, for like other faith traditions, the way Catholics image their faith is the way of life they want to see enacted.

The human body is a veritable treasure trove for Roman Catholic asceticism. Catholics parcel the body parts of saints because they believe that through the body parts we approximate the holiness of the saints. This appreciation for relics stems from the early Christian practice to gather in worship in the catacombs where early Christians could celebrate the hope of resurrection in the "presence" of those who, having gone before them, were actually with the risen Christ.

Underlying Roman Catholic appreciation for the incarnation, passion, and resurrection is a deep belief in the extraordinary suppleness of the human body. This feature is exemplified in the Virgin Mary who conceived through the Holy Spirit and bore Jesus as the incarnate one. The malleability of the body holds enormous repercussions for incorporation of mechanical devices, for just as Catholics believe that in Christ's body he assumed all of humanity, similarly we see in his body the *full incorporation* of anything human (Bynum, 1992). Jesus is nailed to the cross. He is portrayed often, especially in the *ecce homo* ("behold the man") motif, with a crown of thorns still intact in his head. Jesus bears in his body the signs of human redemption.

The ability of the human body to incorporate is astonishingly celebrated in the long-standing history of Catholic iconography.[1] For centuries Catholics have seen human bodies affixed with an array of metal and wooden objects. These objects are not evil; they are instead transformed precisely by being incorporated into the glorified body of the saint. The body's ability to incorporate the divinity as it does in Jesus implies it is quite able to incorporate the inanimate as well.

Catholic asceticism provides a context, then, for implanted hearts, lost limbs, and metal hands. Moreover, incorporation of the mechanical could encompass both therapeutic and enhancement purposes. The capacity of the body for incorporation implies there are no warrants *in se* for prohibiting the possible enhancement of ourselves by incorporating non-human elements into our bodies. Certainly the virtues of justice or self-care, for instance, might prohibit some enhancement procedures simply because they could give a person too much power over others or they might actually cause disproportionate harm to the agent. But in itself the line between the therapeutic and the enhancing is ineffectual.

5.4.2.2 Jewish Thought

Jewish tradition rests on an interpretive, discursive and casuistic tradition in which the events and advances in technology and science are understood against a horizon

---

[1] For instance, Catholics are able to recognize a saint by the instruments of their martyrdom, as exemplified in the martyrdom of Catherine of Alexandria who, while being tortured to death on the wheel, touched the wheel and miraculously broke it. The emperor subsequently beheaded her and so Catherine is always portrayed in Catholic iconography with a broken wheel and sword in hand. Catherine is hardly the only saint with such identifying instruments. Lawrence, a martyred 3rd century deacon, is always dressed in his dalmatics, a sign of his diaconate; he stands holding the gridiron on which he was roasted.

Some saints bear their body parts that they lost in their martyrdom. Lucy, whose eyes were gouged out during her martyrdom is portrayed holding her eyes on a plate, her offering to God. Lucy is often accompanied by Agnes, whose breasts are also on a plate. Not surprisingly, Lucy and Agnes are respectively patron saints of persons suffering from ailments of the eyes and breasts. Perhaps the most remarkable iconographic portrayals are those of Saint Bartholomew, who was flayed alive. In a brilliant depiction in the Sistine chapel, Michelangelo shows the saint in glory praising the risen Christ. Bartholomew holds in one hand the scalpel, in the other his complete epidermis, and the face bears the painter's own visage.

Still other saints are depicted as actually incorporating within their body the instruments of their torture. Stephen, the first Christian martyr who was stoned to death, is seen with rocks stuck on his head. Peter Martyr is depicted with a hatchet in his skull. The celebrated Sebastian bears a half a dozen arrows in his torso. The symbolic importance of these portrayals can only be understood by looking at the iconography of Jesus.

of covenantal law. One of the sources for this framework is embedded in the structure of covenantal life and its relation to the body: The circumcision of every male infant on the eighth day of life has long been understood as a graphic recognition that nature is not normative (Feldman, 1986; Maimonides, 1969) for if nature was normative, such an act would be impermissible. Rather, this intervention on the male body as a covenantal symbol was a statement of human faith and human alteration against powerful instinctual response.

Medicine is a religious vocation in Judaism: its object is to maintain the health of the body so that the adherent can pursue knowledge of God and of the holiness of character modeled by the divine. Jewish ethics is grounded in an affirmation of the embodied existence of persons and the integrity of bodily life (Ramsey, 1970, 187) to such an extent that, as expressed by physician and author Sherwin Nuland, "Jews are in effect the people of the body" (Nuland, 2005, 18).

The central Jewish commitment to a duty to healing of the body and the preserving of life has led to the consideration and ultimate acceptance of technological interventions in the body. David Golinken notes that debates within Jewish tradition about how technology as such should be understood and used traditionally have led to their enthusiastic use in every aspect that would heal or enhance human welfare (Golinken, 2000). An early debate on eyeglasses was concerned primarily about whether they were a part of the body, or an object carried, for the purpose of Shabbat, when such carrying is prohibited. Golinken notes that the ruling recognized that since the purpose of glasses is to restore sight, glasses were then understood as more like a part of the body than like matter foreign to the body.

Similarly, scholars Elliott Dorff and Sherwin Nuland, drawing on medieval commentator Moses Maimonides, contend the core moral gesture of healing in Judaism sets aside all but four commandments (prohibitions of murder, adultery, incest, and idolatry) of the Torah (Dorff, 1986; Nuland, 2005). The twin warrants of *tikkam olam* (healing the incomplete world) and *pekkuah nefesh* (the priority of saving life) frame Jewish consideration of scientific developments and their application in incorporation technologies. Jewish bioethicist Fred Rosner draws on these warrants in an argument that implantation of porcine heart valves does not violate traditional Jewish proscriptions against the consumption of pork (Rosner, 1999).

5.4.2.3 Islam

The status of human embodiment is significant for Islamic thinking because the human body is a sign of the unique creative power of God (*Qur'an* 15:26). A person is endowed with a body as a trust from Allah. This trust provides constructive direction for Muslims to further the purposes of Allah on earth and also generates restrictions on bodily actions, including prohibitions of suicide, torture of one's own body, bodily desecration, and homicide. The embodied sign of the divine means that the boundaries of the physical body should be respected.

Respect for the embodied self places limits on, and requires justification for, medical healing: The justification often delineated in Islamic scholarship is the overriding responsibility to save human life. Medical advances have often been justified within the Islamic community by a Qur'anic passage that affirms: "… whoso saveth the life of one, it shall be as if he saved the life of all humankind" (*Qur'an* 5:32).

Nonetheless, Shi'a scholar Abdulaziz Sachedina observes that technological advancements in medicine have posed a crisis for Islamic scholars and jurists (Sachedina, 2004, 1337), as the authoritative sources of Islamic teaching, including the *Qur'an*, the *sunna*, and the *shari'a* do not directly address such questions. In this respect, Islam is no less vulnerable to the critique Brasher makes of agrarian-pastoralist traditions. However, these foundational sources do set out basic religious-ethical principles—bodily integrity, the imperative to save life, non-harm, ameliorating pain and suffering—that provide a context for Islamic considerations on incorporated mechanical devices. For example, Dr. Gamel I. Serour, medical center director at Al-Azhar University (Cairo) suggests Islam can appropriate technological advancements through its imperative to relieve human suffering: "…the use of biotechnology is welcomed and encouraged by Islamic scholars. All these devices (e.g., pacemakers, joint replacements) aim at alleviating the suffering of human beings, which is encouraged in Islamic *shari'a*" (Serour, 2003).

The limit case for these principles in much contemporary Islamic discussion is the issue of organ transplantation. Although solid organ transplants are not representative of the incorporation of a mechanical device, the method of reasoning by analogy in Islamic discussion of transplantation may exemplify patterns in Islamic approaches regarding incorporation technologies.

Solid organ transplantation has become gradually accepted and practiced in Islamic countries. Some scholars have argued that organ donation exemplifies the obligatory *Qur'anic* norm to "save the life of another," and can override concerns about desecration of the body and immediate burial of the corpse. A government policy entitled "Religious Aspect of Organ Transplant" issued in Saudi Arabia in 1982 pronounced: "The [Senior Ulama] board unanimously resolved the permissibility to remove an organ, or a part thereof from a Muslim or a Dhimmi living person and graft it onto him, should the need arise, should there be no risk in the removal and should the transplantation seem likely successful." And in 1987, the Council of Arab Ministers of Health developed the "Unified Arab Draft Law on Human Organ Transplants," which states: "Specialist physicians may perform surgical operations to transplant organs from a living or dead person for the purpose of maintaining life, according to the conditions and procedures laid down in this law" (as cited in Daar, 1991).

What has proved most controversial in debates and enactment of transplant procedures is not the donation process as such, but rather disputes over the legitimacy of brain death in Islamic teaching. The Iranian parliament in 1995 rejected brain death as a condition for organ transplantation, on the grounds that it may appear that the imperative to save life was fulfilled only by taking life (Sachedina, 2005).

According to Sachedina, the use of technology in medical treatment is guided by an Islamic ethical principle: "No harm, no harassment" (Sachedina, 2003). Nonetheless, the argument for bodily wholeness and intactness can be reinforced not only by customary practice but as well in eschatological teaching. Sachedina writes that at the time of the resurrection, "all parts of the human body will have to account for the actions of the person whose bodily organs they formed" (Sachedina, 2004, 1332).

The ambivalence expressed in Islamic traditions and practices regarding organ transplantation reflect priorities or conflicts in the values and norms of the tradition. By contrast, the representative incorporation technologies we have described pertaining to internal organ functioning, or neuromuscular control for the alleviation of disability, would likely not prompt such reservations because they would not risk violation of the norms of bodily integrity, no harm, saving a life, and alleviation of suffering. Unlike the case with organ transplantation, most incorporation technologies do not involve a donor body whose integrity may be compromised by a non-therapeutic intervention, nor is there risk of harm to the donor. In fact, stress on the significance of bodily integrity may make incorporation technologies religiously preferable. Muslim scholar Ebrahim Moosa reports that recent authoritative edicts, or *fatwas*, from some religious leaders in Pakistan and in Saudi Arabia have expressed a preference for mechanical, prosthetic devices rather than transplants (Moosa, 2004).

Summation: None of the religious traditions described above reduce the body or the person to a machine in the manner that Brooks does. However, it does not follow that a non-mechanistic paradigm regarding the body issues in an abundance of prohibitions or restrictions on the incorporation of mechanical interventions. Indeed, religious diversity is pronounced, reflecting themes of ambivalence, appropriation, and occasionally resistance. The Asian traditions display perspectives of ambiguity and ambivalence about the body; these patterns of practice and thought about the body provide grounds for resistance, ambivalence and appropriation of mechanical devices in the body. Islamic teaching, while still evolving on organ transplantation, may find mechanical interventions preferable in some circumstances. Roman Catholic teachings that affirm the body as a constituent of human identity, and that portray the body as capable of incorporating the human, the divine, and the inanimate, are suggestive of a moral posture more amenable to both therapeutic and enhancement objectives of incorporation technologies. Similarly in Jewish discussion, the overriding importance of the mandate of healing provides grounds for acceptance and appropriation of incorporation technologies.

5.4.3 Religious Moral Diversity

Traditions of religious ethics tend to differentiate a morally significant act into three aspects that correspond with features of human action: the *motivation* or *character* of a moral agent (for example, compassion or egoism); the *rule* or *method* (for example, a Buddhist precept, a Jewish law, or a Christian commandment) followed by the agent in performing the action; and the *results* sought by the agent. These three aspects of moral assessment cannot be separated from each other, yet both religious and secular traditions customarily stress the significance of one feature more than the others.

Roman Catholic teaching affirms that persons must move beyond asking whether they are acting out of religiously acceptable or "good" moral motivations to whether their deliberations about particular courses of actions are specifically morally right ones. Nonetheless, a moral community that proceeds forward through dynamic periods of change invariably falls back on the virtues. The virtues of justice, prudence, charity, compassion and others have given content not simply to what courses of actions moral agents should take, but also what kinds of persons they should become. Perfecting both the agent and the action, virtues guide communities as they venture into new terrain and are integrally related to understandings of human identity.

Judaism is shaped by the complexities of duty (Feldman, 1968). The person guided by legal considerations is affirmed in this by a covenantal conviction that if the commandments are kept—both individually and by the community—good outcomes are promised. This covenant further assumes that acts within the complex interpretive legal network are the core element of shaping the *middot*, or good character of the person. The pragmatic insight that a praxis of duty shapes a character to duty presupposes that intuitive moral judgment is insufficient for moral conduct. Motivation toward the individual's singular sense of the good is shaped by yearning, context or passion. In fact, a core prayer cautions against following one's own "eyes" or internal desires, instead of carefully following the laws of the covenant as interpreted in one's discursive community. Personal agency is always subject to the set of obligations of a covenantal community.

Similar differences in moral emphasis can be discerned in Eastern religious traditions. In early Hinduism, the Brahminical understanding of karma emphasized the importance of following detailed procedures (rules) regulating religious ritual; understandably, however, those persons who paid for the rituals were more interested in the outcome (results) than procedural forms. As a response to this emphasis, the teachings of the Buddha shifted the focus of moral action for his followers to motivation.

The foregoing reflections illustrate a fundamental point about the character of normative religious ethics: the diversity of religious teaching and community defy any stereotypical "either/or" schema in understanding normative religious arguments and conclusions.

1. It is both possible and likely, given the nature of religious diversity, for religious traditions to arrive at different patterns of normative conclusions on incorporation technologies in general, or on particular technological applications.
2. Even when religious traditions agree on their moral conclusions, they might diverge in the content or method of reasoning for such conclusions. The opposition of conservative Protestant Christians to some incorporation technologies (described below) focuses primarily on the prospect for undesirable results given the tradition's emphasis on the fallibility (sin) of human nature; although ends or results are important in Buddhist (the cessation of *dukkha*) and Catholic

(human flourishing) ethics, these traditions also place substantive moral emphasis on motivations and virtues. In Jewish thought, violations of the law create questions about use of technology, but the tradition also directs moral attention to agency and to how virtue can be taught and encouraged.
3. Even when religious traditions seem to agree about which feature of moral action to emphasize, they may disagree about moral conclusions. Buddhist teaching may express ambivalence and skepticism about a narrow economic and utilitarian "progress" achieved through technology, and hence draw some moral lines between the permissible and the forbidden. A teleology of progress as affirmed in Catholic teaching may alternatively lead to skepticism about drawing such lines.

5.5 Consciousness, Identity, and Machine

It is difficult to introduce a bodily invasive medical technology without evoking some moral anguish about technology transforming or altering human nature or identity. This has been true of transplantation technologies, reproductive technologies, genetic technologies, and is also a concern about incorporation technologies. Some authors have (unsympathetically) referred to this as the "essentialist" objection to such technologies (Maguire and McGee, 1999, 10), in that it presumes some unchanging essential human self.

As the history of these other technologies suggests, their application in medicine has not wrought changes in human biology; indeed, they rely on human biology for efficacy. The changes that have occurred instead involve cognitive shifts in imagination and possibility: we are now aware of the life-prolonging potential of organ transplants, alternative methods and familial/social arrangements for human reproduction, or decisions confronted by families and policy makers with regard to genetic interventions.

Given the current state of the science, the incorporation devices that impinge on bodily processes seem similarly innocuous with respect to altering human identity. By contrast, those devices—neuronal control of external objects, nanotechnology, or a computer substrate that houses a disembodied consciousness—have generated responses from some religious traditions, no doubt because they seem to impinge more directly on a fundamental aspect of human identity, mind and consciousness. Speculative visions have been articulated in which humans rely on mechanical enhancements to alter the evolutionary course; incorporation methods that begin with the body, and then are directed towards impingements on memory and consciousness, finally culminate in humans transcending finitude and evolving into a *post-human* status. In this respect, incorporation devices are the entering wedge in a scientific and social project to alter human nature and identity.

In discussing these (currently) speculative prospects, we draw here on representative discussions from conservative Protestant Christian thinkers in their disputes with transhumanist philosophy, as well as on Buddhist and Daoist understandings.

5.5.1 Transhumanism

While current applications of incorporation devices are almost exclusively therapeutic, theological writer Antje Jackelen has observed that the objectives of prospective research in robotics, nanotechnology, and bionics will involve not only "repairing" our physical being, but also "correcting nature" and "improving or even overcoming nature" (Jackelen, 2003, 292). This latter aspiration, the use of technology to improve and enhance the human situation, is the core concern of transhumanist philosophy.

Transhumanism has been described by a leading philosophical advocate, Oxford philosopher Nick Bostrom, as "the intellectual and cultural movement that affirms the possibility and desirability of fundamentally improving the human condition through applied reason, especially by using technology to eliminate aging and greatly enhance human intellectual, physical, and psychological capacities" (Hook, 2004, 2017). These improvements in the human self will involve "biological and neurological augmentation through means such as neurochemical enhancers, computers and electronic networks, intelligent agents, critical and creative thinking skills, meditation and visualization techniques, accelerated learning strategies, and applied cognitive psychology" (Transhumanist Declaration, 2003). Such technologies will increase our "health-span" (as distinct from "life" span), extend both physical and intellectual capacities, and provide greater control over mental states and emotions (Bostrom, 2004b). Research and application of incorporation technologies and on-going advances are thus a necessary, though not sufficient, part of transhumanist aspirations. In this respect, the transhumanist position draws on, but does not defend, the erasure of the therapy-enhancement distinction in medicine.

Transhumanism shares with Judaism especially a basic assumption that human nature is not given or fixed, but malleable and incomplete. According to Bostrom, "human nature (is) a work-in-progress, a half-baked beginning that we can learn to remold in desirable ways" (Bostrom, 2004b). The human species has not, in its current form, reached the culmination of its development; rather, humanity has tapped but a miniscule portion of the experiences, thoughts, and feelings that transhumanism contends are possible. We are currently living in the era of the "transhuman," a being with moderately enhanced capacities. With technological enhancements looming, however, we are on the cusp of making a "transition" to becoming "posthuman," that is, a being with radical enhancements.

The conviction of transhumanist thought is that use of various technologies can unlock fuller human potentiality and extend the self in both space and time. The "humanness" of the "posthuman" is not clearly explicated either personally or collectively, but this does not seem particularly worrisome in transhumanism because human nature and identity are held to be constantly in flux and it is simply not possible to define or locate an essential human identity. Moreover, transhumanist philosophy derives human dignity from a forward-looking assessment of our potentiality and destiny, rather than a backwards-looking derivation from our causality and origins (Bostrom, 2004a).

In this context, the "natural" simply cannot be a moral or normative guide. Rather, transhumanism advocates reliance on both "humane values and personal aspirations," including morphological freedom (a concept that refers to the broad discretion individuals ought to have over which technologies to employ for themselves) and reproductive freedom (a concept referring to the autonomous decisions of parents to determine which reproductive technologies they will use in having children). The expectation of transhumanism is that there will be diverse uses of these freedoms that will produce a "continuum" of individuals with different modifications and enhancements. It is compatible with the transhumanist affirmation of freedom that some persons may refrain from using these technologies altogether. Nonetheless, transhumanism gives voice to a value of "wide access" (a principle of equal opportunity) to defend the claim that "everybody should have the opportunity to become posthuman." The limitation to these freedoms draws directly from Mill's "harm principle," that is, that no incursions into liberty are permissible unless such freedoms inflict harm on another person (Gray, 2001, 26–28).

An evident problem with the human condition for transhumanism is precisely our embodied nature, and especially the limitations imposed by our biology. Indeed, "embodiment in a biological substrate is … an accident of history rather than an inevitability of life" (Hayles, 1999, 2). Embodied experience makes us subject to the limitations of finitude, especially aging, disease, and death. In principle, the transhumanist contends these restrictions ought to be susceptible to technological control; the boundaries of bodily life are permeable and fluid: The embodied person is capable of being "seamlessly articulated with intelligent machines. In the posthuman, there are no essential differences or absolute demarcations between bodily existence and computer simulation, cybernetic mechanism and biological organism, robot technology and human goals" (Hayles, 1999, 3).

In this respect, transhumanist philosophy exemplifies an Enlightenment aspiration to rely on the use of medicine and technology to relieve the human condition of biological necessity and suffering (McKenny, 1997). A core priority is to extend the "health span" to enable us to develop our latent potential and emerge over time as "posthuman." Moreover, if technology can liberate us from the burdens of embodiment, it is possible that we will experience higher levels of sensory experience and responsiveness and that our minds will be free to "discover philosophical and scientific truths." Indeed, "our ideals may well be located outside the space of modes of being that are accessible to us with our biological constitution" (Bostrom, 2004b). The Platonic or Cartesian ideal of disembodied consciousness is a transhumanist aspiration to be realized through technology.

Transhumanism may seem to be an erudite philosophy, but its popular manifestations are played out in various cultural forms through the language and image of the *cyborg*. The word "cyborg" was coined in 1960 by NASA researchers Manfred Clynes and Nathan Kline in a discussion of what characteristics would be needed to improve the capacities of human astronauts for space flight. The term cyborg comes from a hybrid of two terms—*cyb*ernetic and *org*anism—that, when joined, refer to the convergence of technology and humanity (Clynes and Kline, 1995, 26–27). The original meaning-context of "cyborg" is an environment unfamiliar, indeed, alien to what has been the experience of human beings historically. While the term "cyborg" is of recent vintage, its themes are not. From the biblical book of Daniel to images in contemporary film, fascination with and fear of a cyborg figure is pervasive. It is no surprise that a cyborg future is the subject of many films in popular culture, such as *Bladerunner*, *Neuromancer*, *A.I.*, *Terminator*, and *The Matrix*. Theologian Anne Kull has observed of this cyborg culture: "We have become cyborgs because our culture's myths have enabled us to define ourselves that way" (Kull, 2002, 285).

In an influential philosophic and cultural analysis of the cyborg image, Donna Haraway has observed that the cyborg figure draws its potency and power precisely by blurring the boundaries of human bodily life that transhumanist philosophy presupposes to overcome through technology. These include first a breakdown in the sharp boundary between human beings and animals: "The cyborg appears in myth precisely where the boundary between human and animal is transgressed" (Haraway, 1991, 152). As described previously in our discussion of Brooks, the Darwinian challenge to anthropocentrism renders this boundary rigid rather than fluid, and once it is breached, human claims to uniqueness vis-à-vis nature and animals become very problematic.

Moreover, other boundaries also seem to dissipate. A second boundary subversion proposed by Harraway is that of human organism and machine. "Late twentieth century machines have made thoroughly ambiguous the difference between natural and artificial, mind and body, self-developing and externally designed and many other distinctions that used to apply to organisms and machines" (Haraway, 1991, 152). In Haraway's view, we not only have no sound basis to differentiate humans from animals, but also no longer valid reasons for distinguishing humans from machines. Finally, a third diminished boundary is that between the physical and non-physical. The miniaturization of machines and computer chips, as exemplified in nanotechnology, has made them, in several respects, invisible. Although we may live with the pretense of inhabiting a stable world, flux, motion, and change rather than a static equilibrium really characterize our reality. Thus, consistent with the philosophy of transhumanism, human identity and integrity turn on our capacity for life experience made possible through affirming an evolving human-machine constructed self.

5.5.2 *Protestant Christian Critiques of Transhumanism*

The evolution of humans through technology into a post-human state has been sharply critiqued by several authors within diverse traditions of Protestant Christianity. Theology professor Brent Waters has engaged transhumanism not from the standpoint of the feasibility of presumed technical advances but rather from the worldview or "metanarrative" offered by transhumanists in relation to Christian tradition. Waters finds some "suggestive resemblances," or points of convergence between the two views, including:

1. The current human condition is far from ideal, and human beings, whether due to finitude or to sin and fallenness, have not realized their potentiality.
2. Human beings require liberation from this condition. The remedy or agent of liberation for transhumanists, technology, is surely different than that for the Protestant Christian, who looks for redemption through faith, scripture, and the grace of Christ.
3. However, whether as "posthuman" or as member of the Messianic kingdom, the traditions share a common eschatological hope for a future where we will experience a qualitatively different form of life and identity.

Specific incorporated mechanical devices have been interpreted by Protestant Christian writers as manifestations of the divine, indeed as "a step toward the kingdom of God" (Jackelen, 2003, 293–294). However, a Christological interpretation provides a ground for the distinction between therapeutic and enhancement purposes. C. Christopher Hook, for example, invites Christians to see in the incorporation of mechanical devices a means that will "help us pursue the mission of compassion modeled for us by Jesus as he healed others. But we must not forget that He only restored that which was lost by illness and the effects of sin. He did not make people smarter, or stronger, or encourage them to pursue an earthly immortality" (Hook, 2002, 67).

Not surprisingly, the nature of human embodiment is a central issue of disagreement. The transhumanist aspiration to improve the self by overcoming the finitude and limits of the body, even to achieve disembodied consciousness, is theologically problematic. Waters echoes the Catholic theology of the body in affirming the Protestant Christian claim that the self is not rescued from the confines of a bodily prison, but rather that the soul is redeemed as an embodied person. The boundaries of bodily integrity cannot be erased, even if interpenetrated with technology. Moreover, "the doctrine of Christ's bodily resurrection serves as a powerful reminder that the body is God's good gift and not something to be despised" (Waters, 20).

Waters acknowledges that our reliance upon machines and technology has substantively shaped and altered human identity. Normatively, however, the dignity and distinctive standing biblically accorded to humans as beings created in "the image of God" should be affirmed by Protestant Christian tradition against the transhumanist aspiration to "remold" persons in their own self-image as "posthumans." The symbol of the person as "image of God" can assume many meanings, including freedom, moral responsibility, rationality, equality, creativity, relationality, embodiment, diversity, and community (Jackelen, 2003, 298–300). What the normative content of the symbol does not permit, however, is for persons to engage in self-absorbed projects of self-extension.

5.5.3 Buddhist Critiques

Although the above dispute has largely been joined between transhumanists and Protestant Christians, perspectives from relevant Buddhist principles provide alternative conceptual tools.

Buddhism in certain respects begins historically with the encounter of the Buddha with precisely those intimations of our finitude and mortality that transhumanism hopes we can ameliorate or overcome. While our physical nature is not normative in Buddhism, there is a striking contrast between the ambition to escape our finitude through technology and the way the young Shakyamuni (the future Buddha) responded to the same finitude.

The most famous Buddhist narrative describes how Shakyamuni's father secluded him from all awareness of illness, old age and death; but when he finally encountered an ill man, an aged man and then a corpse, the shock motivated him to leave his home and embark on a spiritual quest to solve this fundamental *dukkha* or "suffering." His *nirvana* awakening is the Buddhist solution, one that has nothing to do with technologies for ameliorating finitude, resisting aging, or promising life extension. To live fully, one must overcome the *fear* of physical decline and death by realizing something about morbidity and mortality that is liberating. These realizations constitute the Buddhist religious journey along the "noble eight-fold path"—right understanding, thought, speech, action, livelihood, effort, mindfulness, concentration—undertaken by Buddhists in the quest as "awakening" or "enlightenment," a consciousness enhanced by religious discipline rather than technology (Rahula, 1974, 45–50).

There are meaningful parallels (as well as divergences) between the transhumanist view that humans currently are exercising a minimal portion of their cognitive capabilities and potentiality and basic Buddhist teachings. The Pali term *dukkha* is the most important concept in Buddhism. The four noble truths, taught by the Buddha in what is believed to have been his first dharma-talk, use the term to summarize his teachings: (life is) *dukkha*; the cause of *dukkha*; the cessation of *dukkha*; and how to end *dukkha* (the eightfold path). English translations of *dukkha* as "suffering" do not do justice to the richness of Buddhist thought about the concept; indeed, Buddhist teachings refer to three main types of *dukkha*. The first, *dukkha-dukkhata*, encompasses everything that we normally think of as suffering, including all physical and mental pain. The second type, *viparinama-dukkhata*, refers to dissatisfaction due to our awareness of impermanence, including the uncomfortable fact of our own mortality. The third type is *sankhara-dukkhata* due to "conditioned states" (*sankhara*) because it is those conditioned states—including habitual ways of thinking, feeling and acting—that compose the sense of self (Pali Cannon: Digha Nikaya III.216; Samyutta Nikaya IV.259; Samyutta Nikaya V.56).

For our purposes, what is most relevant is that the cessation of *dukkha* described in the third noble truth has never been understood to involve economic or technological amelioration. A genuine cessation requires an awakening (the literal meaning of "Buddha") that includes realizing the constructed nature of the (sense of) self and its cravings. Buddhist scholar and practitioner David Loy is not convinced that incorporation devices, whether detached from or supported by transhumanism, might assist this process of awakening. Loy maintains that in some circumstances, embedded mechanical devices might remove the physical hindrances that make it more difficult for some people to be mindful; such amelioration that could then be very significant from a Buddhist perspective (Loy, 1996). A core question for

Buddhist moral evaluations, then, is whether incorporation devices encourage or discourage, support or interfere with, our abilities and inclinations to follow the path to awakening.

As with the transhumanist worldview then, Buddhism affirms that the body, and our finitude, can be an obstruction to achieving our full mental capacities and comprehension. Moreover, in Buddhism our present physical identity is not normative. People are not individuals in a strict sense, but ever-changing processes that interact with other processes, constructed beings that are constantly being reconstructed. One's *dukkha* or suffering is due, most of all, to the delusion of self, or, another way of expressing the same thing, to the deceptive sense of duality between oneself (one's self) "inside" and the world "outside."

However, a profound difference emerges regarding the method and context of human improvement: the transhumanist reliance on overcoming finitude and constructing the self-in-becoming by technological means is, from a Buddhist view, misplaced because finitude is not "the" human problem for Buddhism. The problem for Buddhism, as articulated in the four noble truths, is our *dukkha*-suffering and the path we undertake to alleviate it. There are, moreover, non-technological methods to fuller mindfulness, such as meditation, practice, and ritual.

Buddhist teaching thereby shares with Protestant Christian perspectives the suspicion that cultural attachment to medical technologies addresses superficial symptoms rather than the substantive questions we must ask about human nature. Buddhist critiques will inquire about the extent to which incorporation of mechanical devices are part of the solution to the human predicament (limited as we are by physical frailty and death), and how much a technologically-dependent way of thinking is part of the problem. The delusion of self is our fundamental *dukkha*, an intrinsic frustration that would seem impervious to any technological solution. Buddhist teaching must thus ask whether cultural interest and scientific research on incorporation of mechanical devices reflects genuine compassion in promoting remedial therapy, or instead manifests an attempt to provide a technological solution to what is basically a spiritual problem. Insofar as it is the latter, altering human nature through technological augmentation becomes part of our problem. Indeed, in place of the serenity and freedom that comes from realizing the true nature of our ungroundedness, preoccupation with new possibilities may be a symptom of compulsiveness, of a *collective lack* that is always focused on the future because it is unable to dwell in the present (Loy, 1996). This compulsive symptom leads us to think our (constructed) self must be continually altered and improved.

5.5.3.1 Daoist Critique of Technical Progress

According to Chinese thought, the endless circulation of the Dao is an open-ended system. The important norm is not efficient functioning (making the body a better machine for achieving the mind's purposes) but *tzu-jan*, the spontaneity of self-creation, which applies to all creatures. The concept of "progress" over time of self or society simply finds no place in such teaching. The Tao is not going anywhere.

The Daoist understanding brings into critical relief the contrary Western belief that *we* are progressively evolving and that our technological progress will bring us to some ultimate culmination. Unless some content can be given to the concept of post-human, the conviction of unending progress seems to warrant a relentless (and resource-draining) pursuit of the unobtainable. As Western tradition recognizes no particular endpoint or *telos* as the goal of technological development, that is, no point where we will know to *stop* and conclude, "we do not need further technologies," emphasis falls by default on generalities like reduction of suffering and disability, and promotion of human well-being, concepts whose definition is fluid and susceptible to technological re-definition. Indian and Chinese worldviews provide different and relevant alternative perspectives and critiques on what really promotes these goals and how to pursue them morally.

Summation: Religious experience involves immersion in a set of rituals, practices, and moral behaviors designed to elicit a transformative response in the character and conduct of the believer within their community. This transformation might be referred to as a "conversion" experience in some traditions, or as an experience of "enlightenment" in other traditions. The common point is an understanding that the self as currently constituted has not achieved full spiritual potential; the self must be improved upon and needs to embark on a journey to overcome sin, ignorance, or epistemic delusion, and be transformed into right relationship with the divine and the universal. The transformation from the self-as-it-is to the self-as-it-ought-to-be also has social implications.

The content of the ultimate vision of self clearly differs between traditions, but what is common and shared is a conviction that the human person can experience liberation from the mortal experience of bodily decline, finitude, and death. This aspiration is no less shared in the aspirations of the Enlightenment, the progressivist paradigm that motivates scientific researchers to seek a scientific version of the holy grail, and in transhumanist philosophy. The central point of contention is whether human improvement is simply a matter of the grace of scientific technology or rather instead has to do with personal character and enacting good works on the religious journey. Medical advancements, including incorporated mechanical devices, need to be seen as a means, even as religious traditions express skepticism that they are increasingly transformed into personal and societal ends.

5.6 The Common Good

Our analysis has indicated that, in general, attitudes among religious traditions towards incorporation of mechanical devices that impinge on bodily processes and organic functioning reflect patterns of acceptance and appropriation, or occasionally, ambivalence and caution. Meanwhile, currently speculative proposals that would transform human nature through mechanical impingement on consciousness commonly invite resistance and rejection. Yet, there is a further question that religious communities believe it important to discuss in the context of proposals for innovative biomedical technologies: to what extent will the diffusion and adoption of incorporated mechanical devices advance the common good of the society?

The question of the common good has not escaped the attention of scientific researchers, even if at times professional language evokes the historical phenomenon of distancing between the scientific process of device development and the political-moral application of the device. For example, Mnyusiwalla and colleagues identify the ethics of equity as the foremost moral consideration in addressing and regulating incorporation technologies such as nanotechnology. They draw particular attention to the ways that digital and genomic technologies as well as embedded mechanical devices may "ultimately be of most value for the poor and sick in the developing world" (Mnyusiwalla et al., 2003, R11). The welfare of those persons on the social margins, and the responsibility of the community to care for such persons, is shared moral ground for many religious traditions.

For example, an implication of the central Buddhist teaching of *paticcasamuppada* or interdependence is that we always need to consider context—in this instance, the wider effects of introducing new technologies. The ramifications of even small actions are infinite; drawing distinctions between personal consequences and social consequences, or between short-and long-term consequences, is artificial because of our underlying existential condition of interdependency.

This means that as new technological advancements are introduced within medicine, moral evaluation must consider the broader medical system within which they are routinized and made accessible. The Buddhist perspective on interdependency requires an analysis of the reciprocal effects of incorporation technologies and the medical system. If, for example, incorporation technologies become available only or primarily to wealthy persons, they become increasingly problematic for Buddhism because of the prospect that such interventions will increase class divisions and social tension.

The arguments of conservative Protestant Christian authors on incorporation devices, including nanomedicine, as well as in opposition to transhumanist aspirations, also invoke considerations of equity and justice. Technology, it is claimed, is not just an agent of human progress but also an expression of human sin and fallibility. In a Protestant assessment of nanotechnology, Hook contends: "We should expect our technologies to reflect this sin (self-interest) and magnify the consequences of error. While some see technology as the means of leveling the playing field, history has shown that technology tends to produce larger disparities between cultures and subcultures" (Hook, 2002, 60). In particular, sin manifests itself historically in social stratification, inequitable access, and injustice.

The same argument is reiterated in a critique of transhumanism: "The pursuit of transhumanist goals could lead to individuals and communities possessing significant differences in the type and extent of biotechnological modifications. One consequence of these disparities will be the likelihood of discrimination..." (Hook, 2004, 2519). Given these historical and contemporary patterns of inequitable distribution of resources, the transhumanist commitment to a value of "wide access" is not reassuring. The prospect of our evolving into post-humans has not been accompanied by a compelling case for the compatibility of social justice (one indicator of collective human improvement) and technological progress (Jackelen, 2003, 294).

The claim of social justice requires equitable distribution of available resources, including resources for health care, necessary to equal opportunity for full participation in society. At a minimum, then, access to incorporation technologies should not be based on ability to pay, or reflect underlying social indices of stratification, such as race, gender, class, ethnicity, age, etc. Moreover, prior to the issue of equitable *provision* is the question of what needs might go *unmet or omitted* if a certain technological advance is pursued in terms of economic resources and private or public research interests. What priorities should the diverse technologies of incorporation to restore function or even enhance capacity have when some 14% of the American population lack health insurance as their way to claim a basic minimum of health care, or when a good share of the world's population lacks access to vaccinations and basic public health measures of sanitation?

Theological ethicist Gene Outka has proposed several criteria, as influenced by the Christian norm of *agape* (neighbor-love), to help guide decision-making among competing health care priorities. These criteria include: (1) frequency of the condition or illness among the population; (2) level of risks of communicability to a wider population; (3) costs of treatment; (4) extent of pain and suffering experienced by those afflicted with the condition; and, (5) prospects for rehabilitation with treatment (Outka, 1974). As delineated in our section on the state of scientific research, research on embedded medical devices is devoted to conditions that afflict thousands if not millions of persons, who experience substantial pain or suffering as a consequence of impairments of sensory or motor control, for example. What is currently unclear at this preliminary stage of investigation are considerations of cost and rehabilitative outcomes. Even if subsequent studies generate treatment successes at reasonable costs—something that has yet to happen with gene transfer research—policy decisions will be necessary regarding the level of social investment in these devices in contrast to other health care needs.

Liberation Theology and the Moral Option for the Poor: In revealed scripture and in community tradition, each of the Abrahamic faith traditions offer repeated prophetic critiques of social institutions and cultures whose pursuit of economic well-being betrays an ethical superficiality revealed in its neglect of those on the social margins. The U.S. Roman Catholic Bishops, for example, have not idealized any current health care system as optimal, but have portrayed the U.S. system as "so inequitable, and the disparities between rich and poor and those with access and those without are so great, that it is clearly unjust" (USCCB, 1993, 99). The moral imperative for the "poor" in these traditions advances a priority claim of the common good that may restrict technological imperatives.

Some recent work in religious bioethics illustrates the significance of the claims of social justice and the common good. In *Health and Liberation*, physician Alastair Campbell maintains that unless society addresses fundamental injustices in health care prioritization and delivery, and "confront(s) these issues of freedom, oppression, and liberation, we will have missed the *central* problem of modern health care ethics" (Campbell, 1995, 2). He relies on distinctive thematic claims of Christian liberation theology including (a) communal identification with the oppressed in society; (b) a critique of reigning ideologies of oppression and power;

and, (c) construction of an alternative liberation praxis in order to re-orient the debate about justice and priorities in health care. These commitments shift the starting point for discussions of health care justice away from economic preoccupation with efficiency and effectiveness; instead, "liberation theology always begins by heeding the cry of those who are discounted, marginalized, furthest from power or influence in a society" (Campbell, 1995, 16).

When the voices of the ill, the suffering, the aging, and the dying have moral priority, the moral agenda for bioethics resists being set by the progress of biomedical science, or by corporate economic interests, or by academic elites. It is instead defined by "those who suffer and for whom there is no quick technological fix" (Campbell, 1995, 17). On this account, the values and goals of health care must be re-aligned to meet basic needs rather than being propelled by technological advances. Ultimately, health care must be understood as a shared good, a common good, rather than a private commodity whose distribution is stratified according to financial resources.

Campbell claims the central deprivation caused by inadequate access to health care is a core philosophical and legal value: a loss of freedom and moral agency. Indeed, influenced by the Kantian ideal of autonomous persons constituting a kingdom of ends, Campbell contends that health is best understood as an aspect of human freedom; part of the imperative of justice for bioethics then is to "allow for forms of personal and social life in which those made vulnerable by incurable disease, permanent disability, or imminent death remain full members of our human community, with opportunities to continue exercising their personal freedom" (Campbell, 1995, 20).

This emphasis on moral agency entails that more is morally required of distributive schemes than diverting resources to the medically indigent. Such an approach tends to assuage conscience and inculcate passivity. The ethical task is instead to empower those who are currently disempowered so they are able to accept responsibility for their own health care and become participants in communal decision-making. The liberation models for this empowering justice are the biblical narratives of Jesus, who through his healing ministry transformed the physically wounded and socially stigmatized into healers.

The issue of just access in equitable distribution of health care is often lauded in bioethics discourse as a commendable sentiment, but is no less critiqued for its impracticality given the current economic, industry, and legal factors that direct health care delivery. Rather than devote scholarly energy to tilting at the windmills of substantive health care reform, the counter-argument of pragmatic rationality affirms that the bioethics community ought to devote its attention to those matters, e.g., institutional policies on end of life care, transplantation procedures, or stem cell research, in which it can have some meaningful influence.

Resignation to the social status quo is not an acceptable option for faith communities, which have a responsibility to witness, in their discourse and in their practices, to alternative values and institutional structures for just health care. To paraphrase neo-orthodox Protestant theologian Reinhold Neibuhr, the insistence of faith communities on justice and liberation in health care can remind the broader health care community and the society that the economic norms of efficiency and utility are not final or ultimate norms (Neibuhr, 1975). For example, authors of the NSF document *Converging Technologies* claim that "technological superiority is the fundamental basis of the economic prosperity and national security of the United States" (Roco and Bainbridge, 2002, 14; Asher et al. 2002). Once this point is established, the analysis proceeds to argue for rapid development of incorporation technologies through that ultimate trump card, "national security." It would be hard to find any stronger knock-down argument for the technological imperative. Hence, without vigilance on situating technological advances within a broader concept of the common good, society is in continual peril of making relative and instrumental norms into ethical absolutes.

The values of justice, the common good, and liberation from oppression witnessed to by faith communities thus provide a standard of indiscriminate criticism against which almost all systems of health care will fall short. At the same time, such values also serve as criteria of discriminate criticism: they enable informed moral assessments of certain care systems and their priorities in delivery as more or less just, fair, and liberating. Our analysis does not imply that research and social investment into incorporation technologies should not be pursued. Clearly, those who experience blindness, deafness, or loss of functionality constitute a community of the socially marginalized. However, our analysis does entail that substantive discussion of ethical questions about justice and the common good have moral priority over cultural embrace of technological advances in biomedicine. The theological lens in such an ethical examination is necessarily concentrated on the poor, the marginalized, and the oppressed not only for the purposes of exercising social compassion, or exorcising personal conscience, but rather as the source for hope and empowerment.

The commitment to the common good can be defended on religious grounds other than Christian liberation theology. As Sachedina has written, the "Third Pillar" of Islamic faith, the alms levy or *zakat*, "has underwritten badly needed health care for those who cannot afford the cost of rising treatment" (Sachedina, 2004, 1333). The claims of justice in Islamic tradition and religious law (*shari'a*) necessarily would raise questions about the social priorities of developing mechanical devices for incorporation in either body or mind; in Islam, these issues surface with respect to transplantation technologies. Similarly, given the possibility of widespread need and impact of some incorporation technologies, justice might be a supporting principle so long as the technologies are focused on the amelioration of various kinds of disabilities that affect vulnerable persons on society's margins.

Summation: For the religious traditions on this study, it is morally problematic for incorporation technologies to be diffused into medicine either solely on grounds of personal desire and ability to pay or on more collective economic or even national security considerations. Justice arguments about access to such technologies must not presume there is a level playing field of access to begin with, but rather must recognize that society and health care systems already erect barriers or marginalize some persons and groups to social invisibility. Moreover, investments in research and application of incorporation technologies may mean further neglect of the poor and marginalized. Religious communities have a profound commitment to the common good, as displayed in claims about ontic interdependency, equity, and liberation of "the poor." This moral imperative may limit religious acceptance of technologies whose social justification is non-medical in nature and whose use may perpetuate existing disparities in health care access.

5.7 Conclusion: Implications for Policy

The theme of the common good in religious traditions leads conceptually to a brief consideration of the implications of our analysis for biomedical policy. In a liberal democratic and pluralistic society premised on governmental neutrality towards religion, it is not appropriate for religious claims to provide the *basis* or *warrant* for the adoption of policy. However, it is consistent with society's commitments to democratic pluralism, fair procedures, and equal opportunity, for the views of religious traditions to be elicited and their values and conclusions given due consideration in the *process* of policy making. This holds as well for policy approaches to the bodily incorporation of mechanical devices.

There are important precedents for policy attentiveness to religious values and appeals. James F. Childress, a biomedical ethicist, has drawn on his experience as commissioner on the National Bioethics Advisory Commission, to develop several rationales for a policy process that is inclusive with respect to religious traditions:

1. Responsible public policy must ensure that policy conclusions can be explained and justified before a broad audience of citizens, including members of religious communities.
2. Strong support or fervent opposition from a religious tradition can be a valuable indicator regarding the political feasibility of a policy position.
3. The embedded social morality is historically dependent on the resources of moral wisdom mediated by religious traditions; although religious grounds are not required for such basic moral norms as liberty, mutual respect, equality, and fairness, religious values can enable understanding and interpretations of this embedded social morality for policy making.
4. The diversity of religious traditions on a given biomedical policy matter, a diversity that is pronounced in our analysis of religious approaches to incorporation technologies, can provide insight into the diverse views held within the political culture as a whole (Childress, 1997).

Our analysis for the most part reveals attitudes of acceptance and appropriation among many religious traditions towards incorporation technologies, with such acceptance perhaps most pronounced among Jewish thought, as based in the mandate to heal and complete creation, and Roman Catholic thinking that emphasizes a dynamic understanding of the self and of nature. The views of Eastern religious traditions, such as Buddhism, express ambivalence towards these technologies, contingent on the degree to which they impede or facilitate spiritual progress, but it cannot be said that there is an attitude of fervent resistance or opposition to incorporation technologies. The one futuristic prospect that does elicit strong religious objections, primarily from, but not exclusive to, conservative Protestant Christianity, is the possibility of using some incorporation technologies, along with other social and scientific developments, to induce a shift to a "post-human" identity.

Policy attentiveness to these religious patterns and values is likely to be most prominent at what can be described as a pre-regulatory phase of policy making. That is, there are certain regulatory agencies, such as the Food and Drug Administration, which are entrusted by society with ensuring that the use of medical devices is safe and effective. The procedures of regulatory agencies for testing, evaluation, and diffusion of a device presume the medical and social value of the device. Religious perspectives, however, most commonly join with other voices—scientific, industry, political, etc.—in informing initial public discussion about whether research testing and practical application of the device ought to proceed in the first place. This pre-regulatory venue for public discussion seems especially salient given the innovative and transformative nature of the incorporation devices discussed in this essay.

In informing public discussion and debate, religious communities have a special responsibility to provide education and deliberative processes for their own adherents. This is not simply because members of a faith community have expressed certain commitments, and the community has assumed certain responsibilities for the believer's religious or spiritual welfare. This responsibility also follows from the fact that religious adherents are citizens in a democratic society; religious communities have a social and political responsibility of educational stewardship towards their adherents who are also citizens regarding the important biomedical issues and advances of the era.

In addition to information and assessments of particular biotechnological prospects, religious communities are well-situated to step back from the immediate issues and raise broader questions about the character of society. There are serious questions posed in religious reflection not merely about funding and development of incorporation technologies but about the regnant paradigms of the day. We draw attention to two such considerations in conclusion.

1. **The Societal Ends of "Progress"**: It is fair to say that, for several reasons, religious traditions would be critical of an uncritical societal embrace of any new technological development in incorporation technologies as a manifestation of scientific "progress." When there are specific ends for a technology, identified diseases for which incorporation technologies could provide alleviation and treatment, religious traditions in our analysis generally support research development and social appropriation. However, when the objectives of the technologies are ill-defined, or open-ended, religious traditions question whether a biomedical technology is being employed in the service of a superficial solution to a non-medical problem. This critique of scientific progress is perhaps expressed most sharply in Buddhist teaching, but it is embedded in both religious tradition and mythic narratives. In short, religious traditions encourage

both social attention and policymaking insistence on a moral and medical justification for the introduction of a technology, rather than accommodating an after-the-fact justificatory approach.

2. **Social Justice and Resource Allocation**: At present, incorporation technologies are expensive and benefit relatively few persons. Their potential as "halfway technologies," that is, technologies that enable persons to live a reasonable quality of life while still experiencing some kind of impairment, seems very promising for substantial numbers of persons, but has yet to be effectively demonstrated in human beings. The research path from the technologies in their current status and in research anticipations for the future could be prohibitively expensive in terms of time, funding and scientific commitment. Ultimately, religious traditions direct social attention, as well as the interests of policy makers, to the issue of whether the common good is being advanced by development of incorporation technologies in general, and specific technologies in particular. This means consideration of what medical needs, including access to basic health care and education that has a significant impact on health outcomes (environment, lifestyle style choices, preventative measures), may be neglected because of an industry or societal commitment to technological remediation. It also means addressing issue of equitable distribution of the technologies that are developed, with special moral priority given to the disadvantaged, impaired, and poor. Not to ask such questions is a moral failing and evades the central issues of contemporary biomedical ethics.

Acknowledgements The authors wish to acknowledge the research and editorial assistance of Dr. Siobhan Baggot, Lauren Clarke, and Sarah Gehrke.

References

Abiocor. www.abiomed.com (accessed November 30, 2004).

Annas, George J. (1988). *Judging Medicine*. Clifton, NJ: Humana.

Asher, R. et al. (2002). "National Security," in Mihail C. Roco and William Sims Bainbridge (eds.), *Converging Technologies for Improving Human Performance*. Arlington, VA: National Science Foundation, 327–330.

ASSIST (2004). *DARPA BAA for* "Advanced Soldier Sensor Information System and Technology," http://www.darpa.mil/ipto/Solicitations/open/04–38_PIP.htm (accessed August 10, 2004).

Baxi, Maulik V. (2003). "First Artificial Pacemaker: A milestone in history of cardiac electrostimulation," *Asian Student Medical Journal*, http://asmj.netfirms.com/article0903.html (accessed June 15, 2006).

BBC News (2004a). "Bionic Legs Give Soldiers A Boost," http://news.bbc.co.uk/2/hi/science/nature/3502194.stm (accessed March 11, 2004).

BBC News (2004b). "A 'Prison Without Bars,'" http://news.bbc.co.uk/2/hi/uk_news/3906625.stm (accessed July 29, 2004).

Bostrom, Nick (2004a). "In Defense of Human Dignity," www.nickbostrom.com (accessed July 2, 2004).

Bostrom, Nick (2004b). "Transhumanist Values," www.nickbostrom.com (accessed July 2, 2004).

Brasher, Brenda E. (1996). "Thoughts on the Status of the Cyborg: On Technological Socialization and Its Link to the Religious Function of Popular Culture," *Journal of the American Academy of Religion* 64(4), 809–830.
Brooks, Rodney A. (2002). *Flesh and Machines: How Robots Will Change Us*. New York: Pantheon Books.
Bultmann, Rudolph (1952). "Soma," *Theology of the New Testament*. London: SCM, Vol. 1, 192–203.
Business Wire, "First Successful Use of Penetrating Microelectrodes in Human Brainstem Restores Some Hearing to Deaf Patient," Jan. 16, 2004. home.businesswire.com (accessed July 10, 2004).
Bynum, Carolyn Walker (1992). *Fragmentation and Redemption*. Cambridge, MA: Zone Books.
Campbell, Alistair V. (1995). *Health as Liberation: Medicine, Theology and the Quest for Justice*. Cleveland, OH: Pilgrim Press.
Carmena, Jose M. et al. (2003). "Learning to Control a Brain-Machine Interface for Reaching and Grasping by Primates," *Public Library of Science: Biology* 1(2), http://biology.plosjournals.org/ (accessed November 10, 2003).
Chabon, Michael (2005). "The Recipe for Life," http://www.michaelchabon.com/golem.html (accessed March 8, 2005)
Childress, James F. (1997). "The Challenges of Public Ethics: Reflections on NBAC's Report," *Hastings Center Report* 27(5), 9–11.
Clynes, Manfred E. and Nathan S. Kline (1995). "Cyborgs and Space," in Chris Habels Grey (ed.), *The Cyborg Handbook*. New York: Routledge, 29–34.
Cohen, Jon (2002). "The Confusing Mix of Hype and Hope," *Science* 295(February 8), 1026.
Crouch, Robert A. (1997). "Letting the Deaf Be Deaf," *Hastings Center Report* 27, 14–21.
Cyberionics (2003). "What is VNS Therapy?," http://www.vnstherapy.com/aboutvnsfs.html (accessed August 15, 2003).
Daar, A.S. (1991). "Organ Donation: World Experience; the Middle East," *Transplant Proceedings* 23(5), 2505–2507.
Daley, Brian (1990). "The Ripening of Salvation," *Communio* 17, 27–49.
DARPA. www.darpa.mil (accessed November 18, 2003).
Daviss, Bennett (2004). "Vision Quest," *The Scientist* 18(9) (May 10), 38.
Dorff, Elliot N. (1986). "The Jewish Tradition," in Ronald L. Numbers and Darrel W. Amundsen (eds.), *Caring and Curing: Health and Medicine in the Western Religious Traditions*. New York: Macmillan, 5–39.
Etter, Delores M. (2002). "Cognitive Readiness: An Important Research Focus for National Security," in Mihail C. Roco and William Sims Bainbridge (eds.), *Converging Technologies for Improving Human Performance*. Arlington, VA: National Science Foundation, 330–337.
Feldman, David M. (1968). *Marital Relations, Birth Control, and Abortion in Jewish Law*. New York: Schoken Books.
Feldman, David M. (1986). *Health and Medicine in the Jewish Tradition*. New York: Oxford University Press.
Feynman, Richard P. (2004). "There's Plenty of Room at the Bottom," http://www.zyvex.com/nanotech/feynman.html (accessed August 20, 2004).
Fox, Renee and Judith Swazey (1992). *Spare Parts: Organ Replacement in American Society*. New York: Oxford University Press.
Freitas, Robert A., Jr. (1999). *Nanomedicine*, Vol. I. Landes Bioscience.
Freitas, Robert A., Jr. (2003). *Nanomedicine*, Vol. II. Landes Bioscience.
Friedmann, Theodore (2005). Remarks at Symposium, "Super Athletes: A Public Dialogue About Genetics and Sports," February 11, Portland, OR.
Fry, Jenny (2000). "The Construction of Cyborg Bodies: Fact, Fantasy and the Cyborg Continuum," www.soapboxgirls.com (accessed, June 20, 2004).
Gallagher, Raphael (2000). "Catholic Medical Ethics: A Tradition Which Progresses," in James Keenan (ed.), *Catholic Ethicists on HIV/AIDS Prevention*. New York: Continuum, 271–281.
Gibson, William (1984). *Neuromancer*. New York: Ace Books.

Gilbert, Walter (1992). "A Vision of the Grail," in Daniel J. Kelves and Leroy Hood (eds.), *The Code of Codes: Scientific and Social Issues in the Human Genome Project*. Cambridge, MA: Harvard University Press.

Goldblatt, Michael (2002). "DARPA's Programs in Enhancing Human Performance," in Mihail C. Roco and William Sims Bainbridge (eds.), *Converging Technologies for Improving Human Performance*. Arlington, VA: National Science Foundation, 339–341.

Golinkin, David (2000). *Responsa in a Moment: Halakhic Responses to Contemporary Issues*. Jerusalem: Schechter Institute of Jewish Studies.

Gray, Chris Hables (2001). *Cyborg Citizen: Politics in the Post-Human Age*. New York: Routledge.

Haraway, Donna J. (1991). *Simians, Cyborgs, and Women: The Reinvention of Nature*. New York: Routledge.

Hayles, N. Katherine (1999). *How We Became Posthuman: Virtual Bodies in Cybernetics, Literature, and Infomatics*. Chicago, IL: University of Chicago Press.

Haseltine, William A. (2003). "Regenerative Medicine: A Future Healing Art," *The Brookings Review* 21(1) (Winter), 38–43.

Hefner, Philip (2003). *Technology and Human Becoming*. Minneapolis, MT: Fortress.

Henderson, Diedtra (2004). "FDA approves lens implant for nearsighted," *The Oregonian* (September 14), A1, A8.

Hewitt, Jacqueline (1994). "Heated Nanoshells Kill Cancer Cells," www.nanotechweb.org (accessed June 30, 2004).

Hook, C. Christopher (2002). "Cybernetics and Nanotechnology," in John F. Kilner, C. Christopher Hook and Diann B. Uustal (eds.), *Cutting-Edge Bioethics: A Christian Exploration of Technologies and Trends*. Grand Rapids, MI: Eerdmans Publishing, 52–68.

Hook, C. Christopher (2004). "Transhumanism and Posthumanism," in Stephen G. Post (ed.), *Encyclopedia of Bioethics*, 3rd ed. New York: Thomson Gale, 2517–2520.

Huang, Gregory (2003). "Mind-Machine Merger," *MIT Technology Review* (May), 38–45.

Idel, Moshe (1990). *Golem: Jewish Magical and Mystical Traditions on the Artificial Anthropoid*. New York: SUNY.

Jackelen, Antje (2003). "The Image of God as *Techno Sapiens*," *Zygon* 37(2), 289–302.

Janofsky, Michael (2004). "Redefining the Front Lines in Reversing War's Toll: High-Tech Limbs Aid Soldiers' Recovery," *The New York Times* (June 21), A13.

Jewett, Robert (1971). *Paul's Anthropological Terms*. Leiden: Brille.

Jonsen, Albert (1998). *The Birth of Bioethics*. New York: Oxford University Press.

Jonsen, Albert (2004). Personal correspondence, June 29.

Jonsen, Albert and Stephen Toulmin (1988). *The Abuse of Casuistry*. Berkeley, CA: University of California Press.

Joy, Bill (2000). "Why the Future Doesn't Need Us," *Wired* 8.04 (April).

Kasper, Walter (1976). *Jesus the Christ*. New York: Paulist.

Keenan, James (1994). "Christian Perspectives on the Human Body," *Theological Studies* 55, 330–346.

Keenan, James (1996). "The Return of Casuistry," *Theological Studies* 57, 123–129.

Keenan, James and Thomas Shannon (eds.) (1995). *The Context of Casuistry*. Washington, DC: Georgetown University Press.

Kelsay, John (1993). *Islam and War: The Gulf War and Beyond*. Louisville, KY: Westminster/John Knox.

Kennedy, P. R., R. A. E. Bakay, M. M. Moore, K. Adams and J. Goldwaithe (2000). "Direct Control of a Computer from the Human Central Nervous System," *IEEE Transactions on Rehabilitation Engineering* 8(2).

Kohrman, Arthur F. (1995). "Chimeras and Odysseys: Toward Understanding the Technological Dependent Child," in John Arras (ed.), *Bringing the Hospital Home: Ethical and Social Implications of High-Tech Home Care*. Baltimore, MD: The Johns Hopkins University Press, 53–64.

Kopfensteiner, Thomas (1995). "Science, Metaphor and Moral Casuistry," in J. Keenan and T. Shannon (eds.), *The Context of Casuistry*. Washinton, DC: Georgetown University Press, 207–220.

Kull, Anne (2002). "Speaking Cyborg: Technoculture and Technonature," *Zygon* 37(2), 279–287.
Kurzweil, Ray (1999). *The Age of Spiritual Machines*. New York: Viking.
Lakoff, George and Mark Johnson (1980). *Metaphors We Live By*. Chicago, IL: University of Chicago Press.
Lappe, Marc (1996). *The Body's Edge: Our Cultural Obsession With Skin*. New York: H. Holt.
Lavine, Marc, Leslie Roberts and Orla Smith (2002). "If I Only Had a ..." *Science* 295 (February 8), 995.
Leites, Edmund (ed.) (1988). *Conscience and Casuistry in Early Modern Europe*. New York: Cambridge University Press.
Loomis, J.M. (2002). "Sensory Replacement and Sensory Substitution," in Mihail C. Roco and William Sims Bainbridge (eds.), *Converging Technologies for Improving Human Performance*. Arlington, VA: National Science Foundation, 213–224.
Loy, David (1996). *Lack and Transcendence: The Problem of Death and Life in Psychotherapy, Existentialism, and Buddhism*. Atlantic Highlands, NJ: Humanities Press.
Maguire, G. Q., Jr. and Ellen McGee (1999). "Implantable Brain Chips?: Time for Debate," *Hastings Center Report* 29(1), 7–13.
Maimonides, Moses (1969). *The Guide for the Complexed*. New York: Dover Publications.
McCarthy, Patrick M. and William A. Smith (2002). "Mechanical Circulatory Support-a Long and Winding Road, *Science* 295(February 8), 998.
McCullagh, Declan (2003a). "RFID Tags: Big Brother in Small Packages," http://news.com.com/2010–1069–980325.html (accessed January 13, 2003).
McCullagh, Declan (2003b). "Special to ZDNet", August 28, http:www.builderau.com.au/program/work/0,39024650,20277856,00.htm (accessed March 7 2005).
McCullough, Larry (2004). Comments at "Altering Nature" conference, November 12.
McGuire, Meredith (1990). "Religion and the Body," *The Journal for the Scientific Study of Religion* 29, 283–296.
McKenny, Gerald P. (1997). *To Relieve the Human Condition: Bioethics, Technology and the Body*. Albany: SUNY.
Medtronic. www.newhopeforparkinsons.com (accessed, July 10, 2003).
Meeks, Wayne (1993). *The Origins of Christian Morality*. New Haven, CT: Yale University Press.
Mehlman, Maxwell J. (2003). *Wondergenes: Genetic Enhancement and the Future of Society*. Bloomington, IN: Indiana University Press.
Meyyappan, Meyya (2004). *Carbon Nanotubes: Science and Applications*. Boca Raton, FL: CRC.
MEMS and Nanotechnology Clearinghouse, "What is MEMS Technology?," http://www.memsnet.org/mems/what-is.html.
Mnyusiwalla, Anisa, Abdallah S. Daar and Peter A. Singer (2003). "'Mind the Gap': Science and Ethics in Nanotechnology," *Nanotechnology* 14, R9–R13.
Moldice. http://www.darpa.mil/dso/thrust/biosci/moldice.htm (accessed August 25 2004).
Moosa, Ebrahim (2004). Personal correspondence, November 24.
Moravec, Hans (1988). *Mind Children: The Future of Robot and Human Intelligence*. Cambridge, MA: Harvard University Press.
Morreim, E. Haavi (2003). Personal correspondence, August 2.
Morreim, E. Haavi (2004). Personal correspondence, November 30.
Mott, Maryann (2005). "Animal-Human Hybrids Sprak Controversy," *National Geographic News*, January 25, http://news.nationalgeographic.com/news/2005/01/0125_050125_chimeras.html (accessed January 27, 2005).
Naam, Ramez (2005). *More Than Human: Embracing the Promise of Biological Enhancement*. New York: Broadway Books.
"Nanomachine," in *Encyclopedia Nano*, http:www.nanoword.net/library/def/Nanomachine.htm.
Neibuhr, R. (1975). "Why the Christian Church is Not Pacifist," in Arthur F. Holmes (ed.), *War and Christian Ethics*. Grand Rapids, MI: Baker Book House, 301–313.
Nelson, James (1992). *Body Theology*. Louisville, KY: Westminster/John Knox.
Newell, J. Philip (2000). *Echo of the Soul: The Sacredness of the Human Body*. Harrisburg, PA: Morehouse Publishing.

Noonan, John T. Jr. (1995). "Development in Moral Doctrine," in J. Keenan and T. Shannon (eds.), *The Context of Casuistry*. Washington, DC: Georgetown University Press, 188–204.
Nuland, Sherwin B. (2005). *Maimonides*. New York: Schocken Books.
Olds, J. and P. Milner (1954). "Positive Reinforcement Produced by Electrical Stimulation of Septal Area and Other Regions of Rat Brain," *Journal of Comparative and Physiological Psychology* 47, 419–427.
Outka, Gene (1974). "Social Justice and Equal Access to Health Care," *Journal of Religious Ethics* 2(Spring), 11–32.
Pali Canon: *Digha Nikaya* III.216; *Samyutta Nikaya* IV.259; *Samyutta Nikaya* V.56
Pardridge, William (2005). "Intravenous RNA Interference of Brain Cancer with Targeted Nanocontainers," 11th Annual Neuro-oncology and Blood-Brain Barrier Disruption Consortium Meeting, March 18, Portland, OR.
Rahula, Walpola (1974). *What the Buddha Taught*. New York: Grove Press.
Ramsey, Paul (1970). *The Patient as Person*. New Haven, CT: Yale University Press.
Rauschecker, J.P. and R.V. Shannon (2002). "Sending Sound to the Brain," *Science* 295(February 8), 1025–1029.
Roco, Mihail C. and William Sims Bainbridge (eds.) (2002). *Converging Technologies for Improving Human Performance*. Arlington, VA: National Science Foundation.
Rosner, Fred (1999). "Pig organs for transplantation into humans: a Jewish view," *Mt. Sinai Journal of Medicine* 66(5–6), 314–319.
Sabbatini, Renato M.E. (1997). "The PET Scan, A New Window into the Brain," http://www.epub.org.br/cm/n01/pet/pet.htm.
Sachedina, Abdulaziz (2003). Personal correspondence, July 15.
Sachedina, Abdulaziz (2004). "Bioethics in Islam," in Stephen G. Post (ed.), *Encyclopedia of Bioethics*, 3rd. ed. New York: Thomson Gale, 1330–1338.
Sachedina, Abdulaziz (2005). "Brain Death in Islamic Jurisprudence," http://www.people.virginia.edu/~aas/isislam.htm (accessed July 18, 2005).
Sandhana, Lakshmi (2003). "Bionic Eyes Benefit the Blind," *Wired News* (July 16), http://www.wired.com
Schmithausen, Lambert (1991). "Buddhism and Nature," Vol. VI of *Studia Philologica Buddhica* Occasional Paper Series. Tokyo: The International Institute for Buddhist Studies, 22–34.
Scholem, Gershom (1965). *On the Kabbalah and its Symbolism*. New York: Schocken Books.
Science (2002). *Bodybuilding: The Bionic Body* (February 8), 295.
Serour, Gamel I. (2003). Personal correspondence, August 6.
Sherwin, Byron L. (2004). *Golems Among Us: How a Jewish Legend Can Help Us Navigate the Biotech Century*. Chicago, IL: Ivan R. Dee.
Stallard, Jim (2004). "Innovations: The Human Body Shop," www.pbs.org (accessed July 12, 2004).
Stelarc (2004). "Third Hand," http://www.stelarc.va.com.au/third/third.html (accessed July 30, 2004).
Stupp, Samuel (2005). "Nanotechnology: State of the Science including New Biological Materials for CNS Regeneration," 11th Annual Neuro-oncology and Blood-Brain Barrier Disruption Consortium Meeting, March 18, Portland, OR.
The Transhumanist Declaration. www.maxmore.com/extprn3.htm (accessed July 16, 2003).
United States Conference of Catholic Bishops (1993). "Resolution on Health Care Reform," *Origins* 23, 98–102.
Vergote, Antoine (1991). "The Body as Understood in Contemporary Thought and Biblical Categories," *Philosophy Today* 35, 93–105.
Warwick, Kevin (2004). *I, Cyborg*. Urbana: University of Illinois Press.
Waters, Brent (unpublished). "From *Imago Dei* to Technosapien? The Theological Challenge of Transhumanism," unpublished manuscript.
Weissert, Will (2004). "Microchips Implanted in Mexican Officials," http://www.msnbc.msn.com/id/5439055/ (accessed July 14, 2004).
Wickelgren, Ingrid (2003). "Tapping the Mind," *Science* 299 (January 24), 496–499.
Wolstenholme, Gordon (ed.) (1963). *Man and His Future*. London: J. and A. Churchill.
Zrenner, Eberhard (2002). "Will Retinal Implants Restore Vision?" *Science* 295(February 8), 1022–1025.

Chapter 6
Medical Devices Policy and the Humanities: Examining Implantable Cardiac Devices

Jeremy Sugarman, Courtney S. Campbell, Paul Citron, Susan Bartlett Foote, and Nancy M. P. King

6.1 Introduction

Medical devices are ubiquitous: eyeglasses and intraocular lenses enhance vision; artificial limbs restore mobility; and implanted pacemakers and defibrillators treat otherwise debilitating or fatal heart rhythms. Despite the critically important role of medical devices in health care, their development, testing, and use raise significant practical and normative issues.

For the most part addressing these issues in the context of medical device policy is quite straightforward. It seems fair to claim that the overwhelming majority of regulators, payers, and patients do not regard most implanted medical devices as significant alterations posing challenges to the integrity of the individual or the species. The prospect or actual failure of an incorporated medical device can at times loom large in patients' perceptions, but in general, most devices have not presented unique moral or ethical challenges relative to other interventions such as those involving genetics or reproduction. Medical device research and development instead provides a somewhat specialized example of the relevance of the humanities in general, and of ethics in particular, to medical research and medical products. In addition, examining this example in detail may challenge the conclusions made in other settings where technology is incorporated into humans and nature is thereby considered to be altered.

Other forms of incorporation may arguably give rise to more profound concerns about self-identity and integrity, primarily when the incorporated material is organic. For example, the incorporation of new genetic material by means of gene transfer continues to evoke questions about enhancement, the integrity of the germ-line, and the production of chimeras. Blood transfusion and bone marrow and organ transplantation have in the past presented questions about the physical and psychological effects on the individual. In addition, technologies still in development, such as xenotransplantation, give rise to similar issues, at times even raising the prospect of "transhumanism." (Best et al., 2006; World Transhumanist Association). Yet, medical device incorporation on the current scale has only hinted at times of such metaphysical concerns—for example, when devices involve the heart. Our focus in this chapter for reasons we articulate below, is on examining the policy process related to developing devices pertaining to cardiac function.

6.2 Bridging Themes

Before analyzing the policy process related to cardiac device development, we consider the context of this discussion. This chapter, and a predecessor chapter that examined scientific and religious perspectives on mechanized medical interventions, originated in response to questions about the normative status of "nature" or the "natural." These questions assumed the following general form: does medical intervention in natural processes of the human organism, including interventions in those processes that bring on disease, decay, and death, infringe or violate nature in such a way that they require moral justification?

The implantation of medical devices in the body, and particularly the implantation of various cardiac devices, obviously involves altering nature and natural processes. In this context, device implantation reflects a medical or professional assessment that nature has failed, or is inadequate, to the point that human life or the quality of human life is threatened, and that medical intervention to prevent or ameliorate natural cardiac dysfunction is justifiable and perhaps obligatory. Thus, while medical intervention in cardiac function still stands in need of moral justification, as reflected in part by the requirements of informed consent, this justification stems not from an infringement of an intrinsic goodness or normativity of the intervention or the natural, but rather because of respect for the will and dignity of the person.

Given the common background question about the normative status of nature, themes developed in the predecessor chapter on scientific and religious perspectives surface and are re-interpreted in the context of the policy considerations we address in this chapter. For example, it is important to explore the effects of implantation of a variety of medical devices on a person's self-identity. This is an especially salient concern in the context of implanting cardiac devices including replacement of an organic heart with an artificial one or even an implantable artificial heart, because the heart is so significantly entwined with conceptions of personhood, and viewed in some traditions and communities as the source of emotions, and even a symbol of the seat of the self.

As Fox and Swazey have illustrated in their studies of solid organ transplantation (Fox and Swazey, 1992), it is not an uncommon response for a transplant recipient of a donated heart to wonder if they are still the same person, or whether they will, in beginning to undertake their former pursuits, assume some of the characteristics (for example, food preferences) of the heart donor. Similar questions about self-identity can occur, and perhaps be intensified, when a machine or implanted device performs all or many functions previously carried out by the heart. Of course, the degree of impact of implantation on self-identity will vary with the nature of the device; a heart valve replacement or implantation of a pacemaker is unlikely to generate the kind of exploration of self that might be anticipated with a replacement of a whole heart by an implantable total artificial heart.

Advances in medical technology over nearly a half-century, including the development of devices like pacemakers and the feasibility of heart transplants, have demonstrated the heart to be a replaceable part of the self. In addition, definitions of death that have in many countries invoked criteria pertaining to brain function, suggest that the ramifications of implantable devices for self-identity are less pronounced with bodily organs, including the heart, than they are with devices that may have an impact on the brain as the seat of higher functions such as consciousness, memory, language, or relationships. The heart retains its significant symbolic potency but its replacement or assistance by a mechanical device does not seem to reach the threshold level of an existential and ethical quandary that is reached when such devices are used to assist brain functioning. This is to say that the question of the status of the self seem to be raised more significantly by medical interventions in the brain. This phenomenon in turn reinforces the claim above that alterations of the natural processes of the heart are not as politically or morally problematic as other kinds of alterations of nature.

A second bridging theme from the earlier chapter that deals with nature and the natural is that of the "appropriation" of medical technologies and mechanical devices for purposes of human benefit. The current chapter illustrates the political and regulatory working out of the theme of appropriation. The themes of "ambivalence" or "resistance" identified previously as representative of some religiously-grounded responses to medical technologies do not seem to find much of a place in a policy paradigm oriented by concerns of regulation, oversight, and efficacy. It is the case that as part of policy oversight, there may arise technical considerations of paramount importance, such as device failure, that prompt re-design or even risky re-implantation of the device. However, such practices presume the medical and political legitimacy and societal appropriation of the device in the first place. This suggests that the themes of "ambivalence" or "resistance" to a technology are more likely to be voiced prior to the initiation of policy oversight, as part of the general societal and ethical discussion of whether research should proceed in developing a specific device.

These reflections suggest that a primary concern of some religious traditions in contemplating implantable medical devices—avoiding a reductionism of the person to merely a mind, with the body assuming the status of a peripheral appendage—is unlikely to be a consequence of the use of implantable cardiac devices. The body is capable of appropriating mechanical devices of various kinds without thereby being transformed into a machine. However, for some, concerns about post-mortem dissection of the body in order to facilitate device evaluation may generate ambivalence or resistance. As long as such evaluations are conducted respectfully and with the purpose of advancing knowledge to bring about health benefits for future patients, many religious concerns about desecration of the body can be accommodated and addressed in approaches to regulatory oversight. It is nonetheless important for policy to express respect for the somatic being of the person.

A third bridging theme concerns matters of justice and the common good. Even though there is broad social acceptance, as well as a policy process to oversee and regulate the appropriation of implanted cardiac devices, a central issue for religious traditions is the whether distribution of these devices will meet the requirements of justice. As illustrated in the subsequent analysis, matters of access are a vital concern of the policy process, though such questions are contextualized (and complicated) by questions of reimbursement. Given the stratification of health care goods in society currently, if provisions of reimbursement drive the availability of a good and determine access, it is likely that some populations who stand to benefit from implantable cardiac devices will be excluded from access.

There are, moreover, two related questions of justice that bridge the two chapters. Prior to the issue of equitable provision of implanted medical devices is the question of what health care needs might go unmet or omitted if a device is tested, approved, and regulated. To provide a concrete illustration, in a society where 45–47 million people do not have their basic health needs met through an insurance plan, is a device like the totally implantable artificial heart the best use of financial resources and scientific research? As intimated by the discussion below, this question of justice has long been in the shadow of research developments regarding the artificial heart. Moreover, several converging reasons influence the development of medical devices, including medical need, scientific knowledge, industry resources, and potential profitability; considerations of the common good and justice do not seem to play a significant role. Thus, a central question is whether or not provisions for equitable access should be a relevant or a peripheral factor in device development. Nonetheless, the strong presumption in favor of considerations of justice in the initial chapter is recognized here to the extent that justice is held to function as a corrective to policies that rely primarily on economic analyses that emphasize efficiency and utility.

A related question is that of procedural justice, or who decides what devices will be introduced and diffused into practice. There are a variety of stakeholders whose interests are represented and negotiated in various ways in the policy process, including regulators, industry, reimbursement systems, clinicians, and patients. The challenge of procedural justice is to ensure that all parties with a stake in the decision about device development are represented. This entails inclusion of communities that historically have been on the margins of the health care system or biomedical research, including women and racial and ethnic minorities, as well as the medically indigent. While the Food and Drug Administration (FDA) certainly cannot be expected to embrace a religious preference for the poor with respect to device development, efforts should be undertaken to ensure that the voices of the socially marginalized are not excluded from the policy process.

6.3 Implantable Heart Devices: Three Categories

There has been an extraordinary set of developments in the past half century related to enhancing, or replacing, impaired cardiac function by means of devices. This is fortuitous because cardiac disease represents the top cause of mortality in developed nations and there is arguably a moral imperative to attenuate its associated morbidity and mortality. Here we consider three categories of devices.

First, are devices focused on correcting abnormal heart rhythms. By 1957, the first external transistorized cardiac pacing device was introduced. A few years later implantable cardiac pacemakers became clinically available. Subsequent research and development led to the creation of the AICD (automatic implantable cardioverter- defibrillator), aimed at preventing sudden cardiac death. Introduction of these devices was associated with considerable debates about the timeliness of regulatory approval, Medicare coverage, private insurance payment, and appropriate indications for use.

Second, are devices designed to improve cardiac blood flow, including prosthetic heart valves to maintain systemic flow when natural valves fail and stents to hold open clogged coronary vessels. The initial use of totally mechanical artificial heart valves were associated with questions related to balancing the benefits of replacement with the risks of surgery, embolization, and prosthetic valve failure. Cardiac stents are used to maintain patency of vessels following angioplasty. Recent work in stent technology involved coating the stent with medications in an attempt to minimize restenosis.

Third, are devices intended to enhance or replace the mechanical function of the heart muscle. During the 1960s, the National Institutes of Health's National Heart, Lung and Blood Institute issued a challenge to develop an artificial heart. Unlike the rather quiet development of pacemakers, there was extensive public debate and discussion about the possibility of using an artificial heart, known as the Jarvik 7. Ultimately in 1982 Dr. Barney Clark became the first recipient of the device. Continued efforts resulted in the development of the LVAD (left ventricular assist device) that was used initially as a bridging device prior to cardiac transplantation when a donor heart is not available. Subsequently, the LVAD has been employed as a "destination" device, providing final treatments for selected patients. In 2001, an implantable device called Abiocor was tested in a small number of patients. These experiments also attracted considerable public interest.

In this chapter, our goal is to examine how bioethics and the humanities can play a role in enhancing understanding about medical device policy, using as examples the development and use of these three categories of implantable cardiac devices. For instance, in device regulation, now-standard bioethical principles provide justificatory and explanatory tools. In the realm of post-approval evaluation for reimbursement purposes, a robust understanding of the assumptions underlying the policy tools currently in use, such as cost effectiveness analysis and cost-benefit analysis is illuminating. The diffusion and use of technologies may involve addressing an array of religious and moral beliefs. After analyzing these examples, in the last section of the chapter we make some conceptual claims about the relationship between the humanities and medical device policies. Our examination focuses on the U.S. regulatory context, but we expect our analysis to be applicable to other policy settings.

6.4 Medical Device Development and Use

Devices are developed for a variety of overlapping reasons including a medical need, the availability of relevant scientific and technological knowledge, the availability of resources to negotiate the development process, and the potential profitability of the device should it be shown to be safe and effective. Thus, a variety of stakeholders play a role, including inventors, supporters (public and private), engineers, manufacturers, clinicians, and patients. Governmental oversight has evolved in large part in reaction to adverse experiences associated with some devices. In the United States, oversight responsibility during device development and testing rests with the FDA (Food and Drug Administration). Decisions regarding whether a particular device will be reimbursed by Medicare fall in the province of the Center for Medicare and Medicaid Services (CMS). Private insurers also make decisions about coverage for their enrollees.

6.4.1 Device Development, Testing, and Approval

6.4.1.1 Oversight

Since 1906, the US government has exercised oversight of foods and drugs, initially through the Department of Agriculture, and then through the Food and Drug Administration. Early device regulation focused on product labeling and adulterations. Because medical devices, including implantable devices, until somewhat recently were an insignificant part of the health care system, there was relatively little attention paid to them. However, as their use became more widespread, oversight expanded. Specifically, the Medical Device Amendments (MDA) were passed in 1976, giving the FDA authority to require pre-market review of devices on the basis of safety and efficacy. Although there have been a variety of amendments to these rules, primarily via administrative rule-making by the agency but occasionally via legislative amendments, the basic structure of regulatory oversight has remained virtually unchanged for 30 years.

Many products are defined as medical devices in the MDA:

> instrument, apparatus, implement, machine, contrivance, implant, in vitro reagent, or other similar or related article…intended for use in the diagnosis of disease or other conditions, or in the cure, mitigation, treatment, or prevention of disease…or… intended to affect the structure or any function of the body of man or other animals, and which does not achieve its primary intended purposes through chemical action within or on the body of man or other animals and which is not dependent on being metabolized for the achievement of its primary intended purpose (United States Code 321(h), 2000; Foote, 1978).

The regulatory framework organizes this disparate list according to the degree of risk posed to patients when the device is used in its intended manner. Class I devices (e.g., tongue depressors, examination gloves, hand-held surgical instruments) are viewed as carrying the lowest risk. Consequently, the regulatory focus for Class I devices is on general standards and controls, such as sterilization, packaging, and labeling requirements. Class II devices (e.g., powered wheelchairs, surgical drapes, external infusion pumps) carry more substantial risk so the regulators may require a certain amount of clinical evaluation, as well as an assurance that they meet performance standards. Class III devices (e.g., implanted therapeutic products like pacemakers and the range of examples of cardiac devices described above) carry substantial potential risk and accordingly require rigorous pre-market testing and approval. This classification system was intended to focus regulatory effort where it is most needed, based on the risk that the specific medical device poses to the public. Our examination in this chapter is limited to Class III implantable devices.

Oversight of clinical research on devices involves several bodies besides the FDA including Institutional Review Boards (IRBs) and Data Monitoring Committees or Data and Safety Monitoring Boards (DMCs or DSMBs). Other, more specialized bodies, both national and local, provide additional review for certain types of devices, such as those involving radiation or genetically altered material.

The committees charged with reviewing human subjects research are called IRBs in the United States, and Research Ethics Committees, Research Ethics Boards, or Ethical Review Committees in much of the rest of the world. These committees (many of which are based at the institutions where device research is conducted, but which may also be independent, free-standing bodies) review proposed and ongoing research to ensure that risks are reasonable in terms of the anticipated benefits, that risks to subjects are minimized, that the selection of subjects is fair, and that informed consent is obtained. These requirements each find ethical justification in the ethical principles of beneficence, justice, and respect for persons (National Commission, 1979). In contrast, DMCs or DSMBs are typically created by research sponsors (whether commercial or governmental). They are charged with assessing the adequacy and appropriateness of trial design from the perspectives of safety and statistics, and with reviewing aggregated emerging data from trials in order to see if there is sufficient evidence of efficacy or a concern about safety.

In general, the oversight of cardiac device research is and has been fairly standard and straightforward. However, as is the case with some other technologies, such as genomics, this history has also been punctuated by topics of interest and controversy. For example, there has been considerable attention focused on the development of the artificial heart and this has been associated with an additional review mechanism.

6.4.1.2 The Total Artificial Heart: An Example

In 1964, when the NIH's total artificial heart (TAH) program began, there was widespread recognition that heart failure was a major cause of morbidity and mortality. Medical treatments were mostly palliative and were expected to remain so. Even with the emergence of heart transplantation as an effective treatment, it was evident that access would be sharply limited due to the shortage of viable donor hearts. Despite this shortage, the TAH program raised ethical questions based primarily on the matter of allocation of scarce health care resources and differences of viewpoint about how best to maximize utility under the circumstances.

The route to a practicable TAH was understood to be long, and its technological complexity foretold of an intensive, expensive treatment, at least in the early years of clinical application. Questions were raised about other uses for the NIH's financial resources that arguably held greater patient benefit: "The crucial question is whether there are other things that could be done with the resources that would yield greater benefit" (D.M. Eddy, cited in NHLBI, 1985, 56).

Other concerns were raised about patients' access to a very costly intervention: Did the TAH fall into the category of basic care to which patients are entitled? What constitutes equitable eligibility criteria and how should they be set? In addition, how would patients' preferences and values be protected after implantation, in view of uncertain implications for quality of life, as well as unknown aspects of device performance and complications? (IOM, 1991).

Congressional hearings were held in the 1980s on the merits and advisability of an implantable total artificial heart; these hearings led to enabling legislation to provide unprecedented federal funding for technology development based at the NIH. External expert boards were impaneled to advise the NIH about progress and prospects of the initiative, and external Congressional audits and assessments of the program were conducted (NHLBI, 1985; OTA, 1982).

6.4.1.3 Concepts Employed in Research Regulation

By the time the FDA acquired the authority to regulate devices it had developed concepts of "safety" and "efficacy" in the drug arena to determine whether a product would be approved for use outside of the research setting. As with drugs, assessments of any new device must balance its benefits favorably against its risks. In considering whether it is ethically appropriate to test a new device in humans, there must be sufficient pre-clinical data to suggest safety, so that participants are not placed at undue risk of harm in the trials that will be used to secure approval. In addition, risks must be minimized in the design and conduct of clinical trials of devices. Further, there must be evidence from pre-clinical research that there is a reasonable likelihood that the device will show efficacy in humans.

Critical questions arise in clinical research when testing is sought for high-risk interventions that are offered for very serious disorders and conditions for which there is no standard effective treatment. The question that must be asked is whether it is ethically acceptable to expose research subjects to greater risks of harm than would otherwise be permitted simply because of the severity of their underlying disease. Although this question exists for drugs and biologics as well, it can be especially acute for cardiac devices in general, and the TAH in particular. The Abiocor totally implantable artificial heart study provides an excellent example. When the only appropriate research subjects are near death, and few if any would decline whatever treatment was available, a placebo comparison arm is arguably inappropriate.

Even so, many ethical challenges are posed by the design, recruitment, and retention for a trial in which subjects might be randomly assigned to two very different experimental arms (an artificial heart versus "best standard treatment") when potential subjects have an extremely short projected lifespan. The very circumstances that make patients appropriate research subjects may increase the likelihood of the so called "therapeutic misconception" about the experimental device—that is, both research subjects and investigators may view the experimental device as the best treatment, in ways that could adversely affect decision making about research participation (Burling, 2002a, b, c; Lidz and Appelbaum, 2003).

Indeed, the urgency of individual patients' needs for safe and effective cardiac devices can make policy decisions about when a device is ready to be made available extremely difficult, both ethically and politically. In addition to the considerations of distributive justice that have been mentioned for the TAH, the principle of utility suggests that once the decision to pursue a technology has been made it is necessary to ensure that the technology as developed has a maximally favorable risk/benefit ratio under the circumstances and in light of the available alternatives. However, utility maximization often stands in tension with the perceived needs of individuals, and the cautious development of optimally safe and effective devices thus may appear at odds with the desires of patients for more treatment options. As will be discussed further below, this tension can affect what is deemed to count as sufficient data to approve a new device, as well as the amount of uncertainty that regulators, policymakers, and patients are willing to accept.

In addition to the problems associated with conducting research for extremely serious conditions is the scientific desire to remove an unsafe or ineffective experimental device in an effort to best understand the causes of failure. In some cases, this may not be feasible until death since the interventions are in a sense permanent. Although this is not unique to cardiac devices, or even to devices (since the effects of some drugs and biologics are not reversible), much stands to be learned from devices that fail if they are examined. The TAH serves as an exemplar for devices that can result in the proximal death of the patient when turned off or removed (Veatch, 2004). Determining how this additional "permanence factor," since it can impinge upon the ability to learn maximally about such devices, should affect the progression from preclinical to clinical studies, the amount of data needed to determine safety and efficacy, and the duration of the clinical trial process pose serious ethical challenges for device development regulation.

Informed consent is also a key component of medical device research. Informed consent finds ethical justification in the ethical principles of autonomy and respect for persons, and in the political principle of liberty. In order to test a new device it is essential to obtain the informed consent of subjects. In device research, like research involving drugs or biologics, the consent process is well-standardized, and consent forms are routinely scrutinized by IRBs. However, some special consent questions have arisen in cardiac device research. Consider, for example, the case of Barney Clark, who received the first artificial heart. In anticipation of the procedure, there was considerable ethical debate about who should keep the key that could be used to turn the device off. Turning off the device was essentially the only way that Dr. Clark could withdraw from the experiment, if he chose to do so. Ultimately, it was decided that Dr. Clark could wear the key around his neck (Annas, 1988, 391–396).

6.4.1.4 Regulatory Flexibility: The Example of "Combination Products"

Cardiac devices provide an example of how rapid technological advances give rise to new and creative regulatory responses. Implanted cardiac devices have historically been mechanical (e.g., heart valves) or electro-mechanical (e.g., cardiac pacemakers, implanted defibrillators). However, an increasing number of devices also incorporate either a drug or biological component in effort to enhance their performance. Such devices, referred to as combination devices, represent an area of rapid development. One example is the drug-eluting coronary stent. The metallic stent provides radial support to prop the coronary artery open; the drug is added in hopes of preventing re-occlusion of the vessel from cell proliferation in the weeks and months following implant of the stent.

Combination products confront regulators with challenges, as innovation creates configurations that don't fit into existing regulatory molds. The regulatory pathway for combination devices is still evolving, as issues of jurisdiction within the agency's branches (device, drug, and biological) are negotiated and systematized. Combination products involve combining the attributes of at least two of three product types—devices, drugs, and biologics. The concept of "combination products" itself derives from the FDA's historical regulatory silos for these three types of products, each of which are regulated through different statutory requirements, implemented by three distinct centers within the FDA. Devices are under the jurisdiction of the Center for Devices and Radiologic Health (CDRH); drugs are regulated by the Center for Drug Evaluation and Research (CDER); and biologics are governed by the Center for Biologic Evaluation and Research (CBER). Significant differences in these three regulatory regimes pose challenges for combination products. Because regulation tends to be reactive, and thus necessarily slow to respond to change, the FDA is evolving to address them. A new, congressionally mandated Center for Combination Products now manages the coordination of the three centers for combination products, although it does not currently have regulatory authority (Foote and Berlin, 2005).

Combination products may pose some unique ethical issues as well. For instance, mechanical devices that incorporate biological tissues raise questions about zoonoses, and therefore about the safety of products to communities and public health, as well as to the individual. How are such concerns addressed in the usual approval process? Is the input of communities elicited and considered? As described in Chapter 4 in this volume, venues for addressing public health concerns have been institutionalized with respect to recombinant DNA research, but are less well developed for xenotransplantation (Secretary's Advisory Committee on Xenotransplantation).

6.4.2 Use of Devices

A variety of complex factors contribute to when and how particular devices are used. In this section we address four main issues: demands for access to new devices balanced against a desire to be protected from adverse events; decision-making regarding reimbursement; the ethics of evaluation; and marketing and clinician education.

6.4.2.1 Access and Protection

While there is ethical justification for the current approach to determining safety and efficacy, the question of what counts as sufficient safety and efficacy for approving a device for use is a matter of ongoing debate. In short, there is a regulatory tension between protection and access. On the one hand is a history of device failures; on the other hand is the desire for rapid access to clinically important device innovations that have not yet met the test of time. For many years, FDA regulators had few incentives to approve products with high risk profiles, regardless of their potential benefits, because the public and the Congress appeared so risk-averse. Congress has frequently passed more restrictive legislation in response to adverse events. One aspect of the historical evolution of FDA policies is that instigated by product failures, such as the Shiley heart valve. In this instance a design deficiency in tandem with insufficient manufacturing process controls caused some valves to fail unpredictably and catastrophically because of the sudden fracture of a critical portion of the valve. Although the risk of fracture was relatively low, estimated to be 0.7% of the 86,000 implanted, the notion of patients wearing a 'ticking time bomb' was unacceptable for many physicians and patients since failure often resulted in death. The decision to perform an elective replacement of a suspect valve that was functioning in an apparently normal manner was complicated by the fact that the surgical procedure to replace a valve carried substantially higher mortality risk than that of possible mechanical failure (Blot, 2005).[1]

---

[1] Although Class III medical devices may be highly reliable, life-enhancing and life-saving failure-free performance is not a reality today and may never be. In addition, the causes of malfunction may not be tied to the device itself, but may be caused by physician error or patient-specific reasons.

"Intrinsic" device failures can be caused by product designs that are insufficiently robust, manufacturing process errors and lapses in control, and just plain random failures that are characteristic of all manufactured products.

"Extrinsic" device failures can result from physician errors. Two examples include: (1) inadvertent cutting of the insulation of pacemaker leads during implantation which later results in improper pacemaker system performance; and (2) the inappropriate use of forceps on a heart valve disc during implantation that results in surface scratches which eventually results in catastrophic failure of the mechanical valve due to disc fracture. Patients can be responsible for device failures as

The pendulum began to swing toward greater access to experimental therapies and less government protection of research in the 1980s. The desire for early access to AIDS drugs and experimental breast cancer therapies brought out advocates and spurred the creation of patient advocacy movements (Dresser, 2001). Some advocates argued that risk-benefit trade-offs should be made by affected individuals rather than by the FDA. In ethical parlance, concerns about protection and access are both matters of justice or fairness. The initial approach viewed justice as protection—the need to protect both research subjects from excess risk and populations from unsafe treatments. The latter approach views justice as access and incorporates mechanisms to accelerate the approval process.

Although awareness of a desire for greater access had been growing, a change in the FDA's leadership in the early 1990s refocused the agency on protection, thereby slowing new product approval. By 1995, there was a strong political backlash against the newly energized consumer protection functions of the FDA which had the effect of sharply slowing new product approvals. Some politicians argued that delays in approval and irrational requirements deprived Americans of needed medical products, (including both medications) and devices which ultimately resulted in the FDA Modernization Act of 1996 (FDAMA).

Some of the controversy around access to innovations focused on the time that the FDA required to evaluate clinical data as well as the extent and scope of the tests it required. The companies requesting approval of their products are responsible for providing the information needed to show safety and efficacy. Often this included massive amounts of data. When the FDA review time is added to the years required to perform the necessary research, arguably valuable time is lost in getting the products to patients. Congress tried to ensure speedy reviews, yet consumer protection advocates insisted that speed not compromise safety evaluations. The FDAMA was passed in 1996 that included provisions to streamline the review process. Since that time the device industry and patient advocates continue to work to improve the process; significant advancements and efficiencies have been accomplished.

The demands for accelerated access relative to risks of premature approval can be illuminated by a controversy in ethical discourse over "identified" and "statistical" lives. The concept of identified lives refers to persons who are known to clinicians and who have an urgent, and perhaps life-saving need for access to a new device or product. Because identified persons in these circumstances often have no other alternatives or remedies, some may be willing to consent to a higher risk and an unknown

---

well. For example, some pacemaker patients manipulate their implanted device, located just under the skin, causing the lead that connects the pacemaker to the heart to twist and sometimes fail. This has been given the name, "twiddler's syndrome."

Whether malfunctions are caused by intrinsic factors or extrinsic, "Manufacturers must report device-related deaths, serious injuries, and malfunctions to the FDA whenever they become aware of information that reasonably suggests that a reportable event occurred (one of their devices has or may have caused or contributed to the event)" (Department of Health and Human Services, 1997). Because of strict and short time requirements for reporting, the root causes of a particular event may not have been determined when notification is sent to the agency.

potential for benefit. By contrast, the concept of statistical lives refers to persons who are alive now, though unknown, and who have no immediate need for a device or treatment, but can be anticipated to benefit in the future. Thus, their interests in an increased prospect of benefit when they do have need for such a device are enhanced through applying the most strict and rigorous standards of risk, safety, and efficacy at the outset (Schelling, 1966; Childress, 1981, 13). Similarly, an identified life can highlight a tragedy arising from the use of a new technology, and have a 'decelerating' effect despite statistical evidence of overall safety, as was the case with the Shiley heart valve.

There are numerous examples in public policy and in society of shaping policy to reflect a preference for known, identified lives over the interests of persons who may be in need in the future. In defending the societal preference to spend more resources on a known person in peril rather than employ less-expensive preventive measures to reduce the risks for statistical persons in the future, legal scholar Guido Calebresi contends: "the (life-saving) event is dramatic; the cost, though great, is unusual; and the effect in reaffirming our belief in the sanctity of life is enormous. The effect of such an act in maintaining the many societal values that depend on the dignity of the individual is worth the cost" (Calabresi, 1970). However, other scholars have argued that what is at stake in this social preference is more a psychological investment in a social myth—we will spare no expense to save a life—than a morally defensible principle such as the sanctity or dignity of life or social utility. Regardless, it is clear that both the contrasts between known and unknown persons and present and future peril are important in policy. Assuming that public policy is ultimately based on achieving net good for the public, the principles of justice and fairness necessitate including statistical lives as relevant in regulatory decision-making.

6.4.2.2 Decisions About Reimbursement

Rapid innovation in medical device technology after World War II coincided with the growth of health insurance in the private sector and the passage of Medicare for the elderly (over 65) and disabled, and Medicaid for low income individuals in 1965. As a result, the market for medical technologies grew. As costs began to rise in the 1970s, insurers began to experiment with new payment methodologies and to question the value of the many new procedures and technologies diffusing into widespread use, especially when their use broadened beyond the populations and conditions studied in the pre-market process.

Public and private insurers have struggled with the development of criteria for medical technology evaluation beyond the FDA requirements. Over the last few decades, as insurers have grown larger and more sophisticated, they have developed internal evaluation programs. Private insurers do not necessarily disclose their processes of evaluation, and many of these programs are not as transparent as Medicare. However, competition and the internet have opened the way to greater public disclosure of private insurers' coverage policies, and some have provided considerable information on their processes, including Blue Cross/Blue Shield, which has had a Technology Evaluation Committee in place for many years.

This section focuses on Medicare's development of coverage and payment policies because Medicare beneficiaries consume the vast majority of cardiac technologies and procedures.

Medicare was designed as a fee-for-service system, where doctors, hospitals, and other providers would submit claims for payment for each item or service performed. As expenditures increased, Medicare deployed a variety of payment methodologies, moving from price setting for every item or service to bundled payments for a particular case (called diagnostic related groups, or DRGs, for hospital payment and RBRVS, or resource-based relative value scale, for payments to physicians). A bundled payment meant there was one price or fee that included numerous items or services provided to treat the diagnosis or condition. Medicare employs an elaborate coding system (Healthcare Common Procedure Coding System [HCPCS] and Common Procedural Terminology [CPT]) to implement their payment systems.

As new technologies and procedures proliferated, Medicare recognized the need to evaluate technologies to determine if they should be offered in the Medicare program. Under the original Medicare statute, Medicare must provide only "reasonable and necessary" items and services. Medicare employs private local contractors to process claims for payment. In the early days of the program, claims were paid by these local contractors using their judgment about whether to allow the claims. There was little attention focused on new technologies at the national level. The first challenge occurred when vanguard surgeons began to perform heart transplantation in the 1970s, a highly risky and expensive procedure at the time. Medicare struggled to develop an infrastructure and mechanisms to evaluate and manage heart transplantation services (Foote, 2002).

Over time, local contractors and the Health Care Financing Administration (HCFA), the predecessor to CMS, developed an in-house, informal method for sharing information about new technologies and deciding when and how to cover them. In the 1990s, CMS made the process of developing coverage policies more formal and transparent. Local contractors continued to develop local policies if no national coverage policies applied, and the processes for their development also were formalized. These changes began to put pressure on device manufacturers to provide information that Medicare could use to evaluate new technologies. They also put downward pressure on price, especially from hospitals because they purchase the lion's share of medical devices, but are reimbursed for the bundle of services under the DRGs.

Medicare has been particularly challenged in its efforts to develop criteria to evaluate technologies and procedures to use to implement the "reasonable and necessary" clause. Medicare currently considers evidence concerning safety and effectiveness, clinical effectiveness, and appropriateness for the particular patient, or medical necessity in a particular case. Medicare was unsuccessful in its effort in 1989 to interpret its authority to include the use of cost-effectiveness as a criterion. There was a firestorm of protest, primarily from industry but also from physicians. Industry concern was based on public assertions criticizing cost-effectiveness models, as well as less public concern that cost effectiveness evaluations might limit sales and market growth. As of now, there is no explicit cost-effectiveness evaluation in Medicare.

6.4.2.3 ICD: An Example

Implantable cardiac defibrillators (ICDs) present an interesting example of the issues that arise in coverage controversies. Implanted cardiac defibrillators shock the heart when potentially lethal heart arrhythmias arise. The issues with ICDs are less about how well they work and more about which patients should be candidates to receive the device. Defining the patient population that can benefit from a particular technology is very challenging. Although this consideration is part of the device development process, for reasons of speed and cost, many clinical trials enroll only a narrow subset of potential users of an experimental technology.

Experimental Class III devices such as the ICD were first considered as a therapy of last resort because of the rigors of implantation and the unknowns of a new technology. Consequently, the patient inclusion criteria of the clinical study are very narrowly defined. If proven safe and effective, the product is labeled accordingly and the manufacturer can only promote it for the approved indications.

Post-market (i.e., after FDA approval of the base technology) ICD studies demonstrated additional at-risk patient populations could benefit from ICD implantation. Indications for use were expanded markedly to include the prophylactic use of devices in patients having certain diagnostically determined cardiac characteristics. This is a typical pattern for Class III devices: indications for use are very narrow initially and broaden with additional studies, additional experience, and technological advancements to the device itself.

To resist the tendency to expand use of ICDs beyond the parameters of the FDA trials that approved them for human use, CMS labored on defining conditions of medical necessity as part of a national coverage discussion. Early coverage policies were quite limited—to patients who had already experienced and survived a life-threatening heart arrhythmia, namely ventricular fibrillation. Manufacturers, however, sought to show that patients with earlier warning signs, including those at risk for serious heart rhythm problems but not yet experiencing life-threatening events, could benefit from an ICD.

CMS coverage decisions do not always align with the results of well regarded post-market clinical studies that resulted in FDA approval for expanded indications for use. In the case of the ICD, CMS initially went against its own advisory board of experts who unanimously recommended coverage for patients having specific high-risk characteristics previously not covered by product labeling or coverage decisions. The agency's interpretation of the data tightened the eligibility criteria over those of the inclusion criteria of the study. CMS beneficiaries who fell outside the narrower definition set by the agency, but within those covered by the product labeling and endorsed by its experts, were effectively denied treatment until a subsequent coverage decision cycle (Ashwath and Sogade, 2005; Centers for Medicare and

Medicaid Services, 2005). In fact, results of this study were sufficiently persuasive that private insurers issued coverage consistent with the study findins. Further, both the American College of Cardiology and American Heart Association amended their physician practice guidelines regarding the new treatment well before CMS' expanded coverage decision.

In addition to issues of appropriate criteria and process, this example also raises questions about the timing of evaluations. The earlier a device manufacturer seeks approval, whether for marketing or for coverage and reimbursement, the more limited is the information available to support such decision-making. It also raises questions of cost/benefit and risk/benefit. ICDs are expensive and invasive technologies. What type of patient is most likely to benefit? The first round of approval by CMS resulted in a limited coverage decision. The agency demanded that device manufacturers conduct additional studies, which are large and expensive to run, before it would consider an expansion of coverage. After the studies were completed, there were conflicts about how much to extend the indications for use on the basis of study data. CMS was alleged to be too conservative in its interpretation, arguably to keep the costs of ICDs to the Medicare program down. Subsequent recommendations included requirements for a registry of patients with ICDs so that the benefits and the risks could be measured in the real world population.

Both Medicare and private insurers place the burden on device manufacturers and advocates of innovative procedures to provide sufficient information to evaluate new technologies. Increasingly, the clinical information they request goes beyond FDA data, focusing on clinical usefulness and necessity rather than safety and efficacy in research settings. Payers may also be suspicious of data derived exclusively from a manufacturer's own work. Academic researchers and private clinical trial organizations are often recruited by manufacturers to design and run extensive, expensive trials in order to provide data that address payers' criteria.

6.4.2.4 Ethics of Evaluation

The evaluation of technology is inherently, and perhaps somewhat obviously, influenced by the humanities, simply because it entails assessing the value of a technology or device. Value considerations identify what counts as a benefit and what counts as a risk or harm, and how particular benefits and harms count in an evaluation method. Value is thereby a normative and narrative concept as well as a measure.

In very general terms, the ethical principles and values we readily identify as having a regulatory role (in pre-marketing research)—beneficence, justice, and respect for persons—overlap substantially with those identified as having a role in (post-marketing) evaluation. However, in addition, the 'humanities toolbox' includes a range of value considerations that, if rendered more explicit, can inform most evaluation measures. Specifically, some literature in philosophy and bioethics addresses values in economic analyses of health (Menzel, 1990; Ubel, 2000).

Standard measurement tools like cost-effectiveness analysis (CEA), cost-benefit analysis (CBA), quality-adjusted life years (QALYs), and cost utility analysis clearly signal the methodological importance of outcome and consequentialist or utility considerations as approaches to meeting the obligations of the ethical principles of beneficence or justice. Driven by concerns about maximizing efficiency, CEA inquires whether more benefit can be obtained for the health care dollars expended, or the same benefit can be obtained with fewer dollars expended. That is, CEA involves comparisons with other forms of health care. The QALY measurement, which combines both morbidity and mortality, is indispensable for this method. In contrast, CBA investigates how the benefits obtained from health care compare with other goods that could have been obtained through the use of the same amount of time, education, and effort, or whether investments in relatively inefficient health care lead to forgoing other improvements in society. As such, CBA examines health care benefits within a larger societal context (Menzel, 1990).

Despite the usefulness of such methods, non-utilitarian considerations put tension on the utilitarian calculus in such a way as to incorporate considerations that impose limits on utility maximization. These non-utilitarian considerations may make possible the favorable evaluation of a technology on the basis of its ability to benefit those who are least well off, whether the disparity is defined economically or in health terms. For example, CEA and QALY methods are often critiqued as disvaluing the health preferences of persons with disabilities. Accordingly, other considerations of justice and the principle of respect for persons are necessary to supplement and remedy this otherwise exclusive orientation to efficiency and utility.

Finally, the process of evaluation is itself ethically significant, especially as those with oversight responsibility may, as noted above, make decisions in conflict with the positions of medical advisory boards or panels of experts in the device field. The evaluation process thereby should integrate central values of procedural justice, including accountability, transparency, and inclusiveness.

6.4.2.5 Marketing and Clinician Education

Unlike many pharmaceutical products, implanted medical devices require considerable physician skill to tailor the device's features for specific patient needs, and to implant or deploy it. Because of the highly specialized nature of many implant technologies, sub-specialty groups have formed within medicine. For example, electrophysiology has emerged as a sub-discipline of cardiology in order to provide expert care to patients who need a pacemaker or defibrillator to chronically manage heart rhythm disorders. In addition, the complex nature of Class III devices requires companies to develop marketing methods very different from those typically employed by the pharmaceutical industry.

Marketing efforts in the device industry can be segmented into two broad areas, product awareness and technology utilization. Awareness is, in turn, directed primarily at three groups: implanting physicians, referring physicians, and increasingly, patients. Awareness-building physician education programs are necessary because potential implant patients make up only a small portion of a typical general medical practice. These physicians, however, act as gatekeepers in the referral process. The fact that most medical device technology evolves rapidly has led industry, often in partnership with specialized medical societies, to take on this educational role. Patient education materials are also familiar aspects of such outreach efforts. Implant companies have provided patient-tailored information materials to patients, prospective patients, and families upon request. With more widespread use of the internet, many companies have developed detailed web sites specially designed for patients, informing them of new technologies and therapeutic options as well as answering questions they may have regarding older technologies. Web sites have also become an important patient resource for device related information.

The highly sophisticated nature of implants and the pace of innovation require that programs and systems are in place to help assure correct implementation of technologies such as pacemakers, defibrillators and ventricular assist devices. Proper device performance depends not only on the intrinsic attributes of the device, but also on extrinsic factors that relate to patient selection and how the device is surgically implanted. These aspects continually evolve. Companies often are the place where new knowledge of what works and what doesn't is concentrated. Knowledge dissemination can take the form of in-house training seminars, best practices newsletters, symposium sponsorship, hands-on workshops, product performance reports, and educational web sites. However, some questions have been raised about the appropriateness of such approaches (Meier, 2006). In addition, many companies maintain technical experts in the field in order to provide highly specialized consultative support to physicians if they request it.

Conflicts of interest can be created by incentives given to physicians for promotion of new technologies and a physician's relationship with a device manufacturer. This is an unsettled area with uncertainty about the appropriate rules and constraints. When is an "expert" really a marketing tool rather than a neutral expert? How should the line be drawn between impermissible marketing tactics and permissible information sharing? What types of disclosures should be made when an "expert" has a financial interest in the technology or the company?

As articulated by Thompson, a "conflict of interest is a set of conditions in which professional judgment concerning a primary interest (such as a patient's welfare or the validity of research) tends to be unduly influenced by a secondary interest (such as financial gain)" (Thompson, 1993). What is needed to fill out this account, of course, are shared conceptions of primary and secondary interests, as well as the conditions in which the influence of the secondary interest seems overriding or dominating. For the secondary interest, it is important to note, is not an illegitimate interest; it is an important facet of professional life that needs to be managed and directed. Insofar as marketing of innovative medical devices to physicians and clinicians is accompanied by various incentives for adoption and usage, without due regard for the welfare of patients in whom the device will be implanted, the possibility of inappropriate professional and industry conflicts of interest must be addressed.

In order to address the ethical challenges involved in promotion of new products, the drug and device industries have developed and promulgated voluntary standards governing promotion of new products directly to clinicians. These standards are intended to avert unethical practices, such as giveaways (e.g., free trips, etc.) and bonuses for use.

The most common approach to managing conflicts of interest is disclosure to patients of the relationship between the physician and the marketing arm of industry. In addition, various professional associations have developed principles, codes, and procedures for the guidance of their members in addressing conflicts of interest. Insofar as the issue is perceived not to be adequately addressed by disclosure and by guidelines of professional associations, regulatory agencies at the state medical board level or institutional level may intervene and establish standards of performance. Nonetheless, bioethics' commentary continues to make appeals to the personal integrity of professionals to supplement agency requirements and, more generally, recommends awareness of the inevitability of such conflicts as part of the contemporary landscape of medicine (Morreim, 2003).

6.4.2.6 Altering Death: Humanities Implications of Device Use

Although decisions about whether to use devices often obviously implicate ethics and the humanities and raise values questions, it is perhaps even more likely for questions and issues to arise along with use.

As described earlier, the advent of the artificial heart raised concerns about the effects of such a fundamental alteration on the person. Historically, the heart represented the seat of personhood, and Barney Clark's wife briefly wondered if he would still love her after he received the artificial heart. By the time the TAH was developed, the American cultural view had firmly ensconced the brain as the essence of the self—a development presumably attributable at least in part to the availability of cardiac devices.

In addition, however, the ubiquity of these devices has changed the experience of death, not only for affected patients and families, but also for their clinicians. During the dying process, modern medical ethics recognizes the patient's right, through his or her own choice or through a decision by a legal representative, to forgo treatment, receive comfort care, and die without invasive interventions. When a patient dies with an ICD in place, however, the device may at times interfere, unexpectedly and distressingly, with a peaceful death (Stein, 2006). If this potential difficulty is anticipated, then another decisionmaking process regarding the device is added to the already complex choreography of dying. And at times, when faced with the request to disable an ICD clinicians may be reluctant to turn off an implanted device that has become integral to the patient's body and physical function (Mueller et al., 2003). Although the ethical issues are identical to the issues implicated by withdrawal of an external device like a ventilator, an incorporated device can appear to be very different to some people, thus giving rise to disagreement.

Related to these perhaps dramatic effects, incorporated devices may also have profound effects at the individual level. As has been noted, the risk of device failure, although low, can be rightly perceived as catastrophic by patients, not only because the consequences of failure may be serious but also because, the devices may be regarded as permanent intrusions. This may be true even when the failure risk for an ICD is lower than the risks of harm, side effects, or failure from a drug or surgical alternative. At the same time, the permanence of some incorporated devices, and perhaps cardiac devices in particular, may make it more difficult for physicians, researchers, and device manufacturers to assess device function and malfunction in an attempt to reduce uncertainty concerning them.

Because these questions arise along with device use, and because they do not always arise in the same way for all device users, they are unlikely to be appropriate policy-level concerns. However, sensitivity on the part of clinicians, researchers, and device manufacturers to the possibility that the incorporation of a device may have special meaning for patients or give rise to special concerns at times (e.g., device failure or death from other causes) would undoubtedly be beneficial. The requisite sensitivity includes, but is not limited to awareness of the ethical issues raised by device inactivation or removal.

6.5 The Humanities and Medical Device Policy

In considering the current approaches to device development and use, it becomes clear that different criteria are used in policy development. As described previously, the central criteria for clinical research and regulatory approval of a medical device are safety and efficacy. Post-approval studies are conducted to evaluate clinical effectiveness. Diffusion reflects patterns of availability and use of devices as determined by information dissemination, as well as reimbursement and coverage.

Philosophical, religious and other bioethics disciplines have historically been explicitly involved in public discourse both at the initiation of device development, or what we have described above as the process of regulatory approval, and before the beginnings of the approval process, when the device is still at the "drawing board" stage. In general, then, intersections between the policy process and the disciplines of bioethics are likely to occur at a "pre-regulatory" stage.

Alternatively, bioethics may also "weigh in" during the post-approval phases of diffusion and evaluation of medical devices. Traditions of bioethics may present distinctive conceptual or analytical tools to assess medical devices. Further, bioethics traditions offer different ways of "seeing the world," including the political world, or the embodied world of the patient, into which the device is implanted. In some circumstances, these distinctive world-views may inhibit device development. For example, early research on a cardiac pacemaker was opposed by some religious voices not on policy grounds of safety and efficacy, but rather because it interfered with divine providence in the determination of life and death (Schechter, 1972). Such world-views are not invariably opposed to device development, but may in other circumstances, such as experimentation with the artificial heart, provide a cultural context of support (Fox and Swazey, 1992).

Similarly, concepts of the sanctity of the body and its intactness in some religious traditions may hinder comprehensive device evaluation where removal of a device post-mortem may be desirable from a scientific perspective. For instance, if one holds that manipulation of the post-mortem body constitutes desecration and sacrilege, the appropriateness of such evaluations are called into question.

Thus, bioethics invites analysis of the assumptions underlying technical and regulatory standards of safety and efficacy, and definitions and evaluations of risks, harms, and benefits. These common standards of regulatory discourse presume the sufficiency and normative nature of "instrumental" rationality, that is, to devise the most effective and efficient means to achieve desired ends (device approval, patient access, etc.). In addition, consider that bioethics discourse frequently draws on a discourse of meaning and "symbolic" rationality, that is, a form of reasoning that emphasizes the expression of values, such as compassion, dignity, or respect for persons, which cannot be reduced to consequentialist analysis. Such considerations may be most important in public discourse in pre-regulatory and post-approval stages of device development, but may be relatively peripheral to the policy approval process itself. However, consequentialist or utilitarian-based methods may find direct application in this area, and indeed provide justification for the regulatory process and specific methods.

The resolution of the policy issues of regulation, evaluation, and diffusion doesn't dispel the bioethical controversies that may arise at each stage of development. As indicated previously, ethical dissent may often be voiced at the pre-regulatory phase of discourse. However, it is conceivable that dissent may also occur after the approval and evaluation of a device has been completed. This is especially likely when matters of life and death are at stake. The policy context of "approval" of a medical device and the moral context of ethical "diffusion" of the same device, and its meaning in the provision of medical care, can be very different.

Finally, because all decisions are made under uncertainty, it is of course necessary to decide how much uncertainty is too much for a device to be approved and used. Just as individual and policy-level evaluations of benefit differ, policy-level determinations may tolerate an amount of uncertainty different from that appropriate to individual decisions about the desirability of a given technology. Uncertainty in individual decisions is usually managed through shared medical decision making and the provision of informed consent. In the regulatory context, a high degree of uncertainty can signal that an experimental technology is not ready for broad patient use. Acknowledging and taking account of uncertainty is an imperative in all aspects of the development and use of technology. At a very basic level, technology evaluation depends on ethical values to help explain and justify actions under uncertainty. It would be tempting, at least to some, to seek mathematically certain determinations about the appropriate allocation and use of new technologies. However, even prodigious amounts of data simply cannot produce certainty. Instead, coming to terms with uncertainty requires flexibility and analytical transparency.

Every approach to uncertainty is value-laden. Value considerations "dictate the manner in which uncertainty as to the potential adverse consequences will be resolved" (Green, 1975). The bioethics debate on uncertainty suggests the need for transparency in these deliberations, both about the value-laden assumptions built into forecasts of the future and about decisions as to which parties in society assume the burden of proof. Does this burden fall on advocates of a device, to show that its speculative benefits will be greater than its potential harms? Or does the burden fall on those who desire to restrict access, to show in advance that the possible harms will be greater than the benefits? As ethicist Daniel Callahan has argued, it is not possible to demonstrate the burden of harm without actually using the device in research or in practice, and once a device has been introduced it is very difficult to undo its introduction. Since the principle of nonmaleficence typically takes moral priority over the duty to benefit, Callahan argues for a very high threshold of benefit (Callahan, 2003).

In the context of technology evaluation, however, when new technologies are introduced into practice, they are inevitably utilized in ways that outpace the data on which their approval was based. How this uncertainty is regarded depends upon the consequences of saying no to an expanded use. "We won't pay for this" is importantly different from "You can't have this"—and both of those answers acquire different moral valences depending on the severity of the health consequences that could result from lack of access.

6.6 Conclusion

It is clear from our analysis of implantable cardiac assist devices that policy, including regulatory approval, evaluation of use, and reimbursement, evolves with and is frequently reactive to technological innovation, device use, and on occasion, device failure. Innovative technologies similarly positions most bioethical analyses in a posture of reaction after the technological advent; while there are certain settled issues in bioethics, such as the necessity of informed consent in circumstances of human experimentation, we have seen from numerous examples of cardiac devices that even this foundational legal and bioethical concept is open to a variety of risk thresholds that can be less or more demanding depending on the patient population and the immediacy of need. Thus, it often seems to be the case that bioethical questions and guidance are contingent on the state of technological innovation.

Yet, a technologically-determined bioethics loses much of its critical edge, and perhaps also much of its policy significance. The fact is, however, that technology does not create ethical problems as much as it provides an occasion for underlying values to come into conflict, thus requiring clarification and deliberation over these values, as well as their prioritization to facilitate sound decision-making.

In considering the general themes of this two-phase study through the window of implanted cardiac devices, we do not find in the development, diffusion, and use of these technologies to alter human biological nature any intrinsic violation of nature itself. Indeed, we contend that their development is an expression of rational human nature as motivated by some core values, including the preservation of human life and knowledge generated by scientific inquiry and research application. Moreover, a central consideration with implanted medical devices—their impact on personal identity—is not, objectively speaking, altered through the use of a cardiac device. A person who receives a TAH, let alone an ICD, is still the same individual. This objective continuity of the person does not, however, preclude changes in subjective perception, or the prospect for emotional, psychological, relational, and symbolic alterations in the sense of self.

Our analysis of the policy process (whether it be FDA regulatory oversight of medical devices, CMS post-approval evaluation of their efficacy and safety, Medicare and other reimbursement and coverage procedures, or education and diffusion) has illustrated repeatedly the numerous *value assumptions* that are integrated within policy deliberation. That is, while policy must address its own set of questions for approval, usage, and payment, these questions are not shorn off from ethical considerations. Although not always explicit, we find more integration and overlap of policy and bioethics than we do separation and compartmentalization. We have observed, for example, that the "gold standard" of policy regulation relies on two criteria, safety and efficacy. If we were to probe further and ask why these two criteria constitute the policy gold standard, at least part of the rationale would be that safety and efficacy are warranted by core ethical principles of beneficence and non-maleficence, which impose moral, political and professional responsibilities to promote patient welfare and refraining from causing injury, harm or death.

Bioethics disciplines can and do provide philosophical justification for policy assumptions; in this respect, bioethics is devoted to what might be called "a task of problem solving." However, we contend that this does not exhaust the implications of bioethics for policy, and indeed, it may not even be the most important contribution of bioethics to policy deliberation.

There is a prior task for which bioethics disciplines are absolutely vital, namely, that of perceiving the problem that requires policy and ethical analysis. Bioethics disciplines thereby expand moral vision and provide a context for the content of policy problems, and of understanding how a "problem" became such in the first place. In addition to tasks of problem-solving, bioethics is no less devoted to what can be designated "tasks of problem-seeing."

For example, our discussion has previously made reference to the common regulatory conflict between the consumer protection functions of federal agencies such as the FDA and expedited access of patients to devices that are medically needed and politically demanded. This conflict is more fully illuminated when it is situated within a context of a societal value preference between "identified" persons and "statistical" lives.

The full array of humanities disciplines can facilitate the task of sharpening moral vision. Bioethics and the policy process are committed to the principle of universalizability, that is, of giving similar treatment to relevantly similar situations. The implementation of this principle in turn requires an historical account of prior technological innovations to illuminate relevant similarities or dissimilarities. In so doing, we employ not only the logic of rational-deductive reasoning, but also an aesthetic of analogical reasoning.

This aesthetic sensibility may also be cultivated by recognizing how narrative, metaphor, and story shape patient and professional experiences. For example, philosopher Larry Churchill has illustrated how patients' apprehension of the approach of impending death—which may or may not be staved off with an implanted cardiac device—is better understood through storied experience and narrative constructive rather than a mechanistic "stages" model (Churchill, 1979). Stories are the formative texts for illuminating the meanings of illness, dying, or death to a patient; thus, a patient's demand for expedited approval of a potential treatment is revealed to be not so much a controversy over device availability or procedures for reimbursement, but the person's experience of a crisis in meaning and identity. Ultimately, if policy and ethical analyses focus solely on the requirement of respect for patient self-determination, they neglect the meaning patients narratively construct about their illness at the risk of dehumanization and objectification.

It seems likely that these tasks of problem-solving and problem-seeing for the humanities are present when other sorts of medical devices to be incorporated are considered, developed, and used. At least in the case of implantable cardiac devices they seem to have some currency, whether or not they are explicit to those engaged in the process. Indeed, translating the languages of medical technology, policy making and bioethics can be difficult, which likely poses an obstacle to a more comprehensive analysis. Further, in different settings other roles for the humanities will emerge, especially when considering the pre-regulatory or "drawing-board" stages. After all, such an early stage is ripe with creativity and imagination, where ideas are tested, examined, and explored. Here, the humanities would seem to play an essential role in helping to explore such matters as the appropriate types of interventions and their potential influence on humanity. Such explorations could conceivably help distinguish among those types of incorporation that should be abandoned outside the realm of science fiction and fuel the development process for those that at least at the outset seem desirable.

References

Annas, George J. (1988). *Judging Medicine*. Clifton, NJ: Humana.
Ashwath, Mahi Lakshmi and Felix O. Sogade (2005). "Ejection Fraction and QRS Width as Predictors of Event Rates in Patients with Implantable Cardioverter Defibrillators," *Southern Medical Journal* 98(5), 513–517.
Best, Robert, George Khushf, and Robin Wilson (2006). "A Sympathetic but Critical Assessment of Nanotechnology Initiatives," *Journal of Law, Medicine, and Ethics* 34(4), 655–657.
Blot, William J. et al. (2005). "Twenty-five Year Experience with the Bjork-Shiley Convexoconcave Heart Valve: A Continuing Clinical Concern," *Circulation* 111(21), 2850–2857.
Burling, Stacey (2002a). "Mechanical-heart Patient Comes to Regret His Life-saving Choice," *Philadelphia Inquirer*, July 14.
Burling, Stacey (2002b)."Life, but at What Cost?" *Philadelphia Inquirer*, September 29.
Burling, Stacey (2002c). "Widow Sues Artificial-heart Maker," *Philadelphia Inquirer*, October 17.

Calabresi, Guido (1970). "Reflections on Medical Experimentation in Humans," in Paul Freund (ed.), *Experimentation on Human Subjects*. New York: G. Braziller.

Callahan, Daniel (2003). *What Price Better Health?: Hazards of the Research Imperative*. Berkeley, CA: University of California Press.

Center for Medicare and Medicaid Services (2004). "Medicare Expands coverage of Implantable Defibrillators to Save Lives and Develop Evidence to maximize Benefits." Press Release, January 27. Available at: www.cms.hhs.gov/apps/media/press/release.asp?Counter=1331. Accessed 15 March 2008.

Childress, James F. (1981). *Priorities in Biomedical Ethics*. Philadelphia, PA: Westminster.

Churchill, Larry R. (1979). "The Human Experience of Dying: The Moral Primacy of Stories over Stages," *Soundings* 62, 24–37.

Department of Health and Human Services (1997). *Medical Device Reporting for Manufacturers*, March. Available online: http://www.fda.gov/cdrh/manual/mdrman.html

Dresser, Rebecca (2001). *When Science Offers Salvation: Patient Advocacy and Research Ethics*. New York: Oxford University Press.

Foote, Susan Bartlett (1978). "Loops and Loopholes: Hazardous Device Regulation under the 1976 Medical Device Amendments," *Ecology Law Quarterly* 7, 101–135.

Foote, Susan Bartlett (2002). "Why Medicare Can't Promulgate a National Coverage Rule: A Case of *Regula Mortis*," *Journal of Health Politics, Policy and Law* 27, 707–730.

Foote, Susan Bartlett and Robert J. Berlin (2005). "Can Regulation be as Innovative as Science and Technology: FDA's Regulation of Combination Products," *Minnesota Journal of Law, Science and Technology* 6(2), 619–644.

Fox, Renee C. and Judith P. Swazey (1992). *Spare Parts: Organ Replacement in American Society*. New York: Oxford University Press.

Green, Harold P. (1975). "The Risk-Benefit Calculus in Safety Determinations," *George Washington Law Review* 43, 791–808.

Institute of Medicine (IOM) (1991). *The Artificial Heart: Prototypes, Policies, and Patients*. Washington, DC: National Academies Press. Available online: http://www.nap.edu/catalog.php?record_id = 1820

Lidz, Charles W. and Paul S. Appelbaum (2003). "The Therapeutic Misconception: Problems and Solutions," *Medical Care* 40(Suppl), V55–V63.

Meier, Barry (2006). "Growing Debate as Doctors Train on New Devices," *New York Times*, August 1.

Menzel, Paul (1990). *Strong Medicine*. New York: Oxford University Press.

Morreim, E. Haavi (2003). "Conflict of Interest," in Stephen G. Post (ed.), *Encyclopedia of Bioethics*, 3rd ed., vol. 1. New York: Macmillan, 503–508.

Mueller, Paul S., C., Christopher, Hook, and David L. Hayes (2003). "Ethical Analysis of Withdrawal of Pacemaker or Implantable Cardiac Defibrillator Support at the End of Life," *Mayo Clinic Proceedings* 78, 959–963.

National Commission for the Protection of Human Subjects in Biomedical and Behavioral Research (1979). *Belmont Report*. Available online: http://www.hhs.gov/ohrp/humansubjects/guidance/belmont.htm

NHLBI (National Heart, Lung, and Blood Institute) (1985). *Artificial Heart and Assist Devices: Directions, Needs, Costs, Societal, and Ethical Issues*. Report of the Working Group on Mechanical Circulatory Support, National Heart, Lung, and Blood Institute. Rockville, MD: NHLBI.

Office of Technology Assessment (OTA) Congress of the United States (1982). *The Implications of Cost-Effectiveness Analysis of Medical Technology*. Case Study #9. The Artificial Heart: Cost Risk, and Benefits.

Schechter, D.C. (1972). "Background of Clinical Cardiac Electrostimulation. IV. Early studies on the feasibility of accelerating heart rate by means of electricity," *New York State Journal of Medicine* 72, 395.

Schelling, Thomas (1966). "The Life You Save May Be Your Own," in Samuel B. Chase, Jr. (ed.), *Problems in Public Expenditure Analysis*. Washington, DC: Brookings Institution, 127–166.

Secretary's Advisory Committee on Xenotransplantation. *http://www4.od.nih.gov/oba/Sacx.htm*

Stein, Rob (2006). "Devices Can Interfere With Peaceful Death," *Washington Post*, December 17, A1.
Thompson, Dennis (1993). "Understanding Financial Conflicts of Interest," *New England Journal of Medicine* 329(8), 573–576.
Ubel, Peter A. (2000). *Pricing Life*. Cambridge, MA: MIT.
United States Code 321(h) (2000).
Veatch, Robert (2004). "The Total Artificial Heart: Is Paying for it Immoral and Stopping it Murder?," *Lahey Clinic Medical Ethics Journal* 11(1).
World Transhumanist Association. http://transhumanism.org/index.php/WTA/about/

Chapter 7
Biodiversity and Biotechnology

Nicholas Agar, David M. Lodge, Gerald P. McKenny, and LaReesa Wolfenbarger

7.1 Introduction

The title of this chapter brings together two of the most critical issues raised by the relation of human beings to the natural world. At first, the rationale for bringing them together may be unclear. The threats human activities pose to biodiversity are among the most urgent concerns of environmentalists, but of course, biotechnology accounts for only a narrow range of these threats. Similarly, of the urgent ethical questions posed by biotechnology only a few are directly relevant to biodiversity. Nevertheless there is significant overlap between ethical concerns connected with biodiversity and those connected with biotechnology. This overlap appears, for example, in efforts to enlist biotechnology in the preservation of biodiversity, but the most poignant concerns arise with the prospect of impact on biodiversity from transgenic organisms. Opponents of genetically modified organisms regularly point to the potential effects of these organisms on biodiversity as a reason for prohibiting or strictly regulating research on and application of transgenics. This chapter attempts to evaluate these claims by a careful consideration of the issues they raise in the context of current scientific research.

The chapter is structured by an analysis of the factual and normative issues at stake. In order to evaluate the claims regarding transgenic research and biodiversity we must ask two preliminary questions. The first preliminary question is factual; it has to do with the potential impact of genetically engineered organisms on biodiversity. There is widespread concern about the effects such organisms may have on biodiversity, but what do we know about the likelihood of such effects? Uncertainty about these effects themselves and about the future direction of biotechnology makes this a difficult question, but we nevertheless attempt to answer it in Section 7.1.2 below in relation to the most likely foreseeable developments in transgenics.

The second preliminary question is normative; it has to do with what is at stake, ethically speaking, in biodiversity. What claims, if any, does biodiversity make on us? How do we evaluate any such claims in relation to other claims, for example the interests of those who seek to introduce transgenic organisms into various ecosystems and those who are affected by such interventions? This question is addressed in Section 7.1.3 below, where we examine claims made on behalf of natural diversity in various religious and other cultural traditions. Consistent with the questions raised in this volume, we will discuss claims that natural diversity has normative significance.

Because one aim of the chapter is to relate the question of transgenic organisms and biodiversity to current scientific research, Section 7.1.4 offers an analysis of one potential transgenic organism discussed in Section 7.1.2, namely Bt maize, to illustrate the concrete ethical issues involved in adopting or not adopting transgenic organisms. While this focus is limited to a single instance of current research, the controversy surrounding Bt maize in Mexico makes it an ideal case to identify what is at stake ethically.

Finally, because arguments regarding the threat transgenic organisms pose to biodiversity often turn on debates over how to assess risk, what factors should be included in risk analysis, and how much consideration to give different factors, Section 7.1.5 carries out two kinds of risk analysis which, in response to concerns voiced by opponents of biotechnology, expand the range of considerations that typically enter into evaluations of transgenic organisms.

Before undertaking this analysis, however, it is important to understand what biodiversity is. This is the task of the first section.

7.1.1 Conceptions of Biodiversity

The term "biodiversity" has a deceptively simple meaning for those in fields related to biology: It is simply "the variety and variability of biological organization" (NRC, 2001). The CBD defines biodiversity similarly as "…the variability among living organisms from all sources…." Although simple, these definitions are extremely broad and encompass levels of biological organization that fall into three categories of diversity: genetic, organismal and ecological (Harper and Hawksworth, 1995a, b). From a biological view, this all-encompassing definition of biodiversity stresses the interconnectedness of biological systems. Assessing and quantifying biodiversity therefore requires understanding genes, populations, species, communities, and ecosystems.

7.1.1.1 Genetic Diversity

By having genetic variation, populations may respond to changing environmental conditions, a component critical for the future persistence and resilience of ecological systems in nature. Loss of genetic variation increases population or species risk of extinction (Reed and Frankham, 2003).

7.1.1.2 Organismal Diversity

The long-term stability of species depends on the dynamics among populations and among species and how these entities respond to environmental perturbations or change. The presence or absence of populations or species within a community or ecosystem may have significant impacts on biodiversity through ecological interactions and ecosystem dynamics (Whitham et al., 2003). For example, predator removal experiments demonstrate the concept of "keystone" predators whose presence or absence affects the diversity and abundance of other species within a community or ecosystem (Navarrete and Menge, 1996). Similarly, the removal or addition of a species or population may affect the function of an ecosystem, including nutrient dynamics and energy flow (Symstad et al., 1998). Lastly, if a susceptible species is rare or has small populations, any mortality or sublethal impacts on its populations may exacerbate an existing high risk of extinction.

Changes in biodiversity may affect ecosystems through at least four factors: (1) the abundance of each species, (2) the composition of species, (3) how a species functions in an ecosystem, and (4) biotic interactions that affect the magnitude and variability of a species' function(s) (Symstad et al., 1998); however, predicting the outcome of changes in biodiversity on these four factors remains unclear and under study (Naeem and Wright, 2003). Studies on the consequences of biodiversity loss concur that a relationship between biodiversity and ecosystem function exists (Chapin et al., 1997; Loreau et al., 2001), indicating that losses in species diversity will have impacts at the ecosystem level of organization.

7.1.1.3 Ecosystem Diversity

Higher levels of organization include associations among species within and among habitat types. These landscape levels of biodiversity include the ecological processes that maintain habitat types (e.g., grasslands, forests) but also include the connections and dynamics between and among habitat types. Ecosystems represent a collection of biological organisms, their dynamics as well as the abiotic features that affect organisms and processes. Ecosystems provide tangible and intangible, short-term and long-term functions as well as services of value to humans and other organisms, including pollination, air and water purification, decomposition, pest control, seed dispersal, nutrient cycling and generation, fertilization and preservation of soils (Daily et al., 1997). Strong associations between distributions of species and particular ecosystems mean that higher ecosystem diversity will translate into higher species diversity.

7.1.2 *Genetically Engineered Organisms and Biodiversity*

Public debates about genetically engineered organisms almost always presuppose some judgments about the likelihood of the effects these organisms will have. It is, of course, difficult to determine these effects due to the novelty of the technologies. In Section 7.1.5 below, we will discuss ways in which the novelty itself, or rather the uncertainty it involves, complicates our assessments of the risks posed by these organisms. Here, we try to answer a preliminary question: what are the potential effects of genetically engineered organisms on biodiversity?

7.1.2.1 Current Status of Genetically Engineered Organisms

We begin with a brief survey of the current status of transgenic organisms. Genetically engineered crops most widely used in the U.S. are corn, cotton and soybeans. Each of these alters the use of inputs (i.e., insecticides, herbicides) that farmers apply to improve crop yield. Three types of GE corn are available in the US. Round-up Ready corn is tolerant to the herbicide glyphosate (Round-up) so that weeds may be controlled by applying a single broad-spectrum herbicide. Bt corn contains a gene(s) from a soil bacterium that provides insect control. Different genes produce variants that will target different insect groups, including Lepidopterans, Dipterans, and Coleopterans. One type of Bt corn targets the Lepidopteran pest, the European corn borer, and other Lepidopterans vary widely in sensitivity to the Bt toxin. Another type of Bt corn, commercialized in 2003, targets the corn rootworm, a soil Coleopteran. Cotton has also been engineered with Bt genes to protect it from bollworm pests (species varies by region). The most widely used genetically engineered crop is Round-up Ready soybeans; in 2003, these represented more than 85% of soybean acreage planted in the U.S.

Products under development range widely in uses (see Table 7.1). The "second wave" of genetically engineered plants has focused on food processing traits, such as altered oil composition. The "third wave" developments are plants producing pharmaceuticals or industrial chemicals. Genetically engineered animals, at this point, are raised under containment (i.e., greenhouse) or confinement (i.e., grown in the field with practices that minimize escape of pollen, seed or plants). The FDA is currently reviewing the food safety and biosafety of Atlantic salmon engineered with a growth hormone.

7.1.2.2 Possible Effects of Biotechnology on Biodiversity

Impacts of biotechnology on biodiversity and ecological processes will vary on a case-by-case basis. It is necessary to keep this in mind, because there is a tendency to think of genetic engineering in monolithic terms as a single phenomenon that will have only one kind of effect, for good or ill, on environments in which transgenic organisms are introduced. In fact, genetic modifications have the potential to have neutral, positive or negative impacts. What species are modified, what phenotypes result, and where these individuals are released represent three factors that will influence what type of impacts on biodiversity a particular biotechnology product will have. Unconfined releases and intentional releases into the environment have the greatest potential to affect biodiversity; however, unintentional releases into the environment (i.e., due to inadequate confinement) could also have impacts.

In Section 7.1.4 we outline explanations of how transgenic maize may affect biodiversity. These effects depend on whether and how organisms interact with transgenic organisms (e.g., feeding, ovipositioning) intentionally released and with any movement of pollen or seeds, or volunteer plants into habitats where the presence of transgenic organisms is not intended. Similarly, how transgenic organisms are used and what practices they replace (e.g., tillage, insecticides) may have implications for biodiversity as well because these practices are likely to alter species composition or abundances. With respect to changes in biodiversity, we consider two alternative scenarios: increase in biodiversity and decrease in biodiversity.

7.1.2.2.1 How Genetic Engineering Could Increase Biodiversity

While seldom the focus of public debates over biotechnology, scientists point out that in some cases transgenic organisms may increase biodiversity. Products from biotechnology may offer environmentally sound alternatives to existing practices and products that cause environmental harm, and the net result may be an increase in organismal biodiversity. For example, Round-up Ready soybeans have dramatically shifted herbicide use for weed management to glyphosate, an herbicide that is not highly toxic to vertebrates, degrades more quickly, and is less likely to contaminate ground water. Similarly, cotton that is engineered with genes from *Bacillus thuringiensis* (Bt) produces a toxin that controls the pink bollworm, a major cotton pest in the southwestern U.S. Positive effects on biodiversity are assumed to be a consequences of Bt cotton adoption, given that the Bt toxins used have a narrow target range and reduce use of broader spectrum insecticides (Ortman, 2001). Initial consequences of reduced pesticides are likely to occur at the population level of non-pest species affected by pesticides. Changes in species composition and abundances of these non-pest species may have higher-level effects on communities and ecosystems and their dynamics.

Biotechnology is also being explored as a tool for conservation biology. Several research teams are working on transgenic methods to confer resistance to the fungal chestnut blight (*Cryphonectria parasitica*) into the American chestnut (*Castanea dentata*) by inserting resistance genes from Chinese chestnut (*C. mollisima*) (Mann and Plummer, 2002).

Biological control has provided an effective alternative to the use of pesticides for the control of many agricultural pests. Recently, transgenesis has been used to enhance the effectiveness of several insect pathogens, including bacteria, a variety of baculoviruses, nematodes, and fungi (Lacey et al., 2001). If these agents further decrease pesticide use, positive effects on biodiversity may occur if the pest populations have negative impacts on non-crop species.

7.1.2.2.2 How Genetic Engineering Could Decrease Biodiversity

Of course, in other cases transgenic organisms may decrease biodiversity; public controversies over biotechnology often focus on this possibility. How might such a reduction in biodiversity happen? Explanations of how transgenic organisms could decrease biodiversity focus on the importance of understanding their toxicity and their ecological aggressiveness. With respect to aggressiveness, decreases

Table 7.1 Biotechnology products under development

GENERAL CATEGORIES AND THEIR ETHICAL IMPLICATIONS	Crop production	Animal production	Food and nutrition	Bio-processing	Medicine
	Short and long-term benefit: Creators/implementors of technology, crop producers	*Short and long-term benefit*: Creators/implementors of technology, crop producers	*Short and long-term benefit*: Creators/implementors of technology, food processors, consumers if foods meet demands	*Short and long-term benefit*: Creators/implementors of technology, crop producers, society	*Short and long-term benefit*: Creators/implementors of technology, individuals needing treatment
	Long-term risks: Society	*Long-term risks*: Society	*Long-term risks*: Society		
	Short-term risks: Producers, society	*Short-term risks*: Producers, society	*Short-term risks*: Producers (if legally accountable), society	*Long-term risks*: Society	*Short and long-term risks*: Individual
	Environmental impacts: Consequences of gene flow or volunteer plants (evolution of resistance, invasiveness), consequences of ecological interactions (unintended, non-target effects)	*Environmental impacts*: Consequences of gene flow or feral populations (invasiveness, unanticipated effects on other organisms), consequences of ecological interactions (unintended, non-target effects), vary with level of confinement/containment	*Environmental impacts*: Consequences of gene flow or feral populations (invasiveness, unanticipated effects on other organisms), consequences of ecological interactions (unintended, non-target effects), vary with level of confinement/containment	*Short-term risks*: Society	*Environmental impacts*: Consequences of gene flow or feral populations (invasiveness, unanticipated effects on other organisms), consequences of ecological interactions (unintended, non-target effects), vary with level of confinement/containment
	Biodiversity x ethical issues: Genetic changes in land races developed by indigenous peoples, altering genetic diversity at centers of origin	*Biodiversity x ethical issues*: Animal welfare issues, consequences of using animals with feral populations	*Biodiversity x ethical issues*: Consequences of using organisms with feral/wild relatives in growing regions	*Environmental impacts*: Consequences of gene flow or feral populations (invasiveness, unanticipated effects on other organisms), consequences of ecological interactions (unintended, non-target effects), vary with level of confinement/containment Ethical issues	*Biodiversity x ethical issues*: Consequences of using organisms with feral/wild relatives in growing regions
				Biodiversity x ethical issues: Consequences of using organisms with feral/wild relatives in growing regions	

APPLICATIONS AND DESIRED RESULTS	**Crop yield** *Output grain and biomass* Photosynthesis, enzymatic regulation, plant structure, flowering, ripening, sprouting Short and long-term benefit: Creators of technology, crop producers Long-term risks: society Short-term risks: producers, society *Grain quality* composition specifications and grade *Selective breeding* reducing the time it takes to develop improved crops *Abiotic stress tolerance* Increase the ability of crops to grow in a geography by increasing tolerance to moisture and drought, heat and cold, saline, and heavy metals (Al, Se, Mn and Ozone) **Pest management** *Disease resistance* fungus (verticilium, fusarium, sclerotinia, grey mould, botrytis, powdery mildew, black sigatoka) Bacteria (bacterial blight) Virus (BYDV, mosaics, leaf curl, spotted wilt, ring spot, feathery mottle, necrotic yellow vein viruses) *Insect & Nematode Resistance* foliar, root, fruit, grain sucking, chewing, piercing *Herbicide tolerance* more environmentally benign (e.g. glyphosate) alternate mode of action (e.g. IM, SU, Glufosinate) *Bio-pesticides*	**Livestock performance** *Feed to gain improvements* High density, more completely balanced feed resulting in more meat per ton of feed *Feed digestibility Carcass quality* **Animal health** Animal fertility and genetics Plant based animal vaccines Pathogen resistance **Aquaculture** Increased growth	**Organoleptics** Sensory quality improved taste, texture, and appearance **Nutrition** Micronutrients Fiber content Protein Vegetable oils Carbohydrates/starch Probiotics Phytochemicals **Shelf life** **Allergens and safety** Reduced allergens, mycotoxins, detection methods for pathogens, toxins	**Food enzymes** Raw material conversion **Food processing** Improved processing Improved food ingredients **Industrial processing** Bioenergy production waste water treatment Bio-catalysts Detergent proteases Bio-polymers Specialty chemicals Fibers	**Pharmaceutical proteins** Production of complex proteins Efficient drug delivery vehicle **Drug discovery and screening** Bioactive molecules Natural products

in biodiversity may arise if transgenic organisms were able to invade habitats beyond their intended use. Changes in biodiversity would arise if these volunteers affected the survival or reproduction of organisms existing in the invaded habitat. Similarly, gene flow of transgenes to wild relatives could lead to similar results if the incorporation of these transgenes into the wild relatives altered their survival or reproduction and produced cascading changes in other species composition or abundances.

Transgenic organisms may directly harm other organisms (i.e., genetically engineered plants like Bt corn that produce insecticides) and decrease diversity by altering species composition or abundances. Lethal or sublethal effects (e.g., effects on development time, reproductive characteristics, morphological characteristics) on non-target populations may scale up to affect organismal or ecosystem biodiversity. If individuals vary in sensitivity to Bt toxin, the loss of Bt-sensitive individuals from the population would alter genetic biodiversity within populations or species.

The practices associated with using genetically engineered organisms could also have effects on biodiversity if these are more harmful to keystone species or ecosystem functions. For example, the Bt corn targeting the European corn borer has a relatively narrow range of effects on other insect taxa. However, because its effects are focused on butterfly and moth larvae, it may have significant impacts on diversity of these taxa along with any species that depend on butterflies and moths as prey or as pollinators. Transgenic organisms or the practices associated with using transgenic organisms could cause the evolution of characteristics (e.g., disease resistance, insect resistance, herbicide resistance) that would increase aggressiveness. Decreases in all scales of biodiversity could result from measures to control organisms with these evolved traits.

It is clear from this survey that the impact of transgenic organisms on biodiversity must be assessed on a case-by-case basis, and also that it is too early to assess this impact even in locations where transgenic organisms have been introduced. The uncertainty involved here will itself be a factor in the assessment of the risk these organisms pose.

7.1.3 Species, Biodiversity and Normative Status

Having addressed the question of the potential effects of transgenic organisms on biodiversity we now turn to our second question: What is at stake, ethically speaking, in biodiversity? Are there any grounds for ethical concern about the impact of biotechnology on biodiversity? As a first step toward answering this question we introduce a distinction from debates over the moral status of nature in environmental ethics. This is the distinction between anthropocentrism and nonanthropocentrism. Anthropocentrists argue that the ethical significance of nonhuman nature lies entirely in its contribution to actual and potential human interests. These interests may be defined broadly (to include, for example, an aesthetic interest in biological diversity) or narrowly (restricting these interests to clearly determined economic interests). From this perspective, any effects of transgenics on biodiversity should be evaluated with respect to the human interests at stake. By contrast, nonanthropocentrists claim that nonhuman nature has ethical significance that is independent of human interests. They argue that natural diversity is good in itself, that human interventions should maintain certain ecosystem states, or that human beings should not interfere with natural processes such as speciation. Of course, there are also mixed theories that attempt to find a place for both kinds of value. It is now common in environmental ethics to redefine the interests of humans to include interests that are not directly human-oriented (as in the case of an aesthetic interest in natural diversity) or to include the interests of nonhumans in a theory that assigns a limited priority to the interests of humans. In both cases a largely anthropocentric theory accommodates nonanthropocentric concerns. Trends in this direction may indicate movement towards a convergence. Such a convergence would recognize that a sharp distinction between anthropocentric and nonanthropocentric can obscure the complex ways in which human interests, broadly conceived, are closely interwoven with natural conditions. This is especially significant in the case of transgenic organisms introduced into regions where indigenous agriculture exhibits a web of relations between human communities, their crops, and the landscape itself, as we will point out in the case of maize in Mexico. Nevertheless it is still useful for the sake of analytical clarity to distinguish anthropocentric and nonanthropocentric kinds of ethical significance.

7.1.3.1 Anthropocentrism and Biodiversity

Let us begin with anthropocentrism. There is a long history in the West of interpreting nature in general and its diversity of forms in particular as created by God in part for human well being. Natural diversity assisted human contemplation of the divine (Aquinas) or met ordinary human needs (Calvin), though in both cases its divine source precluded its value from being exhausted by its utility for humans. Modern science and technology have greatly expanded the value natural diversity has for human beings while removing the restraints on human utility imposed by the theistic reference. The human-regarding value of biodiversity is now both extensive and well known. Agriculture requires continual introduction of new genetic characteristics from wild strains to create new hybrids capable of resisting disease organisms. Approximately 25% of pharmaceuticals, including many drugs for the treatment of cancer, hypertension and cardiac disease, come from vascular plants. Meanwhile, plants, fungi, and bacteria remove toxins from the air, water and soil. On purely anthropocentric grounds, then, there seems to be a strong case for preserving biodiversity. However, this case often turns out to be difficult to make in practice. The reason is that in many cases in which an activity undertaken for human benefit is likely to have a negative effect on biodiversity, the human benefits of the species or ecosystem(s) threatened by the activity are unknown or only partially known, while the human benefits of destroying or disrupting the species or ecosystem(s) are reasonably well known. For example, the introduction of a transgenic organism into an ecosystem shaped by centuries of traditional farming may involve clearly identifiable benefits for certain parties while many of the benefits the altered ecosystem will no longer be able to provide are uncertain or unknown. In such a case the prevailing forms of cost-benefit analysis will likely favor the intervention over the preservation of natural diversity. Nor should we assume that the argument would go against the introduction of the transgenic organism even if all the relevant values were well known and the balance of these values fell on the side of preserving biodiversity. This is because, as we saw above, some human interventions into nature may actually increase biodiversity.

Anthropocentric considerations are not limited to economic benefits. The moral evaluation of an intervention may also require one to weigh the social or cultural value of an ecosystem to a human group and to determine which human beings or groups benefit from and are harmed by the intervention. Along the lines suggested by the mixed theories noted above, anthropocentric concerns may go even further than this to include a kind of value in biodiversity that is in some significant sense not human-oriented. Some anthropocentrists distinguish between human interests promoted by the abuse of nature and human interests that require the ongoing existence of flourishing ecosystems. One example is Elliot Sober, who argues that natural diversity may have aesthetic value even when the instrumental value of species or whole ecosystems is unknown or insufficient to outweigh what is gained for humans by destroying the ecosystem or reducing its diversity (Sober, 1986). A second example is Bryan Norton, who uses the notions of *demand value* and *transformative value* to help capture this distinction (Norton, 1987). Something has demand value if it satisfies actual human needs. Many of the human needs cited by advocates of transgenic organisms fall into this category. Things with transformative value may not satisfy actual preferences but instead give us reasons to change our preferences. Norton's example is a ticket for a classical music concert for a rock-music-loving teenager. An intact nature may have this second kind of value—it may open up new possibilities of experience and preference-formation for us. One question, then, is whether the enhancement of biodiversity through biotechnology succeeds in promoting nature's transformative value and whether, in the opposite scenario, the decrease in biodiversity inhibits this kind of value.

It is clear, then, that any evaluation of the anthropocentric value at stake in the introduction of a transgenic organism is sure to be complex. There is a growing tendency in the field of environmental ethics to account for this complexity by identifying and weighing noneconomic as well as economic values in the ethical evaluation of interventions into ecosystems. This tendency is exhibited in the analysis carried out in Section 7.1.4.

7.1.3.2 Nonanthropocentrism and Biodiversity

Let us turn now to the nonanthropocentric values at stake in biodiversity. For many contemporary environmentalists these are the most crucial values at stake. They believe that anthropocentrism makes inappropriate concessions to a way of thinking about the world that places too much emphasis on humans. However, these environmentalists are not the first to value natural diversity on nonanthropocentric grounds. Arguments for the latter have a long pedigree in western religious and philosophical traditions of thought.

These traditions were strongly influenced by the principle of plenitude, which formed part of the notion of the "great chain of being" described by Arthur Lovejoy in a classic work (Lovejoy, 1936). First articulated by Plato in the *Timaeus* (41b–c), the principle of plenitude holds that the divine purpose in creating the physical universe was the realization of as many forms as possible. Two implications of this principle are significant for our purposes. First, the principle of plenitude set the terms for one of the major ways in which the question of nonanthropocentric value was raised in the premodern world. According to this principle the universe is complete and therefore good. Parts of the universe contribute to the goodness of the whole. As Augustine and Thomas Aquinas pointed out, it follows that those parts of the universe that have no apparent ordering to the good of human beings—created things that do not fulfill human needs or desires and that may be harmful or simply annoying to humans—are nevertheless good because they contribute to the goodness of the whole (Augustine, *City of God* XI.16, XII.4; Aquinas, *S.Th.* I.72.1).

The principle of plenitude appears in these instances to have direct relevance for contemporary debates. Augustine's distinction between what is good for us and what is good for the whole rests on a more fundamental distinction between an order of nature, correlative to reason, and an order of use, correlative to need or desire. While the order of nature is hierarchical—among other things it assigns a higher value to sensate species over insensate species—it nevertheless recognizes nonanthropocentric value in all species, whether higher or lower. For Augustine it is therefore wrong to deny the value of any created form or to wish it absent from nature. So enduring was this view that as late as 1653 John Bulwer could speak of a dispute over "whether, if it were possible for man to do so, it were lawful for him to destroy any one species of God's creatures, though it were but the species of toads and spiders, because this were taking away one link of God's chain, one note of his harmony" (Thomas, 1983).

It would be premature, however, to draw direct conclusions from these beliefs for contemporary concerns over biodiversity. One reason is alluded to in Bulwer's "if it were possible": Ancient, medieval and early modern thinkers did not believe that it was possible for species to be destroyed. Their determination to establish the goodness of all natural forms was not to urge humans to preserves these forms but rather to defend the goodness of the whole created order, and thus the character of its creator, against an objection based on the apparent uselessness or harmfulness to humans of some species. Augustine, for example, worried that the failure of the created world to conform to the orders of use or desire would lead some to impugn the goodness of the Creator. Moses Maimonides put the principle of plenitude to use in order to refute the belief that the universe or its parts exhibit a divine purposelessness. This belief, Maimonides argued, is due in part to the false belief that the world exists only for the sake of human beings. Rather, because God brings into existence all that is possible—i.e., all that the divine wisdom decrees—the universe is necessary in its whole and in each of its parts (*Guide* III.25). These affirmations of the value of each species, then, were not articulated in ethical terms but in defense of the goodness or purposiveness of God.

This is not to say that these ancient and medieval views have no implications for today's debates. Aquinas argued that the plenitude of species is necessary for the representation of the divine goodness, which could not have been represented by one natural form alone (*S.Th.* 72.1). Contemporary Roman Catholic theologians often appeal directly to this text to argue for the preservation of species. Another debate related to the principle of plenitude may have contemporary implications. Ancient and medieval thinkers disputed over whether divine providence extends, in the case of nonhumans, to each individual or only to species. Maimonides endorsed a view he ascribed to Alexander of Aphrodisias, an Aristotelian, but that was also common among early neo-Platonists, namely that in the sublunary universe divine providence extends to individuals only in the case of humans; for all other sublunary creatures divine providence operates only on species. While there are no direct appeals to this view in contemporary environmental ethics, an echo of it is heard in theories that ascribe intrinsic value not to species themselves but to speciation as a process and derive from this value a prima facie duty not to interfere with speciation (Rolston, 1985). However, any direct appeal from these views to contemporary issues would have to account for what evolutionary biology tells us about extinction as a component of speciation.

Perhaps the significance of this brief discussion lies not in any alleged continuity between these ancient and medieval views and those found in contemporary environmental ethics but rather in the point that in the premodern world the question of non-anthropocentric value was raised primarily with respect to the diversity of species, or more precisely (since what we would designate as organic and inorganic forms are both included), of corporeal forms. The value of these forms, in turn, is secured by their necessary place in a divinely created and governed order. In the end this notion of a natural order in which species exist as distinct kinds may exercise the most influence on current responses to biotechnology. From this perspective transgenic organisms are ethically suspect not only because they may eliminate some forms but also because they generate forms that nature itself would not generate. Occasionally this attribution of ethical status to natural kinds is expressed, however uneasily, in relation to post-Darwinian biology and to developments in conservation biology and agricultural biotechnology. Damien Keown (1995) and Lambert Schmithausen (1997) both attempt to show how these beliefs operate in Theravada Buddhism while related efforts are found in statements on biotechnology adopted by the United Methodist Church (1992) and the National Council of Churches (1986). These efforts exhibit two tensions. One tension is between the traditional conception of nature as an ordered system and post-Darwinian conceptions. The other tension, found in the assignment of ethical status to species as distinct forms, is between one conception that grounds the normative significance of species in a natural order and another that grounds this significance in the integrity of natural kinds or natural processes (McKenny, 2001). In all cases, explicit ethical concern for species remains at the margins of the interface of religious communities with biotechnology.

However, implicitly religious attitudes about species and their ethical status pervade popular responses to biotechnology (Nelkin and Lindee, 1995).

Contemporary environmental ethics takes account of ecological and evolutionary sciences. For contemporary nonanthropocentrists what is at stake in biodiversity is not the goodness of a metaphysical order but the value of the natural state of an ecosystem or the continuation of an evolutionary process. Biotechnology raises ethical problems insofar as it threatens to alter an ecosystem, to disturb its stability, or to interfere with or destroy the process of speciation. In all of these respects nature, or what is natural, serves as a norm. These nonanthropocentrists face three challenges. The first is philosophical: they must show how a natural condition, what "is," possesses normative status, directing an "ought" to human intervention. The second challenge is scientific: to determine what constitutes the natural condition that is to be maintained or respected. The third challenge is also scientific: to determine which human interventions violate and which do not violate the relevant natural condition. The most simple nonanthropocentric principle holds that human beings should leave ecosystems untouched or should restore them to a state prior to human interference. Biocentrists such as Paul Taylor and some deep ecologists subscribe to this principle (Taylor, 1986; Sessions, 1991) but it is problematic for two reasons. One is that it presupposes that all human activity is unnatural; it places humans altogether outside of nature. The other reason is that it is unrealistic; every ecosystem studied by ecologists bears at least some traces of anthropogenic change. For these reasons many nonanthropocentrists hold to a position originally formulated by Aldo Leopold, whose "land ethic" culminated in the principle that defined right and wrong interventions into nature in terms of their effects on the integrity, stability, and beauty of ecosystems (Leopold, 1949). Leopold has near-canonical status among environmentalists who lean towards deep ecology and other similar movements but critics have pointed out numerous problems with his famous principle (Sagoff, 1999; Shrader-Frechette, 1998). First, there is the problem of determining exactly what counts as an ecosystem for the purposes of the criteria of integrity, stability and beauty. Second, Leopold presupposes an ecology in which, left to themselves, ecosystems maintain a kind of homeostatic balance. This "ecology of stability" has come under attack in recent years; most ecologists now subscribe to an "ecology of flux" in which change—even radical change—is considered normal (indeed, conditions of flux and higher levels of disturbance generally tend to increase biodiversity). But if ecosystems are in constant flux it is not clear what it would mean to preserve or restore the stability of an ecosystem—even if we could decide what features of the ecosystem count in determinations of stability. Third, the notion of the integrity of an ecosystem is ambiguous. Does integrity refer to health or the ability to cope with stress? It is difficult to know precisely what ecosystem states correspond to these features. In short, all attempts to determine what counts as integrity or stability rely on extra-scientific judgments of value. This is certainly also the case with Leopold's third criterion, beauty. Finally, Leopold does not adequately address the "is-ought" problem noted above.

In response to these criticisms Baird Callicott, the foremost contemporary defender of the land ethic, has pointed to a less prominent principle articulated by

Leopold. This principle distinguishes between anthropogenic and nonanthropogenic rates of ecosystem change, arguing that human activities that increase the natural (nonanthropogenic) rate of species extinction in an ecosystem are at least prima facie wrong (Callicott). Callicott therefore preserves Leopold's fundamental conviction that nature itself supplies a norm for evaluating human interactions with nature while also accepting the ecology of flux and the anthropomorphic principle, i.e., the fact that every known ecosystem is to some degree the product of anthropogenic change. Other things equal, Callicott's principle would presumably permit human-induced extinction (or diversification) so long as the rate of extinction (or diversification) did not exceed that which would have occurred without human intervention. However, Callicott gives no convincing reason why a natural rate of change has the normative status he assigns to it. Moreover, this principle would seem to prohibit not only some forms of biotechnology-based agriculture but much of traditional agriculture as well, for in many instances the latter quite clearly induces extinction at a faster rate than would have otherwise occurred.

With respect to nonanthropocentric value, then, the situation appears to be as follows. Even if the historically influential principle of plenitude could be formulated as an ethical norm, it does not take into account natural extinction and therefore leaves us with an inapplicable requirement to preserve every species. Leopold's principle recognizes natural extinction and therefore requires only that we uphold the integrity, stability and beauty of ecosystems. But these criteria are also inapplicable because they are either vague or make false assumptions about the nature of ecosystems. Finally, Callicott's principle accounts for the ecology of flux and for the anthropomorphic principle but its normative status is uncertain and in any case many currently accepted methods of agriculture appear to violate it just as clearly as many uses of transgenic organisms would.

7.1.3.3 Relevance for Transgenic Organisms: The Analogy with Nonindigenous Species

The relevance for transgenic organisms of these debates in environmental ethics can be illuminated by considering a similar case, one that also involves the introduction of organisms into an environment. This is the case of nonindigenous species. A nonindigenous species can be simply defined as a species that has been introduced to an area beyond the previous range of this species. Synonyms for "nonindigenous" include "exotic," "alien," and "non-native." In the public policy arena, the species of concern have been introduced by humans, either intentionally or unintentionally. Nonindigenous species that cause ecological or economic harm are referred to as invasive; reducing their occurrence and impact has become a major recent focus of scholarly research and environmental management. Such efforts have recently engendered considerable controversy between environmentalists and others, and some debate among ecologists. These controversies expose some difficulties of building normative claims on nature, and have been reviewed by Lodge and Shrader-Frechette (2003). The controversies are relevant in the present context because many of the concerns raised by nonindigenous species are also raised by transgenic organisms. In both cases the concerns arise from the prospect that one or more kinds of organism recently introduced into an ecosystem will cause extraordinary changes in that ecosystem.

A brief survey of debates over how to manage nonindigenous species quickly brings to the surface the challenges faced by any effort to manage transgenic organisms. These debates indicate the problems involved in the various ways of treating nature or the natural as a norm. We begin with Flannery (2001, 345–347), who suggests that lions and elephants could be reintroduced into North America to replace those that disappeared 13,000 years ago. The lions were apparently the same species that exist on the plains of Africa today. The elephants would replace the ecologically similar mammoths and mastodons that once were part of a speciose herbivorous megafauna. Without these herbivores, with which North American native plants co-evolved, some native plants can barely reproduce and cannot thrive (Barlow, 2001). Turning away from those species that became extinct and instead thinking about species that arrived in North America, Flannery (2001, 141) notes that "had the creosote bush arrived [from Argentina] last century rather than 10,000 years ago, it would doubtless be proclaimed the most noxious weed ever to have invaded North America." The perspective that emerges from these and many other observations about the biogeographic history of life on earth is that the supposed balance of nature is much more complicated than previously thought (Pickett et al., 1992) and that "…biological invasions are natural and, more important, necessary for the persistence of life" (Botkin, 2001). Even if the arrival of humans was responsible for the extinction of the North American megafauna in the last few thousands of years, extinctions and invasions of biota characterized earth long before humans existed (Flannery, 2001). Even on the scale of years and decades, species ranges change (Lodge, 1993). And as Botkin (2001) points out, invasions of new habitats allow the long-term persistence of species, as populations in old habitats are extirpated in the face of environmental change. Thus it is true that species invasions are natural and that the very definition of "nonindigenous" sometimes hinges on what time frame is being considered. These points indicate the difficulties faced by those who want to eliminate nonindigenous species from ecosystems on the grounds that they are unnatural. The problem arises with the very effort to determine what is natural. This, of course, creates problems for any ethical evaluation based on assigning normative significance to what is natural, including the case of transgenic organisms.

The case against nonindigenous species is also difficult to make under Callicott's distinction between natural and anthropogenic rates of ecosystem change. It is undeniable that in recent centuries human influence has increased far more dramatically than that of any other species. The human-induced rate not only of species extinctions but also of species invasions has increased exponentially, in concert with the exponential growth of the human population over the last few hundred years. In addition, in more recent decades, global human travel and commerce have increased disproportionately relative to the increase in the sheer number of humans. Combined, these factors have produced burgeoning rates of nonindigenous species in every ecosystem that has been monitored (e.g., Cohen and Carlton, 1998). Although species invasions are natural, both the rate of their occurrence and the distances traversed by species now exceed by orders of magnitude the rates and distances of only a few hundred years ago (Cohen and Carlton, 1998; Williamson, 1996). The first question the appeal to the difference between natural and anthropogenic rates of change must face is the existence of rational disagreement about the temporal benchmarks for ecological conservation or restoration. However, the US National Park Service's pre-European benchmark is not as "arbitrary" as Botkin (2001) suggests. The arrival of Europeans in North America marked an ecologically significant time, given the rapid increase in human population, travel, and commerce. It was the beginning of an enormous increase in rate of arrival of nonindigenous species. Clearly, any such benchmarks would differ for other continents, and whether the Park Service's benchmark is appropriate for other US agencies and applications should be the topic of scientifically informed public policy discussions. But even if, as this suggests, a nonarbitrary benchmark can be found, the question of why it is ethically significant remains. The argument that it is unnatural begs the question and also seems to remove humans from nature.

In sum, whichever conception of the natural we presuppose, there are insuperable difficulties involved in identifying one state of affairs as natural and another as unnatural. We have already noted that transgenic organisms may have different effects on ecosystems, depending on the case. The analogy with nonindigenous species suggests that it is not possible to identify one set of effects as natural and another as unnatural. It follows that even if one argues for the normative significance of what is natural, it is unclear that the latter could be made applicable to transgenic organisms unless one considers these organisms themselves as unnatural.

The problems posed by these efforts to treat nature itself as a criterion lead many environmentalists and ecologists to appeal to the harms nonindigenous species inflict on ecosystems. Could these efforts supply a model for transgenic organisms? There are numerous difficulties with this appeal. There is, first, the obvious problem that people disagree over whether a certain change is a harm or a benefit, and this kind of judgment is often not distinguished from empirical difficulties involved. For example, the *New York Times* published an article entitled "Alien Species Often Fit In Fine, Some Scientists Contend" (Derr, 2001), which followed by a few pages an article documenting the devastating impact on California forests of nonindigenous oak blight fungus (Woodsen, 2001). The confusion and tension about nonindigenous species exhibited in this single issue of a newspaper follows from the complexity of a situation in which science, different ways of ordering and weighing value, conflicting principles of environmental ethics, and public policy have begun to intersect strongly on this issue. Normative judgments are made about whether the invasive species-induced changes described by ecologists are good or bad. Sometimes normative judgments are made and reported by scientists themselves, without drawing a distinction between the changes in the natural world that they have documented, on the one hand, and the judgments they make about the acceptability of such changes and the grounds for those judgments, on the other hand. Different people, of course, will make such judgments differently, or at least weigh them differently against competing goals, and scientists are just as entitled to make such judgments as anyone is. But any effort to characterize any or all nonindigenous species as good or bad must be recognized as a normative judgment that should not be confused with scientific judgments, as is frequently done (Rosenzweig, 2001; Slobodkin, 2001). Indeed, the very definition of "invasive species" offered above—hinging as it does on "harm"—depends on normative judgments, which is why so many misunderstandings have developed around nonindigenous species.

These difficulties with the effort to evaluate invasive species, and by extension transgenic organisms, on the basis of harms and benefits do yield a helpful result, albeit a largely negative one. The unavoidability of normative judgments combined with their irreducibility to scientific judgments indicates that policy development on invasive species and transgenic organisms should depend on much more than scientific expertise. If normative judgments are inevitable, an important aspect of the ethical evaluation of nonindigenous species—and, by analogy, of transgenic organisms—will be to determine which values enter into such judgments. Frequently, such judgments are made on narrowly anthropocentric grounds for which immediate economic interests—and those of a narrowly defined group—are considered. A major purpose of Section 7.1.5 below is to develop and apply a model that will take account of the broader interests at stake in transgenic organisms.

7.1.3.4 Conclusion

The conclusion we draw from this survey is that nonanthopocentric theories have thus far provided an inadequate basis for the ethical evaluation of transgenic organisms while anthropocentric theories have tended to treat the relevant interests too narrowly. The goal of this chapter is to provide a method of analysis that will take account of a broad range of human interests, including human interests which regard nature in more than narrowly economic terms. In order to identify these interests we will consider a case that has generated considerable controversy, namely that of Bt maize.

7.1.4 An Ethical Analysis of Bt Maize

7.1.4.1 Ethical Issues Associated with Genetic Engineering and Biodiversity

Environmental effects of transgenic organisms and their consequences for biodiversity may affect the value of biodiversity for society or for specific groups of society. Genetic diversity of crops and their wild relatives is often valued by societies generally as well as by indigenous cultures more specifically. Geographical areas where crop domestication originated (hereafter, centers of origin) are highly valued by crop breeders for functioning as a reservoir for traits that may be needed in the future. For example, in response to environmental conditions that have detrimental effects on crops (i.e., new strain of a virus, need to expand the growing range of a crop), new varieties may be bred with diverse, varying genetic backgrounds and screened to produce plants that will survive or reproduce. In contrast to transgenesis, this process for creating new crop variants does not rely on known gene sequences and known gene function but rather the factors exerting environmental selection on crops and existing genetic variation are the driving forces for producing new crop varieties. Transgenesis is a very powerful tool for creating variation given existing information about gene sequences and gene function. However, high genetic diversity at centers of origin provides a reservoir of unknown genes and gene function. Introgression of the genetic background of transgenes (and their accompanying genetic backgrounds) into the existing genetic structure at a center of origin will alter the potential for using genetic diversity for plant improvement, a risk that society will bear.

To the extent that cultures in these centers of origin have evolved in concert with crop domestication, crop varieties unique to particular geographical locations will also have cultural importance so that alterations to genetic diversity may have more than ecological consequences for a region. Moreover, cultural importance may also attach to a landscape itself and the pattern of interaction between a human community and a landscape that is established over centuries of traditional farming. Finally, traditional farming may provide a livelihood for members of a community that is severely disrupted by the introduction of biotechnological agriculture, resulting in unemployment or displacement. The interweaving of human interests and natural conditions in this brief identification of the relevant issues is an indication of how misleading it is, in practice if not in theory, to treat anthropocentrism and nonanthropocentrism as two sharply opposed positions.

To explore ethical issues raised by genetically engineered organisms, we focus on the current debate over whether planting of transgenic maize should be allowed in Mexico. Mexico is the global center of maize and the origin of strains grown commercially around the world. Furthermore, indigenous cultures evolved with maize not only as their main food source but also as a key component of their religious life. The centrality of maize to these cultures is evident by the deities of maize developed in Mesoamerica, in sculptures and paintings featuring maize and in representations of maize with deities of earth or rain (Taube, 1996).

To date, the Mexican government officially has banned the planting of GE corn in Mexico, but imports of GE corn seed intended for consumption are not regulated. Imported maize is a mixture of transgenic (approximately 30% of total) and non-transgenic grain. Imported maize is largely used as feed, but a portion is distributed to rural areas for use as food (Serratos-Hernández et al., 2004). In spite of the ban on planting, molecular evidence indicates the presence of transgenes in landraces of maize in Mexico (Quist and Chapela, 2001). What processes explain their presence is hotly debated (Metz and Fütterer, 2001).

The effects of planting transgenic maize on local landraces and on crop diversity will have repercussions for the global community that depends on maize as a food source and also on the local communities whose ancestry and culture is intricately linked with local landraces of maize and their respective agricultural practices.

7.1.4.2 Global Effects of Losing Genetic Diversity

The global community relies on corn as a crop for food and feed. Maize is among the three major world crops (with rice and wheat) with a production of more than 275 million tons in 2003. Maize has a long history of genetic modification and domestication. It is estimated that the first domestication of maize began some 7,000–10,000 years ago in south-central or south-western Mexico. Introduction of maize into Europe is thought to have occurred as a result of colonization. As a result of this long history of farmer breeding and selection of maize for local conditions, hundreds of landraces have developed in Mexico as well as in Europe, Africa and other areas of North America.

In Mexico, additional genetic diversity is available for maize breeding because of the presence of wild relatives. Maize is a domesticated form of teosinte, a conclusion supported by molecular similarity. Despite profound difference in ear morphology and plant grown forms, interbreeding between maize and teosinte produces highly fertile F1 hybrids (first generation offspring of interbreeding). Hybrids occur at low levels. All forms of maize freely cross pollinate forming fertile hybrids; therefore, field-to-field gene flow due to pollen movement readily occurs. The breeding biology of maize, teosinte, and their relatives satisfies one of the necessary conditions under which a transgene may spread.

The potential for introgression of genes and for genetic alteration of landraces occurs with the introduction of any new variety of maize, transgenic or non-transgenic. Is the introduction of transgenic maize quantitatively or qualitatively distinct from non-transgenic maize? No consensus exists on the answer. On one side, new maize varieties have been introduced for centuries, suggesting that the risks from transgenic techniques are similar to the dynamics of new introductions. Indeed, teosinte and maize have remained morphologically distinct. However, *in situ* conservation and germplasm preservation of landraces is a high priority in order to slow the trend of losing landraces and their genetic diversity. Any acceleration to these existing trends could produce quantitative changes in genetic diversity available for future improvements.

Transgenic crops may accelerate the rate of gene transfer and introgression for two reasons. First, given the tremendous resources for developing and marketing biotechnology, adoption rates of transgenic crops may be more rapid than other introduced varieties. Adoption rates of transgenic varieties in the United States and Argentina, two of the major agricultural markets using transgenic crops, have been rapid, but how these compare to other new varieties is not well documented. Higher adoption rates would increase the absolute acreage available for pollen flow, subsequently increasing the number of transgenic hybrids between maize varieties and between maize and teosinte.

Second, quantitative differences exist in the pool of available genetic information to transfer and these may also impact rates of gene transfer and introgression. Historically the evolution of maize varieties was constrained by genetic diversity available between interbreeding plants. Information from the phenotypic expressions of existing maize varieties and their interbreeding relatives has been an integral part of the planning for improvement. In contrast, transgenesis is limited by knowledge of gene sequence and gene function; no requirement of interbreeding is necessary; and the phenotypic expression of genes from plants with homologous genetic backgrounds is not a prerequisite to producing a new variety. The release from the constraint of using genetic diversity from interbreeding relatives creates novelty in transferring genes among more distantly related taxa. Stated another way, genes interacting with a novel genetic background will increase the uncertainty for predicting the outcome of gene transfer.

7.1.4.3 Local Effects of Losing Biodiversity

The use of current transgenic maize products in the center of origin of maize could alter the species composition and structure of Mexican agroecosystems, causing local indigenous groups to bear the risks of introducing transgenic maize. Of particular significance, non-target effects on other species of plants and animals that are within maize fields could alter the nutritional base on which these cultures depend. If local landraces have religious, ceremonial or other cultural significance, the transfer of transgenes (or any other genes) would constitute a harm. If transgenic maize is adopted rapidly in adjacent communities or areas, the probability of gene flow and this harm increase.

At centers of origin where high genetic diversity occurs, local landraces developed through traditional knowledge and used by indigenous peoples are co-adapted to local habitats. Introduction of genetically engineered varieties that cross-pollinate or interbreed with local landraces may disrupt and, under some scenarios might even obliterate, these genetic-habitat associations that have refined agricultural practices over hundreds to thousands of years. These consequences may occur intentionally if farmers in an area choose to use transgenic varieties or unintentionally if pollen transfer occurs from other planted areas. Furthermore, the known consequences of introducing a transgenic variety are weighted toward short-term benefits; whereas uncertainty exists over the costs, environmentally, socially and economically. Traditional knowledge remains a fundamental component of land management in some areas of Mexico and is often the basis for sustainable farming systems in developing countries. Often, the dynamics and goals of traditional farming systems differ significantly from those that have evolved for large-scale production.

Agricultural practices at the center of maize's origin differ from those in the U.S. and Northern Mexico, where high production monocultures of crops exist. More than 75% of maize in Mexico is produced by peasants who use rather traditional farming systems (Turrent-Fernández et al. in OECD, 2003). Maize is typically sown with a scattered spatial arrangement across the corn field that provides opportunities for establishing other associated crops as beans and squash. Furthermore, this relatively open spatial arrangement of the crop plants allows for the persistence of populations of a large number of other species in maize fields (Vieyra-Odilón and Vibrans, 2001).

In traditional rural areas of Mexico these other species within maize fields are not "weeds" as they would be considered in North America and Europe. In Mexican traditional agricultural regions, even if yield reductions of the main crops occur (see Vieyra-Odilón and Vibrans, 2001), their possible detrimental effects on crop yields can usually be compensated by cash income after being sold in near markets, because of diet enrichment through direct consumption by the farmer's family, or because of other household uses (Nations and Nigh, 1980; Bye, 1981, 1993, Vieyra-Odilón and Vibrans, 2001). Therefore, the relationships of Mexican peasants with the full suite of species in a crop field may be quite complex, and they collectively represent a rich genetic resource on which selection towards domestication may take place.

7.1.5 Risk Analysis of Transgenic Organisms

7.1.5.1 Tensions in the Assessment of Societal Risks

Recent literature on how best to assess the risks of environmental and other hazards to society exhibits considerable tension exists on at least three points: (1) which considerations qualify as legitimate, rational components in an attempt to estimate the magnitude of a hazard; (2) whether apparently incommensurable components of a hazard must be reduced to a common currency to properly inform decision making (and if so, what currency is appropriate); and (3) to what extent options (and costs) of alternatives for managing a hazard should be considered simultaneously (or at least iteratively) with assessing the hazard.

First, in many traditional risk assessments, variables of risk include only obvious, direct, easily quantified damages, such as the number of human deaths caused, the concentration of a putative toxin that kills a certain percentage of an indicator species in a laboratory beaker in a specified short amount of time, or the estimated dollar losses caused by a hazard. However, recent "psychometric" analyses of risk perception have made it increasingly clear that many people perceive risk very differently from the way that scientists and technocrats usually measure risk, and that different groups of people (e.g., different sex, race, nationality) often differ in evaluating a hazard (Douglas and Wildavsky, 1982; National Research Council, 1996; Slovik, 1999). There is disagreement over what to make of these differences. One response (e.g., Sunstein, 2002) is to discount them and relegate to scientists and technocrats the authority to decide on a set of common, appropriate metrics of risk. Others (e.g., Shrader-Frechette, 2004) argue that the traditional approach circumscribes too greatly the variables included in a hazard assessment, and that the concerns of lay people must be taken seriously as components of a hazard assessment. The commonly held fears of novelty, of "unnaturalness," and the considerations of whether new organisms are quantitatively or qualitatively different perhaps fall into this area of tension when considering how risk assessment of transgenic organisms should be accomplished.

Second, the appeal of using a limited set of easily quantified metrics in a calculation of risk is easy to understand: it makes it easier to calculate an aggregate value for risk when, as is usually the case, a hazard has multiple types of impacts of different things. Considerable progress has been made on identifying a wider variety of environmental metrics that should be assessed for effects of transgenic crops, for example (Andow and Hillbeck, 2004). But, having quantified these various essential responses (e.g., potential impacts of transgenic maize on native teosinte species, on native insect species, and the distribution of income to farmers), it will remain difficult, of course, to sum different responses quantified in very different units. As a recent exchange over the health effects of eating farmed salmon illustrates (Hites et al., 2004, and 7/23 letters afterward to *Science*), it can be exceedingly difficult to incorporate a comprehensive set of risks and trade-offs of risks, even within a narrowly technical set of health risk factors. Yet in any decision-making process, different impacts will be aggregated, whether explicitly or implicitly, and the more units that can be made commensurable, the more explicit and transparent a decision can be. Thus there is increasing effort on the development and application of "non-market valuation" methods to convert other units to dollars, e.g., aesthetic values, genetic diversity, traditional lifestyle. These methods allow the explicit inclusion of values that were once more-or-less excluded from hazard assessment, but non-market valuation is not sufficiently advanced for all relevant considerations to be monetized. Thus the tension over the need to adjudicate incommensurable measures of hazards will remain for years to come, and will be important in the assessment of transgenic organisms.

Third, assessments of hazards are made to inform societal decisions about the management of the risks imposed. In initial manifestations of environmental regulation in the US in the 1970s, target values for hazard reduction were set on the basis of the hazard assessment alone, with little if any consideration devoted to the costs of different levels of reductions in risk (Sunstein, 2002). This approach can lead to very cost-ineffective government regulatory actions. Thus, more recently, the executive branch in the US has urged governmental agencies to consider the costs as well as the benefits of risk reduction before issuing new regulations. Sunstein (2002) argues that such a cost-benefit approach is the only rational way to form such policies. In addressing the previous two tension points above, Sunstein (2002) advocates that cost-benefit analyses be fully comprehensive—including all the costs and benefits of any contemplated action, including "qualitative" considerations. Nevertheless, others (Shrader-Frechette, 2004) are suspicious of this approach, arguing that an exclusive cost-benefit approach ignores other ethical imperatives such as whether a risk is suffered voluntarily (including issues of informed consent), the social distribution of costs and benefits, etc. Thus, tension remains about when in a risk analysis it is appropriate to consider the costs of achieving certain levels of risk reduction, and whether, in fact, a strictly cost-benefit approach can adequately subsume all the important considerations in the assessment and management of risks, including the risks of transgenic organisms.

From these tension points we may draw two conclusions regarding risk assessment of transgenic organisms. First, these tensions indicate the high degree of uncertainty involved in any assessment of societal risks. This uncertainty is especially great in the context of transgenic organisms. Second, there is unsettled controversy in risk assessment over the range of factors that should be decisive in such analyses and whose evaluations count. Drawing on the example of maize in Mexico, we have pointed out a wide range of factors raised by the possible effects of transgenic organisms on biodiversity, so these controversies are directly relevant to the topic of this chapter. We will now consider these two matters in succession.

7.1.5.2 Uncertainty in Risk Assessment

Because of the frequency with which it is invoked by opponents of biotechnology, any discussion of uncertainty and risk assessment must begin by considering the Precautionary Principle. The Wingspread Statement, a widely quoted version of the Principle, specifies that

> When an activity raises threats of harm to human health or the environment, precautionary measures should be taken even if some cause and effect relationships are not fully established scientifically. In this context the proponent of an activity, rather than the public should bear the burden of proof. The process of applying the Precautionary Principle must be open, informed and democratic, and must include potentially affected parties. It must also involve an examination of the full range of alternatives, including no action (Science and Environmental Health Network, 1998).

The history of human intervention in the environment provides ample anecdotal support for the Precautionary Principle. Often in the past we have produced short-term benefits by rearranging nature only to find that their worth is more than cancelled by longer-term costs. To cite just two examples taken from a vast catalogue of cases, the sheep carcasses that seemed to provide a cheap source of feed for cattle led to variant Creutzfeldt-Jacob Disease in humans; gorse brought from Scotland to New Zealand to serve as natural fencing material spread throughout natural ecosystems and farmland without regard to the intentions of its importers. Genetic engineers change nature at a more fundamental level than has been possible for humans before and therefore their opponents fear consequences correlatively worse than the ill-effects of earlier changes.

Some interpretative looseness surrounding its key terms has led to intense debate about the principle and its application to transgenics. There are two points of controversy. The first focuses on the principle's requirement for democratic evaluation of new activities or technologies. Since this concern applies to risk assessment generally, it will be deferred to the subsection below, where we consider what factors and whose evaluations of those factors should count in risk assessment. The second point of controversy concerns the Precautionary Principle in particular, namely, its placement of the burden of proof. The most obvious way in which the Precautionary Principle supports opposition to transgenic organisms lies here: The principle demands proof that a new activity causes no harm while those who oppose the activity do not have to show anything. For example, the latter have only to suggest the possibility that the accidental transfer of pest-resistance genes from a crop to a weed species will devastate natural ecosystems. Once this possibility is on the table, biotechnologists must prove that they will not make superweeds. But they must do more than this. Their opponents are under no obligation to have imaginations sufficiently capacious for all of the ways in which technological innovation might lead to disaster. Who could have predicted that introducing sheep carcasses into cattle feed would lead to variant Creutzfeldt-Jacob Disease, or that rats escaping from ships transporting settlers to New Zealand would not only survive but devastate ground nesting birds. There are many ways in which the genetic modification might lead to ecological collapse. Those who deploy the Principle against transgenic organisms seek reassurance both in respect of ways they have imagined and ways they have not.

Gordon Graham points out an interesting structural similarity between the Precautionary Principle and Pascal's Wager argument for the rationality of belief in God. According to the Wager argument so long as we think that it is possible that God exists then the infinite reward of correctly believing makes it rational to try one's hardest to bring it about that one believes. The penalties that are the concern of the Precautionary Principle are very great. Transgenic organisms are presented as having the potential to bring about global ecological collapse. We should avoid practices that risk such a terrible outcome (Graham, 2002, 126–130).

According to one widely made objection the Wager argument ignores the potential benefits of not believing in God. It is not impossible that there is a supreme being who inflicts an infinite punishment upon all and only those who believe in a supreme being (Royal Commission on Genetic Modification, 2002). The mere possibility of such a being may leave belief and nonbelief prudentially on a par. There is an analogous response to the Precautionary Principle. According to this response proponents of the Principle ignore the costs of inaction. Consider the following environmental justifications of transgenic organisms. One of the threats to the environment comes from the increasing amount of land used for agricultural purposes. By making land that is currently under cultivation more productive, transgenics potentially reduce the environmental burden of agriculture. There are environmental defenses of Bt and roundup ready crops. Farmers planting these varieties may be able to reduce their use of environmentally destructive pesticides and herbicides. The ongoing conversion of natural ecosystems into agricultural land and increasing dowsing of this land with herbicides and pesticides might lead to global ecological collapse. Transgenics may prevent this outcome. Opponents of transgenics legitimately challenge these environmental justifications; as we saw above, there is considerable uncertainty over the effects of transgenics. But the point is that it is not entirely impossible that the widespread use of transgenics will prevent rather than bring about ecological collapse.

Those who object to the Wager argument that there might be a supreme being who punishes all and only believers think that the prudential arguments for and against belief cancel each other out. There is an important difference between the Wager argument and the Precautionary Principle, one that may allow us to find a difference between the prudential case for intervening in nature and leaving nature as it is. The Wager posits an infinite benefit potentially available to believers in

God. If the reward for correctly believing in God is infinite then it does not matter how probable His/Her existence is, so long as the probability is not zero. A tiny probability multiplied by an infinite reward produces an infinite expected gain. Although the penalty associated with ecological collapse is certainly very great, we can be confident that it is finite: the disutility associated with global ecological collapse is the sum of a very large collection of finite disutilities represented by the loss of individual species, habitats, and so on. The large magnitude of this loss may mean that somewhat unlikely paths to this outcome have a substantial expected disutility. But there is a point on the spectrum of probabilities at which it would be decision-theoretically irrational to be deterred by the threat of global ecological collapse. This is because multiplying a large disutility with a sufficiently small probability leads a small expected utility.

If the Precautionary Principle is subject to these practical and conceptual problems, how should we then address uncertainty in risk assessment? All experts on risk analysis agree that most single risk factors are inherently probabilistic (not on-off), and that when trade-offs among risks are considered, probabilities multiply (e.g., herbicide use with conventional maize vs. risks to native species from transgenic Roundup Ready maize). In addition to the uncertainty which is inherent in probabilistic outcomes, a large component of epistemic uncertainty is added to risk analysis of transgenic organisms because the organisms are novel. To what extent do living products of biotechnology change the uncertainties in estimates of environmental risk (as a result of any qualitative differences between biotechnological methods/products vs. natural processes/products)? We begin with another comparison with nonindigenous species. Although neither the occurrence nor societal responses (e.g., prevention or control attempts) to nonindigenous species necessarily involve biotechnology, the closest analogue to environmental risk analysis of transgenic organisms is risk assessment of nonindigenous species. Parallels in risk assessment exist because for both nonindigenous species and transgenics, the goal is to estimate the potential impact on the environment of organisms that are different from organisms that have previously existed. In the case of transgenics, the concern is often with organisms that are different from any organisms that have ever existed *anywhere* (a qualitative difference). For nonindigenous species, the concern is with organisms that are different from those that have existed previously *at a given location* (a quantitative difference). Thus, in both cases, the central challenge in risk assessment is to estimate the potential impact of a novel interaction between an organism and an environment (which includes the physical and chemical milieu, other species, etc.).

For nonindigenous species, the novelty resides in placing a known species in an environment new to the species. For transgenic organisms, the novelty always resides in the organism (by virtue of its novel genome), and sometimes also resides in the environment (if the organism is introduced in an ecosystem where even its unmodified precursor organism has not previously existed). In other words, the risk posed by nonindigenous species results from a previously non-existent interaction between two knowns (the species and the environment); in contrast, the risk posed by transgenic organisms results from a previously non-existent interaction between a known (the environment) and an unknown (the transgenic). The estimated risk posed by a transgenic organism, therefore, will generally contain more epistemic uncertainty than the risk posed by a nonindigenous species.[1] We know of no reason to expect that the potential environmental harm of transgenics, on average, is greater than the threat from nonindigenous species, but there is reason to expect that the uncertainty in estimates of risk from transgenics will be greater than that for nonindigenous species.

This greater uncertainty resides in the extent to which history can be a guide to future interactions. Risk assessments of nonindigenous species can be informed by past or current studies of the characteristics and processes involving the species and environments in question. The extrapolation involved in the risk assessment then focuses on the interaction between species and environment, which has not previously existed. Risk assessments of transgenic organisms can be informed by studies of the environment in question, but extrapolation is involved both in any reliance on studies of the precursor organism (which is genetically different from the organism being assessed), and in estimating the likely organism-ecosystem interaction. Finally, then, if overall risk is defined to include the bounds of uncertainty around a point estimate of risk (say, the probability of bad things happening if a transgenic organism escapes into the environment), then overall risk must, on average, be greater for transgenics than for nonindigenous species.

This perspective is sobering because the practice of risk assessment of invasive species is itself in its infancy, with methods still very much under development, with wide bounds around any estimate of probabilities of harm, and with some contention about what are appropriate interpretations of existing results.

With respect to the magnitude of uncertainty in any risk assessment, strong feedback may exist to risk management. That is, uncertainty is both a component of the estimation of risk from any or all variables, but for many people it is also a driver of their perception of risk. For some people (including some cultures or nationalities), the estimation of the magnitude of any hazard increases as the uncertainty of that hazard increases. This may explain, for example, the differences among countries in the reception of transgenic foods in recent years. At this point, a question arises: Do different religious, philosophical, or cultural traditions assess and respond to environmental risk differently? Is risk averseness greater in some traditions than others? Can some traditions respond better than others to probabilistic risk assessments, including the bounds around any probability? To bring this question into focus, consider the following two hypothetical results that are consistent with the argument above: If nonindigenous species X is released, the probability

---

[1] However, as an anonymous reviewer points out, epistemic uncertainty will not always be greater in the case of transgenic organisms. Reintroduction of an organism with a single genetic change into its original ecosystem would like involve less epistemic uncertainty than introduction of a nonmodified species into an ecosystem that is very different from its original one. To use the reviewer's example, it seems likely that there was greater epistemic uncertainty in the introduction of sheep to Australia than there is in the introduction of Bt maize where non-Bt maize was originally grown.

of a bad environmental result is estimated to be 20%, but the true probability could lie anywhere between 15% and 25%. If transgenic organism Y is released, the probability of a bad environmental result is also estimated to be 20%, but the true probability could lie anywhere between 5% and 35%. Another way to say this is that while the point estimate of risk is the same for both organisms (20%), the range of probability for something bad happening is 15–25% for the nonindigenous species, and 5–35% for the transgenic organism. Would different traditions evaluate the same point estimates of risk differently? Would different traditions evaluate the uncertainty around risk estimates differently? And if the answer to these questions is affirmative, what role, if any, should be given to these different assessments in matters involving regulatory policy? This latter question brings us to the topic of the next subsection.

7.1.5.3 What (and Whom) to Consider in a Risk Analysis of Transgenic Organisms

Given the general background above on risk analysis and the immediately preceding remarks on uncertainty, we now want to suggest what general sets of considerations should be included in a full risk analysis of transgenic organisms. Two brief points may be made with regard to the question of "what." First, we have pointed out that no satisfactory way has yet been found to carry out the kind of "nonmarket valuation" that is necessary to account for noneconomic factors, including those we identified with respect to the impact of transgenic maize in Mexico. Second, we have also pointed out Shrader-Frechette's argument that risk assessment restricted to cost-benefit analysis fails to consider ethically important issues including whether a risk is suffered voluntarily or not, and who bears the costs and reaps the benefits. These factors are essential to any ethical evaluation of transgenics, and the inability of standard forms of risk assessment to account adequately for these factors involves a serious deficit in our capacity to judge transgenic organisms ethically.

However, it may be possible to offset some of our incapacity if we can decide whose assessments of risk should count, and how. This brings us to the second question, the question of "who?" It is indisputable that the interests of those affected by a new technology should be considered when deciding whether to introduce it; a morally satisfactory analysis will not ignore the interests of any affected parties. However, the question remains about how we take these interests into account. There is a spectrum of ways in which we might approach the risks posed by a new genetic technology.

According to one view we should leave the analysis of risk up to scientific experts. Shrader-Frechette identifies a range of reasons for giving nonscientists a role in risk assessments. She argues that scientists are not immune to errors in probabilistic reasoning that can lead to faulty assessment of risk. Her most important point is that the choices about new technologies involve values. A novel technology potentially affects not only health and safety, but also "human autonomy, consent, distributive equity, equal opportunity, future generations, civil liberties, social stability and so on" (Shrader-Frechette, 1995, 117). Experts on the science of transgenics lack the distinctively moral expertise to understand differential effects on these values.

While Shrader-Frechette argues only that nonscientists should be given some role, it is possible to extrapolate from the writings of Leon Kass an argument for the exclusion of scientist experts altogether from decisions about technologies that impact on core human values. A recurrent theme of Kass's writings on the new reproductive technologies is a rejection of bioethical expertise. We can apply analogous reasoning to the risk assessments of scientific experts. Consider Kass's evaluation of the morality of human cloning. The starting point of what has come to be called the "yuck" argument is the queasiness that typically accompanies contemplation of the possibility of cloning humans. Kass defends this unease as "the emotional expression of deep wisdom, beyond reason's power to fully articulate it." Kass explains, "We are repelled by the prospect of cloning human beings [...] because we intuit and feel, immediately and without argument, the violation of things that we rightfully hold dear" (Kass, 1997, 20). Some consequentialist bioethicists contend that making somatic cell nuclear transfer safe leaves no decisive argument against its use for reproductive purposes (Brock, 1998). Kass counters that we should not view the lack of such an argument as providing support for cloning. Rather we should see it as the expression of rationality's impotence when faced with an issue that bears on human existence in such a fundamental way. Instinctual disgust is the only reliable guide. It stands in no need of bioethical validation.

Risk assessors are, like many academic bioethicists, preoccupied with consequences. Their focus on potential consequences in itself takes a moral step. An argument of the kind offered by Kass against human cloning may allow laypeople to recognise that their values are under threat, regardless of what scientific experts say. If successful this argument would show that they do not need scientific experts to educate their moral intuitions about the environmental badness of genetic technologies.

The Report of the New Zealand Royal Commission on Genetic Modification gathered together a heterogeneous set of concerns about transgenics. At first glance some of these concerns seem to be premised on serious scientific misunderstandings. One contributor to the Report expresses the fear that the introduction of human genes into species that are normally part of the human diet will lead to acts that are morally indistinguishable from cannibalism (Royal Commission on Genetic Modification, 2002, 34). This fear seems, initially, to be premised on a confusion about what genetic engineers do. The contributor's example involves the hypothetical introduction of a human gene into a tomato. The charge that the genetic modifiers of the tomato are encouraging cannibalism depends on a mental image depicting genetic engineers as essentially introducing human flesh into the tomato. But this misconceives of genetic engineering. Genetic engineers manipulate a mechanism of inheritance that is shared by all living things on this planet. Unmodified tomatoes already share many letters of DNA with human beings. Those who think that the introduction of a human gene into a tomato pushes it past some threshold of similarity that leaves those who consume it open to the allegation of cannibalism should be especially cautious of beef or lamb. These are much more genetically similar to humans that even tomatoes with human transgenes. Kass's argument points toward a different way to think about this objection. While the reasoning behind it may not succeed as a summary of the science of transgenics, the objection may constitute a distinctive moral concern about introducing the transgene by virtue of expressing "deep wisdom, beyond reason's power to fully articulate it." Kass's argument against cloning turns on our inchoate sense of what it is to be human. The fear that introducing human transgenes into food species might lead to acts of cannibalism was expressed by a Maori contributor to the Report. Maori people feel themselves to be the guardians of the natural habitat of Aotearoa/New Zealand. Their arguments against its genetic modification might be interpreted as appealing to an inchoate sense of the way nature should be, combined with the recognition that transgenics transgress these norms.

The two extreme positions just described rely on evaluative monisms. One evaluative monism would permit scientific experts to evaluate transgenics without regard to public concerns. The other monistic position grants scientific expertise no role; the risk assessments of experts are trumped by the occasionally inchoate fears of ordinary people. The sensible middle course recommended by Shrader-Frechette would enable us to avoid these extremes. Rejecting these extreme monisms requires that we find a way to make a dialogue between scientists and laypeople. This should enable us to integrate these two different ways of cognizing genetic technologies. The following analysis offers grounds for confidence that we can, contra the Kassian argument just sketched, give a consequentialist defence of the suspicion that indigenous peoples feel when confronting proposals to modify nature over which they claim guardianship. Since, as we have also noted, risk assessors speak the language of consequentialism, this may help resolve at least some of the limitations of risk assessment indicated above, even for those who believe that consequences are not the only ethical consideration at stake here.

The finitude of the utilities involved in the evaluation of environmental changes means that probabilities matter. In the following paragraphs we take up the suggestion, indicated above, that different peoples may stand in different positions in respect of the potential costs and benefits of environmental changes consequent on the introduction of transgenics. We present this argument by way of a discussion of one of the concepts used by Maori to describe their proper relationship with their natural environment and to express some of their concerns about genetic modification (Klein, 2000). This is the concept of tapu. We should exercise caution when trying to use Western moral notions to capture the meaning of Maori concepts. The authors of the Report of the Royal Commission on Genetic Modification express optimism on this score, saying that "tapu" is "easy to express as [it has] been adopted into English to fill a void" (Royal Commission on Genetic Modification, 2002, 19). But this optimism is misplaced. "Tapu" is part of a network of moral understandings introduced to the young by way of mythical narratives grounded in a worldview that differs markedly from those with which Westerners are familiar

(Patterson, 1994; Klein, 2000). The adoption of the word "tapu" into New Zealand English does not show that speakers of New Zealand English grasp the moral point that native speakers of Maori use it to make, any more than does the casual use of the words "Yahweh" or "hara kiri" by English speakers indicate that their significance to native speakers of Hebrew or Japanese is fully grasped. The Maori "tapu" is related to the Tongan "tabu," imported into English as "taboo" by Captain Cook. It is sometimes translated as "forbidden," sometimes as "restricted," sometimes as "sacred," and sometimes as "set apart" (Barlow, 1991). None of these English terms fully captures the concept's meaning. It is best approached by way of its claimed implications for action. An object's tapu is supposed to limit the ways in which humans can interact with or use it; those who transgress tapu risk some form of retribution. Although the tapu of some objects can be removed by rituals, thus freeing them up for use by humans, this freeing up is never of the unconditional, all-encompassing variety so as to permit the kinds of uses of natural objects often preferred by Westerners. The removal of tapu is often partial or temporary, sometimes demanding compensation (Royal Commission on Genetic Modification, 2002, 35). John Patterson emphasizes a general presupposition in favour of the tapu of natural objects. He notes that the term functions to remind us "of the need to respect the environment—that the world is not ours" (Patterson, 1994, 392). The Report of Royal Commission on Genetic Modification was careful not to give the impression of a monolithic Maori view about genetic modification: Maori were not all of one mind on the moral implications of the new genetic technologies. But it did identify a concern shared by many Maori that genetic engineers interact with nature in more extensive and intimate ways than would be licensed by earlier removals of tapu (Royal Commission on Genetic Modification, 2002, 34–37). Some Maori put this concern in especially strong language, pointing to a belief in the interrelatedness of all life to argue that transgenic technologies transgress the tapu against incest (Royal Commission on Genetic Modification, 2002, 36–37).

The optimism of the Report about the expressibility of tapu in English notwithstanding, we should hold back from a complete analysis of the concept, instead undertaking a less ambitious task. The analysis that follows is at best a partial one, aspiring to capture only one among the multiplicity of concerns that Maori indicate with the word "tapu." It takes as its starting point the history of the Maori relationship with the environment of Aotearoa/New Zealand presented in Geoff Park's *Nga Uruora: Ecology and History in a New Zealand Landscape* (1995). Maori talk of nature's tapu appears to both reflect and validate what Park presents their environmental conservatism.

Park proposes that there was a discontinuity in the relationship of Maori with the natural environment of New Zealand. When Maori first arrived in New Zealand they had a dramatic impact on natural ecosystems, driving bird species including the Moa and Haast eagle to extinction within a few hundred years, and removing much of New Zealand's forest cover (Flannery 2002). Park argues that Maori went on to form an altogether different relationship with nature after this initial period. They came to weigh less heavily on the natural environments by transforming themselves into "edge-of-the-land dwellers," people who lived almost exclusively at "[r]iver mouths, estuaries, dune lakes, lagoons and islets" (Park, 1995, 312). Park contrasts this manner of interacting with nature with the approach taken by Europeans who combined ignorance of the indigenous ecosystems with a desire to remake New Zealand in England's image.

We suggest that at least part of the moral content of 'tapu' indicates the prudence of the environmental conservatism characteristic of the second stage of Maori inhabitation of New Zealand. This semantic element of the concept makes sense in the context of a people who have an established, stable relationship with their natural environment. We should not overinterpret the point about Maori being in a stable relationship with their natural environment. Environmental ethicists are sometimes tempted by an ecological romanticism that sharply distinguishes the balance and harmony attributed to ecosystems outside of the range of human influence from the disturbed states of those that humans meddle with (Rolston, 1988). Some of these romantics credit to indigenous peoples an intuitive understanding of the significance of ecological harmony and how best to limit their activities so as to preserve it. This romanticism relies on outmoded ecological theory. As we have pointed out, ecosystems exist in a state of continual disturbance. Throughout their existence in New Zealand, Maori have set snares for birds, removed forest cover, taken shellfish from coastal waters and otherwise stood in the way of ecosystems finding any theorised point of harmony and balance. Park's contrast between the activities of the newly arrived Europeans and the centuries-established Maori cannot be that the former disturbed local ecosystems while the latter did not, rather it must be made in terms of the kinds of disturbances. Not all disturbances are alike in their ecological consequences; certain types of disturbance are conducive to, or at least compatible with ecological flourishing, however this is considered, while certain are not.

Jared Diamond's recent book provides further cases of the manner of transition in relationships with nature that Park claims occurred with Maori. To take just one of Diamond's examples, the people who first settled Iceland not only disturbed its natural ecosystems but drove them to the brink of collapse. Walrus colonies were eliminated, sea birds depleted, and seals hunted beyond sustainability. The island lost almost all of its forest cover (Diamond, 2004, 200–201). After this initial period, however, Icelanders evolved a more sustainable pattern of disturbance. Diamond explains that this was accompanied by a change in attitude toward nature that permitted a more benign pattern of disturbance to replace one that was inimical to ecological flourishing. Icelanders became instinctual conservatives, according to Diamond, "conditioned by their long history of experience to conclude that, whatever change they tried to make, it was much more likely to make things worse than better" (Diamond, 2004, 202).

We are now in a position to say how the Maori claim to a special relationship with New Zealand's indigenous life forms might be a premise in an argument against their genetic modification. Genetic modification aims directly at the core of the nature over which they claim guardianship. Maori unease about proposals to modify the genomes of indigenous species may reflect a recognition of such changes as disturbances inimical to ecological flourishing as they conceive it. If so, they are right to use the concept of tapu to signal this course of action as especially dangerous. While this conclusion does not itself resolve the question of scientific expertise and non-scientific input, it does show how the two different forms of assessment may be able to share a common currency.

7.2 Conclusion

Debates regarding the threat transgenic organisms pose to biodiversity often pit opponents of biotechnology whose positions are based on assumptions that are seldom submitted to critical examination against proponents who operate with models of risk that systematically exclude certain factors, especially those having to do with cultures and ways of life. In this chapter we have attempted to show how both of these problems can be avoided. After defining biodiversity we began by showing what kind of analysis is necessary to identify the potential effects, both positive and negative, of transgenic organisms on biodiversity. Next we clarified the various normative positions to which both opponents and proponents tend to appeal, arguing that a position which defines anthropocentric interests broadly enough to include nature-regarding matters is best equipped to evaluate transgenic organisms ethically. This conclusion provides the basis for an analysis of the ethical issues at stake in Bt maize and also for our efforts to expand the range of factors involved in assessment of the risks posed by transgenic organisms. Both the example of Bt maize and the two examples of risk assessment (one addressing the question of uncertainty, the other the question of whose and what interests are included in risk assessment) are illustrative, but they demonstrate that it is possible to evaluate transgenic organisms in an ethically defensible way that does not either preclude their legitimacy from the start or exclude the factors that raise serious ethical questions.

References

Andow, D.A., and A. Hilbeck (2004). "Science-based risk assessment for nontarget effects of transgenic crops," *BioScience* 54, 637–649.
Augustine. *City of God Against the Pagans*.
Barlow, C. (1991). *Tikanga Whakaaro—Key Concepts in Maori Culture*. Auckland: Oxford University Press.
Barlow, C. (2001). "Ghost stories from the ice age," *Natural History*, 110(9): 62–67.
Botkin, D.B. (2001). "The Naturalness of Biological Invasions," *Western North American Naturalist* 61, 261–266.
Brock, D. (1998). "Cloning Human Beings: An Assessment of the Ethical Issues Pro and Con," in M. Nussbaum and C. Sunstein (eds.), *Clones and Clones*. New York: W.W. Norton.
Bye, R.A. (1981). "Quelites: Ethnoecology of Edible Greens—Past, Present and Future," *Journal of Ethnobiology* 1, 109–123.
Bye, R.A. (1993). "The Role of Humans in the Diversification of Plants in Mexico," in Ramamoorthy, T.P., R. Bye, A. Lot, and J. Fa (eds.), *Biological Diversity of Mexico: Origins and Distribution*. New York: Oxford University Press, 707–731.

Callicott, B. (1998). "The Land Ethic Dynamized," *Reflections*, Special Issue No. 3. Available online: http://www.orst.edu/dept/philosophy/pese/reflections/Reflections 98/callicott.

Chapin, F.S. III, B.H. Walker, R.J. Hobbs, D.U. Hooper, J.H. Lawton, O.E. Sala, and D. Tilman (1997). "Biotic Control Over the Functioning of Ecosystems," *Science* 277, 500–504.

Cohen, A.N., and J.T. Carlton (1998). "Accelerating Invasion Rate in a Highly Invaded Estuary," *Science* 279, 555–558.

Daily, G.C., S. Alexander, P.R. Ehrlich, L. Goulder, J. Lubchenco, P.A. Matson, H.A. Mooney, S. Postel, S.H. Schneider, D. Tilman, and G.M. Woodwell (1997). "Ecosystem Services: Benefits Supplied to Human Societies by Natural Ecosystems," *Issues in Ecology* 2. Washington, DC: Ecological Society of America, ISSN 1092-8987.

Derr, M. (2001). "Alien Species Often Fit in Fine, Some Scientists Contend," *New York Times*, D4, Sept. 4.

Diamond, J. (2004). *Collapse: How Societies Choose to Fail or Succeed*. New York: Viking.

Douglas, M., and A. Wildavsky (1982). *Risk and Culture: An Essay on the Selection of Technical and Environmental Dangers*. Berkeley, CA: University of California Press.

Flannery, T. (2001). *The Eternal Frontier*. Melbourne: Text Publishing.

Flannery, T. (2002). *The Future Eaters: An Ecological History of the Australasian Lands and People*. New York: Grove Press.

Graham, G. (2002). *Genes: A Philosophical Inquiry*. London: Routledge.

Harper, J.L., and D.L. Hawksworth (1995). "Biodiversity - measurement and estimation – preface," *Philosophical Transactions of the Royal Society of London. Series B: Biological Sciences* 345, 5–12.

Harper, J.L., and D.L. Hawksworth (1995). "Preface. Biodiversity: Measurement and Estimation," in Hawskworth, D.L. (ed.), *Biodiversity Measurement and Estimation*. Oxford: The Royal Society, Alden.

Hites, R.A., J.A. Foran, D.O. Carpenter, et al. (2004). "Global assessment of organic contaminants in farmed salmon," *Science* 303, 226–229.

Kass, L. (1997). "The Wisdom of Repugnance," *The New Republic*, June 2.

Keown, D. (1995). *Buddhism and Bioethics*. New York: St. Martin's Press.

Klein, U. (2000). "Belief-views on Nature: Western Environmental Ethics and Maori World-views," *Archives of the New Zealand Centre for Environmental Law* 4.

Lacey, L.A., R. Frutos, H.K. Kaya, and P. Vail (2001). "Insect Pathogens as Biological Control Agents: Do They Have a Future?" *Biological Control* 21, 230–248.

Leopold, A. (1949). "The Land Ethic," in A. Leopold (ed.), *A Sand County Almanac and Sketches Here and There*. New York: Oxford University Press, 201–226

Lodge, D.M. (1993). "Biological Invasions: Lessons for Ecology," *Trends in Ecology and Evolution* 8, 133–137.

Lodge, D.M., and K. Shrader-Frechette (2003). "Nonindigenous Species: Ecological Explanation, Environmental Ethics, and Public Policy," *Conservation Biology* 17, 1–8.

Loreau, M., S. Naeem, P. Inchausti, J. Bengtsson, J.P. Grime, A. Hector, and D.U. Hooper (2001). "Biodiversity and Ecosystem Functioning: Current Knowledge and Future Challenge," *Science* 294, 804–808.

Lovejoy, A.O. (1936). *The Great Chain of Being: A Study of the History of an Idea*. Cambridge, MA: Harvard University Press.

Maimonides, Moses (Moshe ben Maimon). *Guide of the Perplexed*.

Mann, C.C., and Mark L. Plummer (2002). "Forest Biotech Edges Out of the Lab," *Science* 295, 1626–1629.

McKenny, G. (2001). "Religion, Biotechnology, and the Integrity of Nature," in M. Hanson (ed.), *Claiming Power Over Life*. Washington, DC: Georgetown University Press, 169–191.

Metz, M., and J. Fütterer (2001). "Suspect Evidence of Transgenic Contamination," *Nature* 416, 600–601.

Naeem, S., and J.P. Wright (2003). "Disentangling Biodiversity Effects on Ecosystem Functioning: Deriving Solutions to a Seemingly Insurmountable Problem," *Ecology Letters* 6, 567–579.

National Council of Churches of Christ in the U.S.A. (1986). *Genetic Science for Human Benefit.* National Council of Churches of Christ in the USA.
National Research Council (1996). *Understanding Risk: Informing Decisions in a Democratic Society.* Washington, DC: National Academy Press.
National Research Council (NRC) (2001). *Environmental Effects of Transgenic Plants: The Scope and Adequacy of Regulation.* Washington, DC: National Academy Press.
Nations, J.D., and R.B. Nigh (1980). "The Evolutionary Potential of Lacandon Maya Sustained-yield Tropical Forest Agriculture," *Journal of Anthropological Research* 36, 1–30.
Navarrete, S.A., and B.A. Menge (1996). "Keystone Predation: Interactive Effects of Two Predators on their Main Prey," *Ecological Monographs* 66, 409–429.
Nelkin, D., and S. Lindee (1995). *The DNA Mystique: The Gene as Cultural Icon.* New York: W.H. Freeman.
Norton, B.G. (1987). *Why Preserve Natural Variety?* Princeton, NJ: Princeton University Press.
OECD (2003). *Consensus Document on the Biology of* Zea mays *subsp.* mays *(Maize).* OECD Environment, Health and Safety Publications, Series on Harmonisation of Regulatory Oversight in Biotechnology No. 27. Paris: Organisation for Economic Co-operation and Development.
Ortman, E.E. (2001). "Letter to the Editor: Transgenic Insecticidal Corn: The Agronomic and Ecological Rationale for its Use," *BioScience* 51, 900–902.
Park, G. (1995). *Nga Uruora: Ecology and History in a New Zealand Landscape.* Wellington: Victoria University Press.
Patterson, J. (1994). "Maori Environmental Virtues," *Environmental Ethics* 16(4), 397–409.
Pickett, S.T.A., V.T. Parker, and P.L. Fiedler (1992). "The New Paradigm in Ecology: Implications for Conservation Biology Above the Species Level," in P.L. Fiedler and S.K. Jain (eds.), *Conservation Biology: The Theory and Practice of Nature Conservation Preservation and Management.* New York: Chapman & Hall, 65–87.
Quist, D., and I.H. Chapela (2001). "Transgenic DNA Introgressed into Traditional Maize Landraces in Oaxaca, Mexico," *Nature* 414, 541–543.
Reed, D.H., and R. Frankham (2003). "Correlation Between Fitness and Genetic Diversity," *Conservation Biology* 17, 230–237.
Rolston, H. (1985). "Duties to Endangered Species," *BioScience* 35, 718–726.
Rolston, H. (1988). *Environmental Ethics: Values in and Duties to the Natural World.* Philadelphia, PA: Temple University Press.
Rosenzweig, M.L. (2001). "The Four Questions: What Does the Introduction of Exotic Species do to Diversity?" *Evolutionary Ecology Research* 3, 361–367.
Royal Commission on Genetic Modification (New Zealand) (2002). "Report of the Royal Commission on Genetic Modification". Available online: http://www.mfe.govt.nz/publications/organisms/royal-commission-gm/index.html
Sagoff, M. (1999). "What's Wrong with Exotic Species?" Available online: http://www.puaf.umd.edu/IPP/fall1999/exotic_species.htm
Schmithausen, L. (1997). "The Early Buddhist Tradition and Ecological Ethics," *Journal of Buddhist Ethics* 4.
Science and Environmental Health Network (1998). "The Wingspread Consensus Statement on the Precautionary Principle". Available online: http://www.sehn.org/wing.html
Serratos-Hernández, J.-A., F. Islas-Gutiérrez, E. Buendía-Rodríguez, and J. Berthaud (2004). "Gene Flow Scenarios with Transgenic Maize in Mexico," *Environmental Biosafety Research* 3, 149–157.
Sessions, W. (1991). "Deep Ecology and Global Ecosystem Protection," in M. Oelschlaeger (ed.), *The Wilderness Condition.* San Francisco, CA: Sierra Club Books.
Shrader-Frechette, K. (1995). "Evaluating the Expertise of Experts," *Risk* 6.
Shrader-Frechette, K. (1998). "Ecological Sense and Environmental Nonsense," *Reflections*, Special Issue No. 3. Available online: http://www.orst.edu/dept/philosophy/pese/reflections/Reflections 98/shrader.
Shrader-Frechette, K. (2004). "Risk and Reason," *Ethics* 114, 376–380.

Slobodkin, L.B. (2001). "The Good, the Bad and the Reified," *Evolutionary Ecology Research* 3, 1–13.
Slovik, P. (1999). "Trust, Emotion, Sex, Politics, and Science: Surveying the Risk-assessment Battlefield," *Risk Analysis* 19, 689–701.
Sober, E. (1986). "Philosophical Problems for Environmentalism," in B. Norton (ed.), *The Preservation of Species*. Princeton, NJ: Princeton University Press, 173–194.
Sunstein, C. (2002). *Risk and Reason: Safety, Law, and the Environment*. Cambridge: Cambridge University Press.
Symstad, A.J., D. Tilman, J. Willson, and J. Knops (1998). "Species Loss and Ecosystem Functioning: Effects of Species Identity and Community Composition," *Oikos* 81, 389–397.
Taube, K. (1996). "The Olmec Maize God: The Face of Corn in Formative Mesoamerica," *RES Anthropology and Aesthetics* 29/30: 39–81.
Taylor, P. (1986). *Respect for Nature: A Theory of Environmental Ethics*. Princeton, NJ: Princeton University Press.
Thomas Aquinas. *Summa Theologiae*.
Thomas, K. (1983). *Man and the Natural World*. London: Pantheon.
United Methodist Church (1992). "Genetic Science Task Force Report to the 1992 General Conference," *Church and Society* 18, 113–123.
Vieyra-Odilón, L., and H. Vibrans (2001). "Weeds as Crops: The Value of Maize Field Weeds in the Valley of Toluca, Mexico," *Economic Botany* 55, 426–443.
Whitham, T.G., W.P. Young, G.D. Martinsen, C.A. Gehring, J.A. Schweitzer, S.M. Shuster, G.M. Wimp, D.D. Fischer, J.K.Bailey, R.L. Lindroth, S. Woolbright, and C.R. Kuske (2003). "Community and Ecosystem Genetics: A Consequence of the Extended Phenotype," *Ecology* 84, 559–573.
Williamson, M. (1996). *Biological Invasions*. London: Chapman & Hall.
Woodsen, M.M. (2001). "If Oak Malady Moves East, Many Trees Could Die," *New York Times*, F3, Sept. 4.

Chapter 8
Swimming Upstream: Regulating Genetically Modified Salmon

Paul A. Lombardo and Ann Bostrom

8.1 Introduction

8.1.1 Biotechnology's Impact on Biodiversity

Humans have been manipulating their food supply for thousands of years, often by consciously breeding both plants and animals for traits that made foodstuffs more plentiful, more convenient to use, and even more nutritious. Most of us have eaten items as common as hybrid corn, steaks from beef cattle bred for lower fat content, or tomatoes specifically bred for tougher skins and more efficient transport. But with the advent of genetic engineering in the last 30 years, it has become possible to create completely novel organisms, whose characteristics have been fixed altered at the molecular level by the introduction of new combinations of genes.

This prospect is troubling to some, who see intrusion into plant and animal genomes as somehow less "natural" than the more traditional methods of horticulture or animal husbandry. Fears also arise at the practical level, when novel varieties of food products—plant or animal—appear to pose a threat to the environment, or alternatively, to people who have survived on the cultivation of established plants or the nurture or capture of existing varieties of animals.

The emergence of these newly designed sources of food has generated a call for additional legal regulation of food production, or at least of some the biotechnologies, such as recombinant DNA technology, used to produce genetically modified organisms. Unlike most of the world, the United States has been slow to adopt such a regulatory scheme. By focusing on the example of genetically modified salmon, this paper explores why that reluctance persists, and how model regulations already in place internationally might be employed in the U.S.

8.1.1.1 An Example: GM Salmon

We have chosen to explore possible (and actual) responses to the introduction of genetically modified (GM) salmon into the food supply as the focal point for our investigation of the theme of biodiversity and biotechnology. Why salmon? Salmon is at the same time a food, a significant link in the economic order, and the first of what might be many marine species—from flounder to lobster to mollusks—that can be efficiently mass-produced via GM technology.

To several indigenous cultures, threatened by the continuing encroachments of industrialization and expanding technologies, salmon is a cultural icon. To these people, salmon symbolizes ties to the land and the water that are in danger of being permanently severed. The image of the fish, leaping free of the constraints of civilization, presents a vision of a state of nature that existed in the distant past. To some tribes, salmon fishing marked the transition from a primarily nomadic culture to a collection of clans who lived by rivers of the Pacific Northwest, marking the seasons by the return of a crucial foodstuff. Harvesting the salmon was community work; curing and storing their meat was a shared responsibility. To these early settlers, salmon provided not only an anchor for economic sustenance, but also represented a spirit that populated their myths or lent a face for totemic images of the divine. Descendants of these "first peoples" can be threatened by the introduction of GM salmon farmed in ocean pens, competing in the market previously dominated by indigenous fishers, potentially displacing wild stocks in existing habitats.

Salmon aquaculture provides us with a concrete example of an activity that is considered by many "un-natural," while others see it both as a humanitarian effort—it promises to generate a reliable and comparatively less expensive food source for millions in need—and a replacement for the dwindling stocks of wild fish in many parts of the world. Fish characteristics—such as the potential for accelerated growth and maturation–are specifically selected for economic and commercial purposes. The critics say this process produces fish of a type that does not occur, absent human intervention, as part of any existing ecosystem.

Salmon also represents food, and in the United States the ready availability and safety of food are axiomatic. For those who fear genetic manipulation of foodstuffs, GM food should be banned, or at least labeled to allow consumer choice. Supporters of a growing reliance on genetically modified, commercially farmed salmon (and other fish) consider these concerns overblown and look to aquaculture as a significant source of food, and in some cases a way to salvage economies traditionally dependent on fishing to generate jobs and incomes.

Opponents of GMO salmon also argue that they pose a threat to the ecology. They see aquaculture as a potential pollutant—both environmentally, via nitrogen pollution of waters surrounding fisheries or introduction of drugs and chemicals into the natural habitat, and genetically, by adding alien strains to the ocean and rivers without proper understanding of their long-term impact. Hatchery-reared salmon compete with wild salmon for food, adding further stress to already dwindling existing stocks, but are less viable in the long run. Even though producers play down the genetic impact of the manufactured fish by noting that they are made sterile to obviate any concerns about excessive reproduction or possible interbreeding, because they are produced using genetic technologies that are inherently suspect to many, the fear of "frankenfish" remains.

Finally, the cultural significance of salmon to indigenous peoples is difficult to exaggerate. The fish's career of birth, travel out to sea and return to a native habitat

8 Swimming Upstream: Regulating Genetically Modified Salmon

to spawn and die represents the life cycle, a process inextricably incorporated into the identity of tribes who traditionally fished for salmon and identified it as a spiritual force. The presence of the salmon as a literal icon—on flags, logos or other symbols of indigenous peoples—is ubiquitous. The salmon thus plays a part in indigenous mythologies; since myth is an inherent part of all human belief systems, it often undergirds policymaking and policies as well, as earlier papers in this volume have noted.

Yet in the face of all these questions about the possible impact of GM salmon on the ecosystem, on the economy, and on indigenous cultures, we still have no unified policy to guide the expansion or restriction of salmon aquaculture. Thus GM salmon provides a prime exemplar through which to consider how biotechnology might be regulated in the US. In this paper, we intend to discuss how state, federal and international law might be marshaled appropriately to take into account competing interests, and how existing policies on biotechnology might be fruitfully employed in the North American context.

A key inquiry in this volume is the attempt to define the role of the "natural", both within belief systems and as that concept might have an impact on policy. Nature is defined in many ways, and within some world views, altering existing biological, ecological and cultural relationships can be seen as "natural" also. The following chart lays out the various ways we discuss "nature" in this paper.

Chart: concepts of "the natural"	Policy implications re: GM salmon
Preservation of existing species and ecosystems	*Avoid any novel technology applications that could increase the likelihood of further species extinction or threaten the current balance among the array of species already present; pay attention to and forestall ecosystem degradation as a result of introducing new GM fish.*
Respecting belief systems of indigenous people	*Protect existing fish stocks and habitat; subsidize traditional aqua-culture by indigenous peoples; block introduction of GM stocks that could pose a threat to economy or cultural habits of "first peoples."*
The Biodiversity Convention's approach: existing biosystems as "natural"	*Protect existing species, along with the habitat/ecosystems and array of genetic diversity; control risks to above that could result in degradation to existing species or biosystems.*
Hobbesian "state of nature;" market based competition in products, technology and ideas	*Decrease regulatory restraints; allow market or other competitive forces to determine which products are distributed and where; e.g., unfettered consumer choice in producing and/or exporting GM salmon.*

8.2 The Existing Regulatory Framework

8.2.1 International

8.2.1.1 The Convention on Biological Diversity

The Convention on Biological Diversity was adopted in 1992[1] under the auspices of the United Nations Environmental Program and is now endorsed by more than 190 countries, but not the United States. The Convention lays out a framework that recognizes the three dimensions of biodiversity: ecosystems, species, and genetic diversity. The Convention's overall goal is to accommodate the dual objectives of environmental conservation and economic development. It focuses on conservation of biodiversity, sustainable use of the components of biodiversity, and sharing the benefits that grow out of using genetic resources (Article I, *Objectives*).

While admitting the sovereign right of nations to exploit internal resources, the Convention simultaneously notes that nations share responsibility for ensuring that activities within their control do not cause damage to the environments of areas that are not within their jurisdiction (Article 3, *Principle*). In addition to undertaking the duty of identifying components of biological diversity, member nations are charged with monitoring conservation, identifying activities that may impact conservation negatively, and maintaining data on all such activities (Article 7, *Identification and Monitoring*).

Member nations agree to establish a system for identifying areas that need to be protected in order to conserve biological diversity, promoting the protection of ecosystems and natural habitats, rehabilitating and restoring degraded ecosystems as well as maintaining viable populations of species in their natural surroundings. Signatories to the Convention commit to take steps to control the risks associated with the use and release of living modified organisms resulting from biotechnology which are likely to have adverse environmental impacts and to prevent the introduction of alien species which threaten ecosystems, habitats or species. Importantly, Convention signatories agree to respect and preserve practices of indigenous and local communities embodying traditional lifestyles relevant for the conservation and sustainable use of biological diversity (Article 8, *In Situ Conservation*).

In the area of genetic resources, the members agree to a principle of cooperation, which would rule out the development of conservation biotechnology for any nation's exclusive use. Research in one nation using genetic resources originally obtained from other nations should be carried out with full participation of those other nations. The benefits—commercial and otherwise—of genetic research should be shared equitably between the nations where research is done and the nation originally providing genetic resources that formed the basis for technology development (Article 15, *Access to Genetic Resources*).

---

[1] http://www.cbd.int/biosafety/protocol.shtml

Signatories agree to provide mutual access to and transfer of genetic technologies, within the boundaries of private intellectual property rights, and to move toward modifying legislation that define such rights, in order to support the objectives of the Convention (Article 16, *Access to and Transfer of Technology*).

The Convention spells out goals for encouraging the safe handling of new biotechnology, particularly in the case of living modified organisms, and lists requirements for disclosure of information about the safety and potential adverse impact of specific organisms to the environment in which they will be released (Article 19, *Handling of Biotechnology and Distribution of its Benefits*). It also commits signatories in developed countries to provide financial assistance to developing countries to help achieve the aims of the Convention (Article 20, *Financial Resources*; Article 21, *Financial Mechanism*). The structure of administration of the Convention includes a critical section requiring disputes among the signatories to be submitted the International Court of Justice (Article 17, *Settlement of Disputes*).

8.2.1.2 The Biosafety Protocol

Within a few years of its creation, the parties to the Convention began working on a legally binding agreement that would address possible risks arising from the creation and use of Genetically Modified Organisms (GMOs). That agreement took shape as the Cartegena Protocol on Biosafety,[2] formally adopted in 2000. The Protocol sets out a comprehensive regulatory system for the transfer, handling and use of GMOs that are subject to movement across national boundaries. Its two major features are an insistence on biosafety and precaution.

Biosafety is a concept that focuses on reducing risks to the environment and/or human health that biotechnology may generate. The "precautionary principle" is a rule for insuring biosafety. It is commonly the case that new technologies are put into use unless some oversight group or the public deems them unsafe and can, in advance, prevent their deployment. When the precautionary principle is in place, any new policy or set of activities that would lead to serious and/or irreversible harm to biodiversity must be demonstrably safe *before* it can be employed. The principle shifts the burden to determine and demonstrate safety from those who oppose specific innovations to those who propose those innovations.

The precautionary principle relies on several ethical justifications (World Commission on the Ethics of Scientific Knowledge and Technologies, 2005, 17–21). In its counsel to move slowly in the face of uncertainty, the precautionary principle incorporates the virtue of prudence, that is, taking care to avoid unnecessary harms. It also represents the moral basis for the legal "duty to care," the obligation to pay attention to activities that have foreseeable negative consequences for others. For those who would undertake activities with potential risk to the environment, it

---

[2] http://www.un.org/millennium/law/cartagena.htm

insists on altruism as a default position. Thus the precautionary principle is an analogue of the ancient Hippocratic prescription in heath care ethics: first do no harm.

Although this perspective on the precautionary principle ignores that in any given situation action entails risk tradeoffs in the face of uncertainty (e.g., special issues of *Journal of Risk Research* (June 2006), and *Human and Ecological Risk Assessment* (2005); see Barry Johnson's editorial in the latter, pp. 1–2, for an introduction), research on the prevalence of situations where there are false positives (early warnings or uncertain evidence of harm turn out to be unfounded) suggests that is so rare that it is not a valid counterargument to the precautionary principle (Hansen et al., 2007).

Among the potential beneficiaries of the precautionary principle are future generations. Those to whom potential harms (the burdens tomorrow traceable to risks undertaken today) would fall benefit from an emphasis on risk avoidance. Intergenerational equity is served when current actors attend to the consequences of their behavior that could have a negative impact on generations yet to be born.

The international regulatory framework provided by the Biosafety Protocol has as its main goal preventing negative effects on biological diversity and human health by living modified organisms—defined as those that possess a novel combination of genetic material created via biotechnology. The definition of "biotechnology" includes *in vitro* nucleic acid techniques, recombinant deoxyribonucleic acid (DNA) and other techniques not typically used in traditional breeding or selection. The Protocol establishes mechanisms to ensure the safe transfer, handling and use of living modified organisms when they are moved over national boundaries. It also specifically clarifies that it will not impact the sovereignty of nations overseas within their jurisdiction under international law or over economic zones defined by continental shelves.

The Protocol requires any nation that plans to export a living modified organism (LMO) to provide notification to the nation of destination. The notification must incorporate details of travel (where it is going, when it will arrive, who will receive it) and technical descriptions of the LMO, such as its genetic composition, technologies used in modification, and how it will be used. An exhaustive risk assessment report must be appended to the notification. The recipient nation may consent to the import, refuse the import, require the exporting nation to abide by existing legal regulations governing conditions for safe passage of such organisms, and/or ask for additional information upon which it may condition its consent.

In the case of LMOs that will be used as food for humans or feed for animals, a different procedure applies. When the producer of such an organism decides to place it on the market or otherwise make it available for use, it must deposit detailed technical descriptions of the product, including information similar to that required for exports and methods for proper safe handling, in a Biosafety Clearing-House. The Clearing-House will make the data available to any potential recipient nation.

When LMOs are unintentionally moved between two countries, the one responsible for the organism must notify the recipient nation and provide all relevant details that would normally have been included in a notification. Additional rules apply to the handling, transport and labeling of LMOs, setting up internal bureaucracies to manage notification, etc., concerning transboundary transport, steps for maintaining the confidentiality of technical information accompanying transfer requests, and the like. The potential recipients of LMOs are specifically empowered by the Protocol to take into account socio-economic considerations concerning the impact of LMOs on biological diversity, particularly with reference to the value of biological diversity to indigenous communities.

8.2.1.2.1 US Resistance to Adoption of the Precautionary Principle

There are several reasons why the United States has refused to adopt the BioSafety Protocol and the precautionary principle it embodies. Some critiques describe the principle as unworkably vague, and thus leading to arbitrary burdens on technological innovation. They say it emphasizes regulation in already regulated areas, where the need for additional restrictions have not been demonstrated. The principle, say critics, discards the technique of cost/benefit analysis in favor of an undetermined and impossible to calculate concern for harms that may never materialize (Merchant, 2002). The precautionary principle shifts the burden of proof, which makes it a powerful policy tool. But as some authors note, it is not an approach to analysis, per se; risk assessment and benefit-cost analyses can be used to implement a precautionary principle (e.g., Farrow, 2004).

The precautionary principle also defies a major premise of those who favor a regime of unfettered market competition. From that position, no principle trumps profit maximization, and no parties in living or future generations take precedence over current shareholders and their interests.

Such concerns also underlie U.S. objections to the Protocol from those fearing encroachments to existing intellectual property rights.

Debates over the meaning and usefulness of the precautionary principle are common in the academic literature. But on the purely political level it is likely that U.S. refusal to endorse the Biosafety Protocol reflects little more than interest group politics. Major industries in the U.S., wishing to preserve the option of extending international markets and continuing to exploit resources unique to other countries, had powerful motives to lobby against adoption of the Protocol.

8.2.2 National State/Federal Litigation

A search of databases recording appellate court decisions involving genetically modified salmon (or other fish) yielded surprisingly few instances when these issues had been addressed. Neither state nor federal courts have dealt with genetic modification very often, though the attention given to the few lawsuits filed in lower courts might suggest otherwise. Several of the more important state and federal cases that raised issues about aquaculture and salmon, along with one Canadian decision, are described below:

1. More than 20 years ago, in a case from Oregon, a multinational corporation engaged in operating a commercial salmon hatchery successfully challenged the propriety of a ballot initiative that included this language: "a person or corporation holding a salmon hatchery permit shall not release fish at nontraditional times, or practice genetic altering or acceleration of the life-cycle of fish that may cause a change in the size, migration pattern, or times of return from the ocean that differs significantly from a natural run in an affected watershed." (*Oregon Aqua-Foods, Inc. v. Paulus*, 296 Or. 469, 676 P.2d 870, Or., 1984) The ballot measure was ruled invalid on technical grounds.
2. Representatives of the aquaculture industry sued challenging a rule published by National Marine Fisheries Service (NMFS) listing only naturally spawning coho salmon as "threatened" species pursuant to the Endangered Species Act (ESA). Since "the NMFS listing decision creates the unusual circumstance of two genetically identical coho salmon swimming side-by-side in the same stream" with only one receiving ESA protection (and not the variety produced in a hatchery), the court called the distinction arbitrary and invalid. (*Alsea Valley Alliance v. Evans*, 161 F.Supp.2d 1154, D.Or., 2001.)
3. Environmental organizations brought a citizen suit alleging that salmon farms released pollutants into water in violation of the Clean Water Act. The United States District Court found the farms liable for polluting Maine waters, and granted injunctive relief. The Court of Appeals upheld the trial decision. (*U.S. Public Interest Research Group v. Atlantic Salmon of Maine, LLC*. 339 F.3d 23, C.A.1 (Me.), 2003.)
4. The Wilderness Society and the Alaska Center for the Environment challenged a decision by the United States Fish and Wildlife Service to permit a sockeye salmon enhancement project at a lake in Alaska, within a designated wilderness area in the Kenai National Wildlife Refuge. They claimed that the project violated the federal Wilderness Act, because it contravened that Act's requirement to preserve the "natural condition" and "wilderness character" of the area, and because it constituted an impermissible "commercial enterprise" within a wilderness area. They also alleged that the Project violated the National Wildlife Refuge System Administration Act of 1966, because it was not "compatible" with the purposes of the Refuge Act. The challenge failed at the trial court, but was eventually sustained by the entire 9th Circuit Court of Appeals. (*Wilderness Soc v. U.S. Fish and Wildlife Service*, 353 F.3d 1051, C.A.9 (Alaska), 2003.)
5. Following the listing of the Gulf of Maine distinct population segment (DPS) of Atlantic salmon as an endangered species, Maine and Maine businesses brought action against the Secretary of Interior, challenging the listing. The Court decided, among other issues, that the listing of Gulf of Maine Atlantic salmon population was supported by the best available scientific evidence, and let the listing stand. (*Maine v. Norton*, 257 F.Supp.2d 357, D.Me., 2003.)
6. An exceedingly complex and lengthy controversy (over 12 years) between the National Wildlife Federation and a division of the National Oceanic and

Atmospheric Administration reached a partial conclusion in May, 2005, with the court invalidating the latest of a series of "biological opinions" issued by the government that attempted to balance the economic, commercial, and recreational interests in the Pacific Northwest, served by the ongoing operations of the Columbia and Snake River dams, with the conservation of salmon species listed under the Endangered Species Act. Among the findings, along the way, were that the size of hatchery-produced salmon posed significant genetic and ecological threats to the remaining natural populations in the species. Invalidation of the agency opinion does not settle the claims made either by wildlife advocates or representatives of tribal fishing interests. (*National Wildlife Federation v. National Marine Fisheries Service*, 481 F.3d 1224 [C.A. 9 (Or.) 2007.]

7. A British Columbia Supreme Court Judge ruled that the provincial Ministry of Agriculture, Food and Fisheries erred in failing to properly consult with the Homalco First Nation regarding the approval of a species amendment to allow the fish-farming arm of Dutch trans-national Nutreco to introduce Atlantic salmon smolts into its pens at the mouth of Bute Inlet. The judge ordered the company to consult with the Homalco First Nation not only about the farming of Atlantic salmon but also about the location of their fish farm. (*Blaney v. British Columbia Minister of Agriculture, Food and Fisheries*, 2005 BCSC 203.)

8.2.2.1 Legislation: State

The first-ever U.S. labeling legislation for genetically engineered food became law in Alaska after being passed unanimously by both the state Senate and House. The Alaska House approved legislation requiring that genetically engineered fish be "conspicuously labeled to identify the fish or fish product as a genetically modified fish or fish product," whether packaged or unpackaged.

In the 2003–2004 legislative sessions sixteen bills were introduced nationally on transgenic fish and aquaculture. In contrast, five similar bills were introduced in 2001–2002. Michigan introduced nine bills which either limit the introduction of transgenic aquaculture species or designate their unauthorized release as criminal.

California legislators passed a bill which makes it illegal to spawn, cultivate, or incubate any transgenic fish in the waters of the Pacific over which the state has jurisdiction.

Florida created a Transgenic Aquatic Species Task force in 2005, which has to date approved only one transgenic fish (the Glofish, discussed below).

Maine passed a bill seeking voluntary labeling of foods, food products or food ingredients as free of, or made without, recombinant DNA technology, genetic engineering, or bioengineering. The state also now prohibits the Department of Natural Resources from issuing a permit for the raising of transgenic fish unless the fish are limited to state waters that do not flow into any other body of water. It imposed a civil violation for any manufacturer or retailer who falsely labels any product such as commercial feed as made without genetic engineering or bioengineering.

Maryland banned raising transgenic species except where they can be confined in 2006.

Michigan now prohibits the release of GM or nonnative organisms including game fish into the state without a permit. The state criminalized the intentional release, importation, or possession of GMOs (especially GM fish, plants, and aquatic organisms) absent a permit, and explicitly prohibited the importation, unless expressly permitted, of GM variants of species under quarantine or GM variants which have the potential to spread disease or harm to livestock or wildlife. It also amended the Michigan Aquaculture Development Act to state that GM variants of aquaculture species are considered distinct species, and prohibits introducing or transporting them unless they are specifically identified on a list of approved aquaculture species or in subsequent rules. The prohibition includes research organisms unless specifically identified in a research permit.

North Dakota established procedures for inspecting, analyzing, and verifying the genetic identity or physical traits of seeds or crops by state Agricultural Commissioners who may prescribe the type of labels to be affixed to the seeds or crops that are inspected and analyzed.

Minnesota, Mississippi, Oregon, Washington and Wisconsin have also passed regulations pertaining to modified organisms or genetically engineered fish specifically.

Several towns in Massachusetts and Vermont have passed resolutions opposing genetically modified organisms, while several states have banned local control of GMOs. By far the most prominent and successful of state legislative actions in recent legislative sessions related to supporting biotechnology (GM specifically) or protecting GM crops and other products by criminalizing vandalism or destruction of those products.

8.2.2.2 Federal Bills

Under international law, countries may claim the right to explore and exploit marine resources as much as 200 miles offshore, free of competition with other nations. Such claims set up "exclusive economic zones." On June 7, 2005, a US Senate bill announced that it would allow aquaculture pens between 1 and 50 miles offshore of US under an "exclusive economic zone." Tuna, salmon, halibut and cod could be farmed there. NOAA drafted the bill, which would remove offshore farming from state oversight.

8.2.2.3 Federal Regulation

On June 16, 2005, NOAA issued its final policy for considering hatchery salmon in making Endangered Species Act listing determinations, putting 131 strains of hatchery fish under the same protection as their wild "cousins."

8.3 The Failure of Policy in the U.S.

In 1986, the U.S. adopted the Coordinated Framework for Regulation of Biotechnology (51 FR 23302).[3] Under the Coordinated Framework, the U.S. Food and Drug Administration (FDA), the U.S. Environmental Protection Agency (EPA), and the U.S. Department of Agriculture's (USDA) Animal and Plant Health Inspection Service (APHIS) are charged with managing biotechnology. "APHIS is responsible for regulating all GE plants, the EPA is responsible for GE microbes and plant-incorporated protectants, and the FDA is responsible for ensuring the safety of food from GE crops and for pharmaceutical chemicals produced in GE crops" (Pew Initiative on Food and Biotechnology, 2006, 9).

This arrangement has been criticized because of the potential for regulatory gaps. In particular, environmental NGOs have noted that the Coordinated Framework involves stretching pre-existing laws to cover biotechnology, resulting in weak environmental protection. NGOs such as the Union of Concerned Scientists have described this arrangement as a policy vacuum, as it may leave environmental and ecological risks out of the picture. Animals not intended to enter the food supply may escape regulation (Bratspies, 2005).

Criticism has come from National Academy committees as well:

> The committee notes a particular concern about the lack of any established regulatory framework for the oversight of scientific research and the commercial application of biotechnology to arthropods. In addition to the potential lack of clarity about regulatory responsibilities and data collection requirements, the committee also notes a concern about the legal and technical capacity of the agencies to address potential hazards, particularly in the environmental area. [...] there is a need for clarity about whether the regulatory agencies consider it within their charge to consider only the direct health and environmental impacts of biotechnology, or also the social or economic impacts of a technology that, in turn, might have an adverse health or environmental impact (National Research Council of the National Academies et al., 2002, 14).

Currently, transgenic animals, including fish, are considered new animal drugs if they will enter the food supply, and so are reviewed by the FDA. Aqua Bounty submitted its request for approval of its transgenic salmon "AquAdvantage" over 6 years ago. AquAdvantage incorporated a portion of a gene from the pout—a bottom dwelling bait fish—that promotes growth year-round; growth hormones in salmon are otherwise light dependent. Hence these transgenic salmon grow at twice the pace of other salmon. Both EPA and APHIS withdrew and left the FDA with regulatory oversight. The FDA claimed oversight over AquAdvantage by designating the genetic modification and growth hormone it produces as a new drug for animals. In December, 2006, Aqua Bounty announced that it had submitted all studies required for approval, and that the FDA had

---

[3] http://usbiotechreg.nbii.gov/Coordinated_Framework_1986_Federal_Register.html

approved a major study and was expediting review. The studies submitted address aspects of food safety, animal health, and product efficacy. The absence of long-term environmental and social impacts in the list of attributes addressed is notable.

A clear example of genetically modified organisms slipping through this federal regulatory gap is the Glofish. The Glofish is a zebra fish with a coral gene that produces a red fluorescing protein. The CEO of the company that sells Glofish told the press that all three federal agencies said the Glofish did not fall under their purview. The state of California banned sales of Glofish under state laws governing transgenic fish (*Nature Biotechnology*, 2004, 22, 11–13). Litigation to compel the FDA to regulate Glofish under federal law has thus far failed.[4] Complicating this picture, recent reviews of FDA oversight in other domains suggest the agency suffers from managerial weaknesses, and in particular, lacks the teeth to regulate post-marketing problems (Institue of Medicine, 2006). In 2001, 41% of a national random sample reported that they trusted the FDA a great deal; this proportion dropped to 29% in late 2006 (Pew Initiative on Food and Biotechnology, 2006). Familiarity with GM foods also declined slightly from 2001, with the majority of US consumers unaware of the prevalence of biotechnology and GM foods in the supermarket (Mellman Group Poll for Pew, Nov. 16, 2006). Nevertheless, a plurality of those polled deemed current federal regulation of GM foods inadequate. Further, transgenic and cloned animals have been judged immoral by a majority of respondents in many U.S. polls, because of the "playing god problem" (Pew Initiative on Food and Biotechnology, 2005a, 14–15; 2005b).

Bills introduced in the U.S. Senate in early 2007 propose the establishment of a post-marketing surveillance center at the FDA. This has probably been proposed with more conventional pharmaceuticals in mind, in response to criticisms of the FDA stemming from withdrawal of drugs like Vioxx from the market. Adequate post-market oversight of biotech foods is also noticeably absent from current FDA priorities and capabilities (Taylor and Tick, 2003). A mechanism for post-market surveillance of transgenic animals might provide an appropriate regulatory structure to monitor latent environmental effects, if staffed with appropriate environmental expertise.

Regulatory oversight of commercial biotechnology applications, even for GM Fish, is hampered by conflicting treatment and communication of confidential business information between state and federal government agencies (Pew Initiative on Food and Biotechnology, 2006). While some have argued that the states should and

---

[4] *International Center for Technology Assessment et al., v. Michael O. Leavitt, DHHS*, Civil Action 04-0062, United States District Court for the District of Columbia, *Memorandum Opinion*, 8 January 2007.

will step in to fill federal regulatory gaps (as illustrated by the increase in state and local regulations described above), state regulators polled in 2006 felt they had insufficient information to be informed of risks of GM crops (Pew Initiative on Food and Biotechnology, 2006).

Regulatory oversight of commercial biotechnology applications is also weakened by the lack of international coordination in financial markets. Over the last few years, many European biotech startups have merged with U.S. companies or moved to the U.S. to take advantage of better financing available in the U.S. However, Aqua Bounty was one of two companies that went public on the Alternative Investment Market the UK in 2006, rather than facing additional scrutiny in US financial markets (NYT, July 12, 2006; http://list.web.net/archives/foodnews/2006-April/000035.html).

8.3.1 What Form Might US Policy Take if the Precautionary Principle and the Provisions of the Biodiversity Convention Governed GM Salmon in US Waters?

Under the Coordinated Framework a 'lead' agency is negotiated for specific cases. Relegating the approval of transgenic animals to the FDA means, for example, that the U.S. has adopted a voluntary approval process, in which the FDA regards the application as confidential during the process, relies on the applicant for safety assessment, and does not provide much transparency. Were the provisions of the Biodiversity Convention to apply, it would lower the likelihood of the government 'opting out' of addressing ecological, social and cultural concerns, and require a greater level of transparency. Canadian regulations provide a relevant counter-example to U.S. biotech regulations. Canadian regulations apply a precautionary approach, and regulate transgenic animals under the Canadian Environmental Protection Act (Kochhar et al., 2005).

8.4 Conclusion: Different View of "the Natural"

Philosophical concepts of nature pervade our understanding and beliefs about policy, as well as the structure of specific policy domains. Samuel Adams and others who signed the Declaration of Independence took to heart Locke's notions of the "natural" human right to life and one's own property (Fradin, 1998). These implicitly relegate nature to human property. Many government acts assume or assert property rights to nature, including the right to assign that property to other entities.

The Convention on Biodiversity does this as well, by recognizing the rights of sovereign nations to exploit their own natural resources.

In contrast to the requirements of the Convention on Biodiversity, however, patenting by multinational corporations—and in the U.S., patenting in agricultural biotechnology has increased dramatically (Barham et al., 2002)—privatizes the rights to exploit patented resources, including indigenous species. Some argue that indigenous peoples should have the same patenting rights as multinationals (Christensen, 1992).

The Supreme Court has said that "a new mineral discovered in the earth or a new plant found in the wild is not patentable.... [...] Such discoveries are 'manifestations of... nature, free to all men.'" (*Diamond v. Chakrabarty*, 447 U.S. 303 (1980); cf. Andrews et al., 2006). However, a live, human-made micro-organism is patentable under the patent statute, because it constitutes either a "manufacture" or a "composition of matter" within the meaning of that statute.[5]

Designating species as endangered also illustrates the role of philosophical concepts of nature in governance. Endangered species designation presumes an identifiable, unchanging, genetically distinct segmentation of species into "evolutionarily significant units" (ESU). In 2001 the U.S. District Court in Oregon ruled to delist all Oregon coast coho salmon from the Endangered Species Act, on the grounds that hatchery-raised salmon and their wild genetic relatives belong to the same evolutionarily significant unit. Scientists later pointed out evidence that survival and behaviors are not comparable across wild and hatchery-raised salmon, and that it can take two generations to see such differences (Myers et al., 2004). These same scientists were science advisors to NOAA, and had authored a prior report for NOAA on the topic. The Union of Concerned Scientists in their update on scientific integrity cites this as an example of the Bush administration excising science from scientific advisory board reports (see http://www.ucsusa.org/scientific_integrity/interference/deleting-scientific-advice-on-endangered-salmon.html). Indeed, NOAA's regulatory action in June of 2005 appears to have ignored this evidence, and the precautionary position of the scientists who attempted to advise the agency.

Ultimately, the battle in the US over genetically modified salmon and other GMOs may turn on whether the concept of nature implied in international agreements like the Biodiversity Convention prevail over other concepts of what is natural. To some, a laissez-faire "free market" is the true state of nature, and business interests must vie with both government and private interests in an economic contest where competition is the proper test of what should survive. In that context, survival alone may be the measure of what is natural.

---

[5] 60 Am. Jur. 2d Patents § 80.

References

Andrews, Lori, Jordan Paradise, Timothy Holbrook, and Danielle Bochneak (2006). "SCIENCE AND LAW: When Patents Threaten Science," *Science* 314(5804), 1395–1396.

Barham, Bradford L., Jeremy D. Foltz, and Kwansoo Kim (2002). "Trends in University Ag-Biotech Patent Production," *Review of Agricultural Economics* 24, 294–308. Available at SSRN: http://ssrn.com/abstract = 371448

Bratspies, Rebecca M. Bratspies (2005). "Glowing in the Dark: How America's First Transgenic Animal Escaped Regulation," *Minnesota Journal of Law, Science and Technology* 6, 457–504.

Christensen, Jon (1992). Letter in *Science*, 257, 1482.

Farrow, S. (2004). "Using Risk Assessment, Benefit-Cost Analysis, and Real Options to Implement a Precautionary Principle," *Risk Analysis* 24(3), 727–735.

Fradin, Dennis Brindell (1998). *Samuel Adams: The Father of American Independence*. New York: Houghton Mifflin.

Hansen, Steffen Foss, Martin P. Krayer von Krauss, and Joel A. Tickner (2007). "Categorizing Mistaken False Positives in Regulation of Human and Environmental Health," *Risk Analysis* 27(1), 255–269.

Institute of Medicine, Committee on the Assessment of the US Drug Safety System Board on Population Health and Public Health Practice. Alina Baciu, Kathleen Stratton, and Sheila P. Burke (eds.) (2006). *The Future of Drug Safety Promoting and Protecting the Health of the Public*. Washington, DC: National Academy Press.

Kochhar, H.P.S., G.A. Gifford, and S. Kahn (2005), "Regulatory and biosafety issues in relation to transgenic animals in food and agriculture, feeds containing genetically modified organisms (GMO) and veterinary biologics," in H.P.S. Makkar and G. Viljoen (eds), *Applications of Gene-based Technologies for Improving Animal Production and Health in Developing Countries*. Dordrecht, The Netherlands: Springer, 479–498.

Mellman Group Poll for Pew (2006). Available online at: http://pewagbiotech.org/research/2006update/2006summary.pdf.

Merchant, Gary E. (2002). "The precautionary principle: an 'unprincipled' approach to biotechnology regulation," *Journal of Risk Research* 4(April), 143–157.

Myers, Ransom A., Simon A. Levin, Russell Lande, Frances C. James, William W. Murdoch, and Robert T. Paine (2004). "Hatcheries and endangered salmon," *Science* 303(5666), 1980.

National Research Council of the National Academies, Committee on Defining Science-based Concerns Associated with Products of Animal Biotechnology, Committee on Agricultural Biotechnology, Health, and the Environment, Board on Agriculture and Natural Resources, Board on Life Sciences, Division on Earth and Life Studies (2002). *Animal Biotechnology: Science-Based Concerns*. Washington, DC: National Academies Press. www.nap.edu

Pew Initiative on Food and Biotechnology (2005). "A Future for Animal Biotechnology." Proceedings from a forum hosted by the University of Illinois at Urbana-Champaign Agricultural Genome Science and Public Policy Program and the Pew Initiative on Food and Biotechnology, December 5, Chicago, IL.

Pew Initiative on Food and Biotechnology (2005b). "Exploring the Moral and Ethical Aspects of Genetically Engineered and Cloned Animals." Summary of a Multi Stakeholder Workshop sponsored by The Pew Initiative on Food and Biotechnology, January 24–26, Rockville, MD.

Pew Initiative on Food and Biotechnology (2006). "Agricultural Biotechnology information disclosure: Accomodating conflicting interests within public access norms." Prof. James T. O'Reilly, College of Law, University of Cincinnati, Cincinnati, OH.

Taylor, Michael R. and Jody S Tick (2003). "Postmarket oversight of biotech foods: Is the system prepared?" A report commissioned by the Pew Initiative on Food and Biotechnology and prepared by Resources for the Future, Washington, DC.

World Commission on the Ethics of Scientific Knowledge and Technologies (2005). *The Precautionary Principle*. Paris: UNESCO.

Index

A

Abiocor implantable device and humanities, 263, 265
Activa therapy, for deep brain stimulation, 213
Adoption, 17, 29, 38, 42, 43, 46
Adult attention-deficit/hyperactivity disorder, 177
Adultery, 38–42, 44, 64, 65, 69, 235
Al-Azhar *fatwa,* on assisted reproduction, 40
Altruism, 326
Alzheimer's disease, 173, 175, 208
    diagnosis of, 185
    human genetic modification as treatment for, 175–176
    patients with symptoms of, 179–180
American
    and Athenian, willing to use ARTs, 77
    ensconced the brain as essence of self, 277
    experiencing form of brain disease, 216
    infertility epidemic in collusion with fertility industry and, 60
    patriarchal cultures, 24
    population lack health insurance, 248
    religion and self-improvement, 144
    report religious service attendance, 103
    Roman Catholics, contraception and fertility treatment, 73
    success in gene therapy trials, 171
    surrogates, morally and emotionally offended, 42
    technology and careful deliberation, 192
    trust on ART policy, 91
Animal and Plant Health Inspection Service (APHIS), 331
Anthropocentrism, 228, 242, 292, 294, 302.
    See also Biodiversity
Antidementia drugs, 183, 189.
    See also Dementia
Appellate court decision, database recording, 327

Artificial insemination (AI), 26, 51, 87, 126
Artificial insemination-donor (AID), 26
Artificial intelligence, 200, 214, 216
Artificial limbs, 200, 201, 206, 209, 213, 259
Assisted reproduction, religion and naturalization of. *See also* Assisted reproductive technologies (ARTs)
    ambiguity and resolution
        ARTs and intercourse, 52–54
        gamete collection and donation, 54–56
        gender and gender roles, 46–50
        natural family, 28–37
        sex selection and eugenics, 56–58
        sexual orientation, singleness, and reproduction, 50–52
        surrogacy, 41–43
        technologies, 25–27
        themes and variations, 24–25
        third-party donation, 37–41
        unused embryos, 43–46
    cultural anthropology and bioethics to investigate, 23
    naturalizing the technology
        divine in and against nature, 74–76
        eagerness or reluctance of traditions, 76
        naturalizing ARTs, 60–74
        pronatalism and religious strategies, 58–60
    patriarchal cultures, 23
    polymorphism of nature, 18–23
    religion, boundaries of, 17–18
    religious thought policymakers, 23
Assisted reproductive technologies (ARTs)
    abortion and to combining PGD with, 46
    birthrate and covert use of, 60
    and chemical mode, 204
    church teaching on, 45
    claim for destruction of traditional religion, 22

337

Assisted reproductive technologies (ARTs) (*cont.*)
  conception of nature X and setting policy, 90
  de facto or *de jure* objections to, 30
  divinely intended nature, Hinduism *vs.*
    Islam or Judaism, 68
  divinization of, 61
  Donum Vitae's analysis of, 33
  ethnographic observations of Traina, 99
  existing policy and regulation of, 87–89
    FDA indirectly regulation and behavior
      of practitioners, 88
    non-governmental regulation, 89
  fatwas permit use of, 64
  federal and state policy on, 93
  Finland's sense of, 21
  in France, law and regulatory surveillance, 102
  and gender identity, 50
  for Greek Orthodox women, 36
  homologous procedures and introduction
    of third-party gametes, 22
  and human dignity, 103
  infertility and, 24
  Iranian Shi'ite muslims *vs.* Sunni
    counterparts, 65
  in Japan, 75
  legislatures, 95–96
  lesbian and gay couples, use of, 28, 52
  limited use of, 33, 35
  for Middle Eastern Muslim women and
    men in infertile marriages, 32
  natural family and, 50–51
  naturalizing of, 60, 76
  and newly-energized identity movements, 70
  normalization or naturalization of, 20
  objection to, 43
  orthodox christianity, moral and vocational
    argument for, 47
  orthodox Jewish attitudes to, 39
  permission to use, 63
  policy
    incorporation of religious conceptions
      of "natural," 87, 90, 102
    NIH for reviewing, 91
    proposals, 77
    religious customs and arguments, 19
  privileging of biological or gestational
    relatedness over, 79
  psychological and spiritual barriers to use
    of, 30
  religion and nature to practitioners and
    patients, 19
  Roman Catholic Church opposing, 73
  Roman Catholic teaching on, 61
  for sex-selection, 57

Auditory brain implant (ABI), 210
Automatic implantable cardioverter-
  defibrillator (AICD), 263

B

Bacillus thuringiensis (Bt). *See also*
    Genetically engineered organisms
  corn target, 288
  maize
    ethical analysis, 301–305
    in Mexico, 286
  sensitive individuals loss, 292
  toxin production, 289
Biodiversity
  and anthropocentrism, 293–294
  and change in ecosystem, 287
  effects of losing, 304–305
  and genetically engineered organisms
    biotechnology, impact, 288–292, 321
    current status, 288
    ethical issues with, 301–302
  meaning of, 286
  and nonanthropocentrism, 294–298
Bioethics
  and accepted standards of practice, 89
  contributions of, 161
  cultural anthropology and, 23
  feminist and, 151
  national bioethics commissions, 168
  pattern of appropriation in, 200
  provide philosophical justification for, 281
  and public policy in 1980s, 162–164
  and religion, 189–194
  in religious doctrine for altering DNA, 163
  role in
    enhancing memory and cognition,
      173–174
    in health policy decisions, 174, 193
    regulation of human DNA transfer
      research, 165
Biological diversity, convention on, 324–325
Biomedicine, 199, 204, 221, 225, 250
Bionics, 202
Biosafety
  Cartegena protocol, 325, 326
  and food safety of Atlantic salmon
    engineered with GH, 288
  refusing to adopt protocol by U.S., 327
Biotechnology. *See also* Biodiversity
  bioethics commissions and, 190
  definition, 326
  and encouragement by Islamic
    scholars, 236

impact on biodiversity, 288–292, 297, 321
policy failure, U. S., 331–333
for purpose of self-improvement, 144
from recombinant DNA to, 165–166
revising RAC and uses of, 167
Bt maize, ethical analysis. *See Bacillus thuringiensis* (Bt)
Buddhism, 18, 21, 45, 78, 109, 114, 166, 218, 231–232, 296
dukkha concept in, 244
perspectives on enhancement, 146–150
pratityasamutpada, for physical world, 114
and Roman Catholicism, 36
treatment to cure disease, 149
Western viewpoint, 30

C

Canadian Environmental Protection Act, 333
Cannibalism, 312, 313
Cardiac implantable devices, 262–263
Castanea dentata, 289
Casuistry, 219, 220, 225–227
Catholic iconography, 234
Celibacy, 33, 55, 59, 72
Central nervous diseases, approach for treatment, 213
Childless marriage, 15, 16
Cholinesterase drugs, 186
Christian theology
 entrance to public arena, 136
 to love and affirm all of God's creation, 133
 perennial problem in, 135
Cloning, 35, 72, 98, 101, 123, 125, 127, 190, 226, 312
Cochlear implants, 209
Cognitive psychology, 240
Combination products concept, 268
Confucianism, 22, 31, 231
Consequentialist based method, 279
Cryphonectria parasitica, 289
Cyborg continuum, 206

D

Death
 altering of, 277–278
 ambiguity of, 226
 apprehension of approach of impending, 282
 brain death and organ transplantation in Muslim bioethics, 140
 Bridging themes and definition of, 260–262

Daoism, alternation of life and, 230
 of extra embryos, 45
 Hindu view of, 229–230
 of Jesse Gelsinger, 165, 170, 171
 of patients from rare viral disease, 119
 putative toxin and, 305
 rituals necessary and passage of father's soul, 68
 Roman Catholicism view of, 233
Defense Advanced Research Projects Agency (DARPA), 214
Dementia. *See also* Antidementia drugs
 amelioration of, 180
 clinical treatment of, 173
Destination device. *See* Left ventricular assist device
Developed nations
 cardiac disease in, 262–263
Direct-to-consumer (DTC) advertising, 184
Disease associated DNA mutations, 163
Diversity. *See* Biodiversity
DNA molecules
 in bacteria, 165
 composition of, 227
 as form of nanotechnology., 225
 in humans, 166
DNA transfer
 clinical protocol for, 169
 and gene for defective OTC protein, 171
 interplay of bioethics and public policy, 162–164
 outcomes of, 170–172
 role of bioethics in regulation of research, 165
Donor technologies
 as marriage saviors, 67
 mut'a, to make egg donation legal, 65

E

Ecosystem
 biodiversity changes, affect, 287
 representation, 287
Efficacy concept, by FDA, 265
Embedded electronic technologies, for internal organs, 211
Embryo donation, 66, 102
Embryo research, 45, 102
Endangered species act (ESA), 328, 329
Enhancement. *See also* Eugenics; Genetic enhancement
 Buddhist perspectives on, 146–150
 feminist perspectives on, 150–155
 vs. therapy, 111–113

Index

Enlightenment, 18, 30, 45, 90, 125, 139, 147, 241, 244
Environmental Protection Agency (EPA), 331
Eternal law, 125
Eugenics, 56, 109–111, 138, 151, 154, 162

F

FDA Modernization Act of 1996 (FDAMA), 270
Fertile married couples, 16
Fertility industry, 60
Food and Drug Administration (FDA), 167, 192, 262
   approval for base technology, 273
   combination products concept, 268
   historical evolution, policies, 269
   medical device amendments (MDA) development, 264
   safety and efficacy concept, 265
Frankenstein, 192

G

Gamete collection and donation, 17, 54–56
Gamete intrafallopian transfer (GIFT), 26
Gay couples, 16
GE corn plantation, in Mexico, 302
Gene modification protocols, 169
Gene therapy patient tracking system, 171
Genetically engineered organisms
   and biodiversity
     current status, 288
     ethical issues, 301–302
Genetically modified (GM) salmon
   in food supply, 321–322
   in U.S. waters, governance, 333
Genetic code, 162
Genetic defects, 16
Genetic diversity, 286, 301. *See also* Biodiversity; Biotechnology
   effects of losing, 303
Genetic enhancement, 15, 109, 110, 116, 124, 129, 132, 136, 151, 155
   Hindu perspectives on, 142–146
   Jewish medical ethics and, 136–139
   memory, clinical scenarios for enhancement of, 178
   technologies for, 174
Gene transfer protocols, 163
Gene transfer rate, transgenic crops acceleration, 303
Germ line therapy, 179
God and procreation of child, 72

Golems, 223–225
Growth hormone, ethics of
   human GH protein therapy, 120
   recombinant DNA technology (rGH) for children, 119–120
   societal ethical concerns, gene therapy, 120–121

H

Hashgacha (divine providence), 63
Healthcare common procedure coding system (HCPCS), 272
Health Research Extension Act of 1985, 178
Heart transplant, 211
Hemophilia, 171
Heterosexual childless marriage, 16
Heterosexual couples, 15
Hinduism
   danger of *pollution*, 230
   encompasses array of traditions, sects, and, 68
   karma and religious ritual, 238
   practice of accomplishing orthodox family formation, 70
   response, for gene therapy, 143
   use of animals for human enhancement, 146
   view of conception, 69
   worldview X over worldview Y, 90
Hip-joint replacement, 203
Hippocratic prescription, ancient, 326
Human breeding, 162
Human dignity, 103, 125, 126, 128, 133, 134, 140, 226, 240
Human DNA transfer. *See* DNA transfer
Human Fertilisation and Embryology Act of 1990, 104
Human Fertilisation and Embryology Authority (HFEA), 104, 105
Human gene therapy, 111, 166, 168
   RAC subcommittee on, 169
Human genetic engineering, 161–162
Human genetic enhancement, 111
Human gene transfer, 167
Human GH protein therapy, 120
Human inheritable genetic modification, 162
Human-machine interactions, 216

I

Implantable cardiac defibrillators (ICDs), 273–274, 277
Implantation impact, on self-identity, 260, 261
Implanted medical electronics, 209–212

Index

Incest
  in Jewish population, 40
  sperm donation and, 41
  transgenic technologies and, 314
Infertility
  affecting men's gender identity and virility, 49–50
  Christian communions intervention, 30
  epidemic, 60
  in Euro-American cultures, 47
  and failure to produce boys, 48
Inherited Genetic Modification (IGM), 112
Institutional Review Boards (IRBs), 265
Insulin-like growth factor I (IGF-1), 116–118
International court of Justice, 325
Intracytoplasmic sperm injection (ICSI), 26–27
  eliminating the randomness of sperm selection, 53
  and epidemic of divorce in Egypt, 79
  in Middle East, 49
  restoring men's masculinity, 50
In vitro fertilization (IVF)
  based on assumptions about importance of autonomy, 151
  Buddhism and, 45
  contemporary Greek Orthodox women and, 36
  Donum Vitae's objections to interventions, 34
  genetic selection in, 112
  Greek Orthodox theologian view, 53
  HFEA's view, 105
  homologous AI, 74
  Iran, donor embryo law and, 66
  lackness in privacy, 55
  and natural family, 56
  needs to remain legal, 96
  for Orthodox Christians, 44
  policies determine basic safety standards of, 93
  in Sunni Muslim, 41
  third party donation, and other procedures, 69
Irish American catholicism, 58
Islam
  to appropriate technological advancements, 236
  fatwa declaration for egg donation, 65
  law, 41, 141
  mandate, procreation and biological inheritance, 40
  marriage and procreation of genetically related children, 31
  medical advancement justified within, 236

reasoning method of *qiya*, 226
traditions to porcine heart valves, 218
view of origin of natural world, 114
zakat and health care, 250
Israeli surrogacy law, 42

J
Jarvik 7, artificial heart, 211, 263
Judaism, 18, 29, 31, 63, 68, 109, 114, 122, 136, 235, 240
Justice, in cardiac device implantation, 261–262

K
Kinship, 25, 32
  genetic relationship to parent-child, 42
  between Jewish ethics and Roman Catholic ethics, 137
  notions of divinely-ordained, 67
  surrogacy and claim for, 97
  understandings of gender and, 77
  Western views of nature and, 76

L
Land ethic principle, 297–298
Left ventricular assist device (LVAD), 211, 263
Legislation, for genetically engineered food (US), 329
Lesbian couples, 16, 50–52, 79
Living modified organism (LMO)
  notification, 326
  recipient nation movement, 326–327

M
Male infertility, 24, 47, 50, 66
Maori
  as guardians of natural habitat of Aotearoa/ New Zealand, 313
  transgenic technologies and incestry, 314
  unease for modify the genomes of indigenous species, 315
Marriage
  ascetic monk and, 147
  Catholic moral theologians, normative relationship of, 74
  and childbearing, 31
  as divinely sanctioned locus for procreation, 72
  gestational vision and family, 67

Marriage (*cont.*)
    Greek Orthodox Church, purposes
        of marriage, 35
    heterosexual childless, 16
    homosexual, 96, 99
    for Middle Eastern Muslim, 32
    now-defunct custom of, 69
    production of male children essential to, 68
    Roman Catholic vision for, 52
    second, 70
    Shi'ite temporary marriage, *mut^{c}a*, 65
Masturbation
    Confucians view, 55
    Jewish law and, 40
    Roman Catholic view, 35
    and surgical procedures, 38
Mechanical devices
    for augmenting human capacities, 203
    common good of society, for advancement
        of, 246–251
        liberation theology and moral option
            for poor, 248–251
    different kinds of embodied incorporations
        of, 205
    purpose of the intervention, 207–209
    religious meaning and implications of
        incorporation of, 201
Medical devices
    current scale, incorporation on, 259
    development
        research regulation, concepts
            employment, 266–268
        total artificial heart (TAH) program,
            265–266
        in United States, 264
    implantation of, 260
        equitable provision of, 262
    policy
        and humanities, 278–280
        for ICDs, 259
    religious traditions in, 261
    use of
        access and protection, 269–271
        decision on reimbursement, 271–273
        evaluation, ethics of, 274–275
        ICDs (*see* Implantable cardiac
            defibrillators)
        marketing and clinician education,
            275–277
Medical device amendments (MDA), 264
Michigan aquaculture development act, 330
Micro-electro-mechanical systems (MEMS), 204
Microsorting, 27, 57
Miscarriage, 74

Monism, 313
Moral discourse, 90, 92. *See also* Assisted
        reproductive technologies (ARTs),
        Policy venues
    thick discourse, 89–91, 99, 101, 104
    thin discourse, 90, 91, 98
Moral law, 125
More Than Human, 208
Motherhood, 17, 22, 24, 40, 45
    Athenian women and, 44, 61
    femininity and, 46
    Greece and Israel, objections to single, 51
    Greece, cultural understandings of, 42
    Greek pronatal policies visions, 60
    identification, womanhood and, 48
    policies to help married women to realize, 79
    primary objections to single, 51
    surrogate, 31

N
Nanotechnology
    applications of, 215–216
    and communication and information
        systems, 226
    for correcting nature and, 240
    to create molecular-sized physical
        machines, 204
    metaphor of DNA as form of, 225
    molecular composition of DNA as form
        of, 227
    Protestant assessment of, 247
    in regenerative medicine, 215
    robotics and, 224
Narcolepsy, 176
Nasab preservation, for Muslims, 41
National Bioethics Commissions, 168
National marine fisheries service (NMFS), 328
National oceanic and atmospheric
        administration, 328–329
National Wildlife Federation, 328–329
National wildlife refuge system administration
        act (1966), 328
Natural concept, 323. *See also* Genetically
        modified (GM) salmon
Natural family, 17, 28
    for ARTs, 50–51
    Buddhism and, 30
    church-sanctioned methods of planning, 53
    divinely-intended, 21
    divinely ordained, 70
    Euro-American, 20
    genetically related, 21
        children and IVF, 96

Index

Judaism and, 63
natural reproductive processes and, 77
for raising children and protecting dignity of children, 35
reductive core of, 78
Naturalism and genetic arguments, 121–124
Natural law, 124–125, 128, 162
Nature
  altering with IGF-1, 116–119
  comprehensive cosmos, 113
  as essence, 113–114
  for *guide* and inspiration, 114
  human genome as gift of, 116
  inclusive of both subject and object, 115
  opposite of culture or artifice, 114
  patenting of
    for engineered bacteria, in *Diamond v. Chakrabarty* case, 166
    by MNCs and in U.S, 334
  result of divine act of creation, 114
  as untamed, 113
Niyoga, 69, 70
Nonanthropocentrism, 292, 294, 302.
  See also Biodiversity
Nonmarket valuation methods, 306.
  See also Transgenic organisms
Non-X-linked SCID, 172
Normative religious resources.
  See also Mechanical devices
  metaphors and casuistry, 219–220
  chimeras, 221–223
  concept of casuistry, 225–227
  golems, 223–225
  metaphorical interpretation, 220–221
  perspectives of appropriation, ambivalence, and resistance, 217–218
  schema in understanding of arguments and conclusions, 238–239
North America
  Europeans arrival in, 300
  lions and elephants introduction, 299
  megafauna extinction, 299

O

Office of recombinant DNA activities (ORDA), 171
Office of technology assessment (OTA), 168
Ooplasm donation, 27, 87
Organic nanostructures, 216
Organismal diversity, 286–287
Orthodox Judaism, 31, 51, 54, 62, 67.
  See also Judaism

P

Pacemaker, 211
Parkinson's disease, 177, 213
Patenting of nature
  for engineered bacteria, in *Diamond v. Chakrabarty* case, 166
  by multinational corporations and in U.S, 334
Patriarchal ideology, 24
Pekkuah nefesh, 224
Pharmaceuticals and cognitive function, 176
Plenitude principle, 295, 296, 298
Policy venues
  and conceptions of nature, 92–98
    courts, 96–97
    executive branch agencies, 92–93
    government commissions, 97–98
    legislatures, 95–96
    professional societies, 94–95
  different appetites for moral discourse, 90
  European, 104–106
Polygamy, 72
Post-Darwinian biology, 296
Post-mortem dissection, body, 261
Precarious social status for women, 16
Precautionary principle
  Hippocratic prescription, ancient, 326
  insuring biosafety, 325
  opposition to transgenic organisms, 307, 308
  US resistance, 327
Preimplantation genetic diagnosis (PGD), 27, 46, 56, 57, 105, 112
Principle of *hishtadlus*, 63
Prodrome syndrome, 174
Pronatalism, 17, 25, 30, 58–60

Q

Quality improvement organizations (QIOs), 185

R

Racial and social stratification, 24
Radio frequency identification chip (RFID), 205
Regenerative medicine, 215
Religion and religious people, 17–18
Religious moral diversity, 237–239
Religious traditions, in medical device, 261, 262
Reproductive technologies, 15
Reproductive tract infections, 16

Retroviral vectors, used for genes transfer, 172
Roman Catholicism, 22, 36, 48, 51, 52, 75, 109, 127, 139, 232
   somatotropin deficiency syndrome, genetic intervention for, 127
   view on therapy and enhancement, 124–128
Royal Commission on genetic modification report, 312, 313

S

Safety concept, by FDA, 265
Schizophrenia, 177
Scientific naturalism, 122
Self-identity, medical device implantation, effect, 260
Severe combined immune deficiency (SCID), 171
Sex determination, 16
Sex selection, 46, 56, 57, 79, 154
Sexual differentiation, 16, 72
Solid organ transplantation, 260
Spinal cord regeneration, 215
Splicing life, 166–167, 191, 193
Subclinical syndrome, 174
Sunni Islamic cultures, 40–41
Surrogacy, 17, 26, 35, 41–43, 51, 52, 63, 102

T

Tapu concept, in New Zealand, 313–315.
   See also Maori
T-cell acute lymphoblastic leukemia (T-ALL cancers), 171
Technological intervention into human body.
   See also Mechanical devices
   Abrahamic traditions
      Islam, 235–237
      Jewish thought, 234–235
      Roman Catholicism, 232–234
   Asian traditions
      Buddhism, 231–232
      China, 230–231
      India, 229–230
   method and impingement intersections, 207
   modes of, 203–205
   research applications, ethical concerns
      consciousness and immortality, 216
      cyber soldier, 214–215
      external devices, neural control of, 212–214
      implanted medical electronics, 209–212
      nanomedicine, 215–216

Theological vision, of covenantal law, 63
Therapeutic misconception, 267
Therapy and enhancement.
   See also Genetic enhancement
   Protestant views on, 128–136
   Roman Catholic views on, 124–128
Tikkum olam, 224
Timaeus, 295
Total artificial heart (TAH), 212
   program, 265–266
Transgenic crops, 303
Transgenic organisms
   decrease in biodiversity, 289, 292
   relevance for, 298–301
   risk analysis
      considerations for, 311–316
      societal risks assessment, tension in, 305–307
      uncertainty in, 307–311
   status of, 288
Transhumanism, 240–242, 259
   Buddhist critiques, 243–245
      Daoist critique, 245–246
   protestant Christian as critiques of, 242–243
Translation problem
   conflation of God's will and nature, 101
   path between religious and secular translation, 100
   principle of autonomy in, 100–101
   true meaning, distortion of, 100

U

United States
   legislation for genetically engineered food, 329
   maize production in, 304
   oversight responsibility for device development in, 264
   policy failure on biotechnology framework, 331–333
   resistance to precautionary principle, 327
Unused embryos, 43–46
   abortions as applications of *ekonomia*, 45
   Buddhist teaching *vs.* Greek teaching, 45–46
   frozen embryos, use of, 46
U.S. Department of Agriculture (USDA), 331

V

Ventral cochlear nucleus, 210

W

Western Europe
  Germany, restrictive nations and policy of ART, 102–103
  policy venues and arguments, 104–106
    European policy *vs.* U.S. policy, 103, 105–106
  regulatory surveillance and heterosexual couples, in France, 102
  religious conceptions of nature into ART policy, 101–104
Women's rights, 16

X

Xenotransplantation, 123, 259, 268
X-linked severe combined immune deficiency, 171

Y

Yahweh and hara kiri words, 314. *See also* Maori

Z

Zygote intrafallopian transfer (ZIFT), 26, 44